ROBOTIC TECHNOLOGY

PRINCIPLES AND PRACTICE

ROBOTIC TECHNOLOGY

PRINCIPLES AND PRACTICE

Werner G. Holzbock, M.S.M.E., P.E.

with a Foreword by
Professor Jack D. Lane

 VAN NOSTRAND REINHOLD COMPANY ———— New York

Copyright © 1986 by Van Nostrand Reinhold Company Inc.
Library of Congress Catalog Card Number 85-15634
ISBN 0-442-23154-7

Manufactured in the United States of America

Published by Van Nostrand Reinhold Company Inc.
135 West 50th Street
New York, New York 10020

Van Nostrand Reinhold Company Limited
Molly Millars Lane
Wokingham, Berkshire RG11 2PY, England

Van Nostrand Reinhold
480 LaTrobe Street
Melbourne, Victoria 3000, Australia

Macmillan of Canada
Division of Gage Publishing Limited
164 Commander Boulevard
Agincourt, Ontario M15 3C7, Canada

16 15 14 13 12 11 10 9 8 7 6 5 4 3 2 1

Library of Congress Cataloging in Publication Data

Holzbock, Werner G.
    Robotic technology, principles and practice.

    Includes bibliographies and index.
    1. Robotics—Handbooks, manuals, etc. I. Title.
TJ211.H65 1986   629.8′92   85-15634
ISBN 0-442-23154-7

In memory of my friend Boris B. Bodelm

FOREWORD

Because of the spectacular growth of the robotics industry and the significance this holds for future industrial growth, there must be textbooks and courses on all levels to teach the technological principles involved. Among the significant publications that have already appeared, none so far has described the structural elements of robots in their totality. This book makes an important contribution to filling the need for such a comprehensive book.

At the GMI Engineering & Management Institute in Flint, Michigan, we have given courses and workshops on robotics for several years. Werner Holzbock was one of my students. He has followed to a large extent the general outline of such a course. His coverage of the subject is excellent. It can be expected that this text will find widespread use in the robotic community and become the basis for the study of the subject in coming years.

JACK D. LANE

Director of Robotics Center at GMI.
Chairman of the Education and
Training Division and Member of
the Technical Council of
Robotics International of SME.

PREFACE

Robotic technology refers to the design and application of robots, and specifically of industrial robots that are used for manufacturing or construction purposes with the intent of handling, processing, assembly, and inspection of materials and parts.

Robotic technology embraces mechanics, hydraulics, pneumatics, electronics, and so on. This is an exceptionally wide spread for any specialized technology. A total treatment of all aspects of these various categories could never be given in a single book, and that is not the object of this book. The aim is rather to describe the principles that control behavior, operation, selection, and development of robotic technology.

Robots are a kind of automation. There are many automatic devices in factories; a bottling plant, for example, is a miracle of interacting, automated mechanical devices. However, they differ from robots in that they are not easily reprogrammable; and, hence, such automation is usually referred to as hard automation as compared with the soft, flexible automation of the robot.

Another kind of industrial automation is a machining center under computer control, which may hardly need human attention. The robot differs from this type of automation too because the robot is capable of working in relation to other production machinery and is capable of bringing about the coordination of various pieces of production equipment in a way that is readily reprogrammable. Thus, it is either the easy programming or the coordinating ability or both that mark the robot as a very particular automation tool. Robots are *not* humanoids. They are practical mechanisms designed to grasp and move objects; that is, they are manipulators. But they are manipulators of a special kind because they can be programmed to operate in a specific manner to execute repetitive tasks. They are designed to move in a variety of ways and to be capable of carrying workpieces as well as tools, such as welding torches and paint spray guns and other implements. In other words, they are flexible enough to be truly multifunctional.

If all this is considered, one arrives at the "official" definition, as formulated by The Robot Institute of America (now the Robotic Industries Association):

A robot is a programmable, multifunctional manipulator designed to move material, parts, tools, or specialized devices, through variable programmed motions for the performance of a variety of tasks.

The definition speaks of a "programmable, multifunctional manipulator." Many manipulators, like the robot, have sliding and rotating joints for the purpose of grasping and moving objects. But it should be well understood that the robot is a *programmable, multifunctional* manipulator, and that not every manipulator can be considered a robot.

Note also the concept of "variety of tasks." If only a single task is involved, a robot is hardly required. There are plenty of mechanisms that fall under the category of hard automation and that can perform the task more efficiently than any robot, even when reprogramming from time to time may be required. As always in life, it is paramount to consider the *simplest* approach for the performance of the specific task or tasks. Only when a variety of tasks is involved should the robot be considered.

H. R. Leep and D. K. Hagan[1] made a random survey of 24 companies and found that fewer than one-third of the robots discussed used more than one computer program, an indication that they were applied in repetitive operations that do not necessarily require a robot.

They also stated that since "robots are not assigned flexible tasks . . . they relieve people from hazardous or repetitive work." This is correct, but it leaves unresolved whether these are tasks for a robot or for a mechanism with suitable linkages to perform this work.

Furthermore, the variety of tasks to be performed determines structure and complexity of robots. Again utmost simplicity—and that spells economy—is the goal.

The judgment about which robot is the most suitable for a given job requires a knowledge of the capabilities of the many ways in which robots can be built, a knowledge of their principle and their transformation into practice. If this knowledge is the task of our endeavors, then the logical first question is: What does a robot consist of? How do the parts work by themselves? And then how do they operate as an integrated whole? It is this kind of analytical approach to the subject matter of robotic technology that is applied in this text.

An attempt has been made to illustrate as many concepts of robotic technology as possible by actual products of a variety of manufacturers. The selection has been arbitrary. Many manufacturers should have been included, and the author extends his apologies to those that were overlooked.

In some subject areas (vision, touch sensors, etc.), developments are in such flux that it was necessary to include some of the ongoing research to give a wider spectrum of the potential offered by these techniques.

Many manufacturers, institutions, and researchers have given their active support to the preparation of this text. The author can only hope to have done justice to all the advanced technology they represent. It is hoped that no misrepresentations have occurred. However, if mistaken statements of published data have been made, they are the author's sole responsibility.

International licensing and agreements between robot manufacturers can lead to confusion: Unimation is now part of Westinghouse, which also is involved with Olivetti, Kawasaki, Mitsubishi, and Kamatsu; General Electric has license agreements with Hitachi, Volkswagen, and DEA; GMF Robotics Corp. is a combined venture of General Motors and Fujitsu Fanuc; IBM is associated with Sankyo Seiki, and so on. Little or no attention has been given in this book to where the different technologies were originally developed. Robotic technology is, at least for the engineer, a welcome example of international coordination and cooperation.

Customary U.S. units have been used throughout this book, wherever feasible. This means inches, pounds, and so forth. In some isolated case, however, use of the metric system seemed preferable.

With regard to terminology, extensive use has been made of the glossary for robotic terms of the National Bureau of Standards,[2] the dictionary of electronics by Graf,[3] and the microcomputer dictionary by Sippl.[4]

REFERENCES

1. Hagan, D. K. A survey of flexible manufacturing methods applied in metalworking industries. M. Eng. thesis, Dept. of Industrial Engineering, University of Louisville, Louisville, KY, 1982.
2. Smith, B. M., et al. A glossary of terms for robotics—revised. SME Paper MS83-914. Dearborn, MI: Society of Manufacturing Engineers, 1983.
3. Graf, R. F., *Modern Dictionary of Electronics*. Indianapolis, IN: Howard W. Sams & Co., 1977.
4. Sippl, C. J. *Microcomputer Dictionary*. Fort Worth, TX: Radio Shack, 1981.

Bloomfield Hills, Michigan   WERNER G. HOLZBOCK

CONTENTS

ROBOTIC TECHNOLOGY

PRINCIPLES AND PRACTICE

Chapter 1
STATICS AND DYNAMICS

Statics is concerned with forces that act on a body and the conditions required to obtain their equilibrium. Robots require hydraulic, pneumatic, or electric forces to change the position of the free end of the arm. These forces are needed because the robot is a mechanical device that offers resistance to change due to inertia and friction. Once a new position is obtained, the equilibrium has to be reestablished.

In this relationship between acting forces and their opposing factors, errors in the response of the robot are produced that must be recognized.

Dynamics also deals with these forces, but it does so with respect to the motion they produce. It establishes the relationship between accelerations (or decelerations) and the forces that cause them. The point of departure of dynamics is Newton's second law, which states that force (F) equals mass (m) times acceleration (a), or in algebraic form:

$$F = ma \qquad (1.1)$$

One particular characteristic of dynamics in robotic technology is that stability often means reducing the acceleration and/or deceleration in order to prevent the alternative of undesirable oscillations of the robot arm.

In either case—statics or dynamics—the concern of robotic technology is the accuracy and fast but stable motion of the robot.

ACCURACY AND ERRORS

Methods used to position the robot will be considered in other chapters. In the present chapter, the characteristic behavior that applies to a mechanical system in obtaining the desired position will be investigated.

The robot is a mechanical system subject to and part of an automatic control system. The accuracy of this system is the conformity of the final position of the robot to the true position that is expected. However, not

1

Fig. 1-1. Block diagram of robot.

only is optimal accuracy required, but also the robot must not be *excessively* qualified for the task at hand. Otherwise, it will unnecessarily burden production costs.

Before going further, some additional clarification of the concept of accuracy is necessary. Figure 1-1 shows the robot as a block with input and output. The robot here is nothing but a transducer changing one form of input into some other form of output. The degree to which the output changes in response to changes in input depends on the accuracy of the robot. When the input calls for a certain position, the transducer (robot) must act to produce this position. The input is the true value; the output, the actual result. The difference between true value and actual result is the inaccuracy or error. There are dynamic and static errors.

Consider a sudden change in input, producing oscillations in the output. After a certain time, the oscillations die out, and the output assumes the proper position as dictated by the input signal. Sometimes, a few oscillations are tolerated in the interest of higher speed, but by and large they are better avoided because usually they delay obtaining the desired result. There are other considerations as well. One is the wear and tear on the robot that can be produced by oscillations. Another is that the robot arm that overshoots its final position may knock into solid obstacles and produce damage.

Oscillations and their resulting overshoots are *dynamic* errors. If there are no oscillations, but the actual result (i.e., the position to which the robot arm moves) differs permanently from the true value of the input, then there is a *static* error. A robot can exhibit both static and dynamic errors.

The following sections consider the various types of static errors and then dynamic errors. Instability of dynamic systems as the most common consequence of dynamic errors is treated under a separate heading.

STATIC ERRORS

Position Accuracy and Repeatability

Accuracy specifications as given by robot manufacturers practically always refer to the robot arm without end effector and without load being carried. Generally, either one diminishes the specified accuracy.

Position accuracy also depends on the deceleration of the motion of the robot arm. For maximum accuracy, the robot has to make the final approach to its programmed position very slowly. Thus, Unimation distinguishes between three levels of accuracy. Accuracy 1 is the highest level. It obtains ±0.050 in. for the 2000 Unimate robot and ±0.080 in. for the 4000 Unimate robot. If this high accuracy level is used in program steps where it is not required, cycle time for these steps will be unnecessarily long. This is of particular importance where the motion from one point to another is composed of several steps, as, for example, when the robot moves around obstacles. Instead of moving directly from point A to point B, it may have to move from A to A' to B and make a 90° turn at A'. Frequently, point A' can be approached with much less accuracy than point B. Hence, a change to accuracy levels 2 and 3 is provided in programming these robots.

Accuracy 2 is an intermediate level of accuracy. Here the arm will not come to a complete stop between program steps, but will bypass point A' at a slight radius and smooth over into the next step. The accuracies in this case are adjustable within the range of 0.05 to 5 in. for rotary motions, 0.05 to 2.50 in. for up-or-down motions, and 0.05 to 1.25 in. for all other motions.

Accuracy 3 operates the same way, but the accuracy ranges are adjustable within about twice the limits of accuracy 2. The transition radius between two steps becomes even larger, and, in changing from one step to the next, the deceleration is correspondingly less.

Position accuracy is a generic term. It comprises repeatability and absolute position accuracy. When a robot is programmed by guiding its arm through a series of points or in a motion pattern that is recorded for subsequent automatic action by the robot, distances and positions need not be quantitatively expressed. All that is expected is that the robot can more or less exactly repeat the position in space. This is repeatability, and is quite different from absolute positioning accuracy. Repeatability of, say, 0.001 in. means merely that the robot can obtain the same position over and over again, and that the deviation from the point that it has been taught by guiding it to it, will never be more than 0.001 in.

Absolute position accuracy becomes the criterion when the robot is to be programmed by computer. It is no longer led by hand like a child to learn its working routine. Programming by computer is comparable to the child's having to walk by itself, by the commands of its own brain.

With a machine tool, programming is easier than with a robot: reference points and bases on the machine tool and workpiece are used, from which measurements are taken. The computer-programmed robot, too, requires such references or points of origin for the coordinates, but it operates in space where it is more difficult to define such a reference. An absolute value is needed that is related to the envelope within which the robot moves. The envelope may be given by a work cell that consists of two or more work stations and the interconnecting materials transport mechanisms and storage buffers. The absolute position accuracy of the robot and its end effector is the spatial relationship to these surroundings. Sensors that can see and feel and feed back their information to the control system are usually needed. Even the human finger cannot blindly find a given point. It requires positional, visual, and tactile feedback to get there. The robot may operate with any or all of them.

To state the main distinction once more: repeatability refers to one type of error, absolute positioning accuracy to another. In one case, it means doing the same thing over and over within a given tolerance. In the other, it means getting it right the first time.

Specifications of robots generally mention repeatability, but rarely do they state absolute positioning accuracy. The reason is that the latter is only important for robots that can be programmed off-line, and they are now a minority. In either case, however, the numerical indication by itself is incomplete. For example, it is imprecise to state a repeatability of ±0.040 in. There are influences such as warm-up period, payload, wear, and speed and length of motion that may influence accuracy. There is also the question of whether or not the stated repeatability refers to the sum of errors of all linear and rotational axes or only to one axis. Finally, the influence that end effector and payload may have should also be stated.

Consider the case of a robot operating in a system of rectangular coordinates. The robot is supposed to move from the point of origin to the position $x = 8.000$ in., $y = 2.000$ in., and $z = 12.000$ in. Let the stated repeatability be ±0.040 in., and let all axes be affected by the same maximum deviation.

The length of motion in a system of rectangular coordinates is given by $x^2 + y^2 + z^2$. In the ideal case, this is 14.560 in. Considering, however, an error in all three axes of +0.040 in., the movement will be 14.621. This means the resulting error will be $14.621 - 14.560 = 0.061$ in. rather than 0.040 in., which means a difference between stated and actual error of over 50%.

Robot manufacturers and/or their associations need to adopt a clear and exact definition of "repeatability." In the meantime, the lack of determination has to be accepted.

Causes of Static Errors

Dead Zone, Static Friction, and Resolution. Free play (backlash) between two gears that are part of a power transmission produces a dead zone in which no output is produced while the input changes. The increase of the input must first be sufficient to lock the gears together before the output gear begins to follow the input gear.

Initially, at zero input, the gear teeth may be just in the center of their free play. This may correspond to the position of the robot arm that coincides with the point of origin 0 in the graph of Fig. 1-2a. When the input signal calls for a certain output, there is no initial response of the output, as shown on the graph, until the input has increased to point k and taken out the free play of the gears. As the input continues increasing to m, which is equivalent to point P, the output lags behind by a fixed value e, which is the error produced by the dead zone.

The same characteristic is repeated when the signal is reversed. It has to decrease sufficiently to take the backlash out. That is, it has to move the input gear from one end where the gear teeth locked at the end of the motion, to the other end where they lock again before they can begin the reverse motion. From then on, output motion again takes place.

By the time the input has returned to 0, the output, however, is only at d. From here on out, for a given signal excursion 0-m-0, the output will

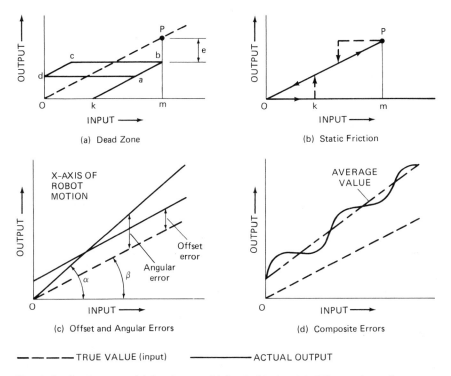

(a) Dead Zone

(b) Static Friction

(c) Offset and Angular Errors

(d) Composite Errors

— — — — TRUE VALUE (input)   ———————— ACTUAL OUTPUT

Fig. 1-2. Static errors. (a) Dead zone. (b) Static friction. (c) Offset and angular errors. (d) Composite errors.

move through *a–b–c–d*, always lagging behind by the error *e*. Although gears have been used here as an example, a dead zone can also be produced by linkages and many other components.

While errors, as illustrated in this and the following graphs, are vastly exaggerated for the purposes of illustration, even small errors can be of grave consequence. This becomes particularly important in dealing with assembly robots.

There is also the influence of the dead zone as well as that of static friction on the dynamic behavior of the robot, as will be shown later.

Static friction, also called stiction and break-loose force, is defined as the force required to initiate sliding or rolling motion between two contacting bodies. Its effect is illustrated in Fig. 1–2b. In order to initiate motion, the input signal has to increase to *k*, at which point it develops enough force to overcome the static friction of the mechanism. Once in motion, the robot arm follows the input signal to point *P*, which is equivalent to the input signal *m*. On reversal of the signal, it again has to decrease sufficiently to overcome the static friction, from which point on the output follows the input all the way to the point of origin 0.

There is actually no error, as far as the graph can show. However, when the input signal calls for a very small motion and thus cannot compensate for static friction, no motion will occur at all. Static friction may also show up in jerky motion of the robot arm, which usually is not desirable, and, as mentioned before, it can affect the dynamic behavior.

Dead zone and static friction determine the resolution sensitivity of the robot. Resolution is the smallest increment of the input that causes an output. Thus in Figs. 1–2a and b, resolution is clearly equivalent to the increase of the input from 0 to *k*.

Offset Errors, Angular Errors, and Linearity.   If the robot consistently assumes a position that differs from the programmed value by some fixed amount, the error is called offset. As shown in Fig. 1–2c, offset changes the point of origin 0 for the robot arm, and the error remains the same independent of the magnitude of the input. It can be expressed by a dimensional number. For example, assuming that the graph represents movement of the robot arm in the *X*-axis only, it can be stated that the robot has an offset in the *X*-axis of, say, 0.004 in.

The other possibility is that the error increases with the input, as shown. Here, angle α, which the actual output forms with the abscissa (the horizontal input line in the graphs), differs from angle β of the true value of the input. The angular error is defined as $\alpha - \beta$. It may be, say, 2° in the *X*-axis of the robot motion. Here, the point of origin, contrary to offset errors, is unaffected.

If actual outputs with either offset or angular errors are lower than the true value of the input, then the error is expressed with a minus sign. Thus, it may be -0.004 in. or $-2°$. Also, there may be combinations of offset and angular errors, and there may be nonlinearities that further complicate the picture. Figure 1–2d illustrates the case of composite errors, combining offset, angular error, and nonlinearity. The dash-dot line is the best straight line that can be drawn through the linearity deviation. Expressing the maximum deviation with respect to this line gives the normal or independent linearity. Offset of this dash-dot line relative to the true values of the input line as well as angularity can be readily determined from the graph.

Measurement of Static Errors

Robot movements can be checked for accuracy. A suitable evaluation system is one that is produced by Selpine, a joint venture company of Selcom and Spine AB of Gothenburg, Sweden. Their Selspot II system is marketed in North America by Selcom Selective Electronics, Inc. It is designed to help keep a robot operating efficiently and accurately to maintain productivity.

The system measures repeatability, trajectory (i.e., the accuracy of following a programmed path), and accuracy of robot movements. It includes camera, light-emitting diodes (LEDs), LED controller, main processing unit, and specialized software.

LEDs are attached to selected points on the robot being monitored. Minute changes in position are transmitted by infrared beams through an optoelectronic detector which picks up the center of the light image and generates output signals that are converted into position information and routed to the controller.

The controller converts the positions into digital or analog output data for real time recording and analysis. Movements are in three dimensions. Access to information is immediate, which permits field testing and correction.

There is also the system proposed by Ranky and Ho[1] for a test method with software in order to determine the accuracy of robots. The system is based on three-dimensional displacement vector analysis to measure positioning and orientation errors of robots. The orientation error of the end effector is measured in three dimensions. The test set-up is shown in Fig. 1–3. Together with the software, it can be used as an off-line test facility. Also, the software can be integrated into robot controllers and executed in real-time in the assembly system, where increased accuracy is an essential requirement. A firmly anchored test rig, which is not shown, is equipped with nine dial gages. They are applied to points P_1 through P_9 of the test

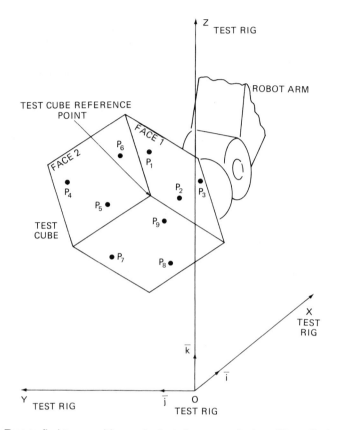

Fig. 1–3. Test to find true position and orientation error of robot. (*From Ranky and Ho, Ref. 1*)

cube which rests against the dial gage probes. The end effector is applied against the reference point of the test cube as shown. Consider face 1 of the test cube. If the indicated points P_1, P_2, and P_3 are measured, the corresponding face normal vectors can be determined. The same can be done with the other two faces. If a sufficient number of such measurements are carried out, then the analysis of these measurements indicates the statistically average error of position and orientation of the robot. This analysis is obtained from the software developed by Ranky for this purpose.

Accuracy and repeatability figures as given by manufacturers generally refer to a specific point in space which is approached at sufficiently slow speed to optimize the results. However, many manufacturing operations require the robot to move a tool in a continuous path—such as welding, deburring and painting. In these cases it is important to know the accuracy and repeatability of the robot's path under working speeds and conditions.

This made the National Bureau of Standards develop a tracking system to provide data on accuracy and repeatability of the robot in motion under actual working conditions. It uses a laser interferometer, a servo-controlled tracking mirror, a similar target mirror mounted on the robot's wrist, and a computer to control the system. Continuous measurements of the position of the target mirror are made while the robot is moving at its normal working pace. The position measurements include not only the three main axes of the motion of the robot but also the pitch and roll of the robot wrist (see Chapter 2). It is expected that such a system will determine the position of a robot in motion with an accuracy of ± 0.0005 in. within the volume of a 9.8-foot cube.

DYNAMIC ERRORS

Joints and Systems

Robot joints are rotational and translational or linear elements that provide elbow, wrist, and other movements. Each degree of freedom of the robot requires its own joint to produce the desired motion. The joints are either directly or remotely driven by actuators that convert electrical, hydraulic, or pneumatic energy into motion.

Figure 1–4 shows two types of robot arms,[2] the so-called articulated arm in (a) and the Cartesian arm in (b). In (a), all the joints are rotational and correspond to angles θ_1 to θ_6. In (b), the first three joints are sliding or prismatic joints. The corresponding linear displacements, d_1, d_2, and d_3, form a rectangular or Cartesian reference frame. In both configurations, the last three linkages, θ_4, θ_5, and θ_6, are usually part of the wrist. Actual robot systems may differ from the two types shown in Fig. 1–4 as far as position of the joints is concerned.

It has been shown that to realize an arbitrary position and orientation of a robot wrist in space, motion in at least six axes (i.e., six degrees of freedom) is needed. Each one of the degrees of freedom requires its own joint. Thus, in both configurations of Fig. 1–4, six joints and, hence, six degrees of freedom are provided. Frequently, the operation of the robot does not require the versatility of six degrees of freedom; and many robots operate with three, four, or five axes. There are other cases where additional degrees of freedom are provided. Thus, a robot with six axes may be able to move on rails and, thereby offer a seventh degree of freedom.

There is also the case in which a robot may go into a so-called condition

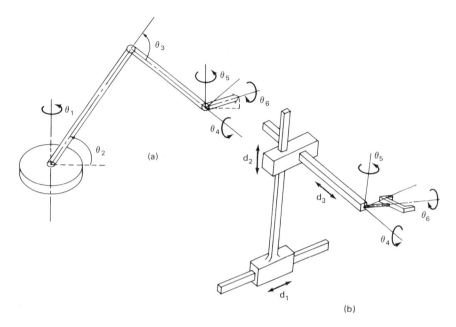

Fig. 1–4. Articulated arm (a) and Cartesian arm (b). (*From Lumelsky, Ref. 2*)

of degeneracy.[3] This means that under specific combinations of position and orientation, it loses one or more degrees of freedom.

Furthermore, in some cases, such as those shown in Fig. 1–5, the end effector has to avoid collisions or wind its way through restrictions that call for contortions of a multiplicity of additional degrees of freedom. In (a), for example, a wall prevents full extension of robot linkages to the left. This may limit the motions to the right. Only additional degrees of freedom would provide the desired flexibility in such cases.

If an actuator were to drive the joint without feedback, its movement would continue until stopped by its inherent mechanical limitations or a limit device. The latter could be a mechanical stop, a cam-operated valve, an electric limit switch, or any equivalent device. However, when this method is used, the motion of any one of the axes is limited to two positions only, although the stops can be changed manually from time to time.

To obtain a higher degree of flexibility, robots are controlled by servo-mechanisms, as illustrated by the block diagram in Fig. 1–6. The feedback element measures the position of the joint or joint actuator continuously. A potentiometer, encoder, or resolver (see Chapter 8) can be used for this purpose.

The signal from the feedback element is tied in with the input signal at

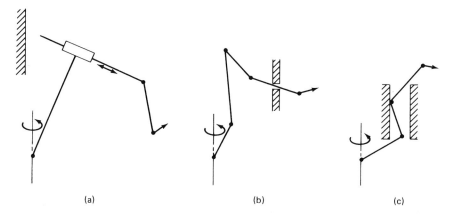

(a)   (b)   (c)

Fig. 1-5. Examples of physical constraints requiring additional degrees of freedom. (*From Lumelsky, Ref. 2*)

the summing point. The difference between input signal and feedback signal becomes the error signal that drives the joint actuator.

The feedback signal opposes the input signal; therefore feedback is considered negative and input, positive. Motion of the joint continues until the absolute magnitude of the feedback signal equals the absolute magnitude of the input signal. At this point, the error signal is zeroed out, and the motion of the joint is stopped.

The velocity of the motion may also be measured and fed back. However, this is done primarily for stability of the system rather than for positioning, and will be discussed under a later heading.

Mass

Any physical system has (a) a mass, (b) damping, and (c) compliance of some sort.

Mass is a measure of inertia. It is considered a property of matter that

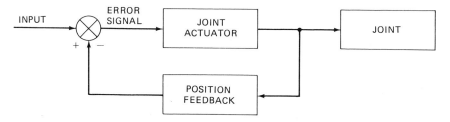

Fig. 1-6. Servo control of robot joint.

resists change of motion. Mass, in accordance with Eq. 1.1, is the ratio of the force applied to a body divided by its acceleration. Thus, the lighter an automobile is, the less its mass, and, hence, the greater its acceleration for a given driving force.

There is a difference between mass and weight. For terrestrial robots it is academic and of significance only in the treatment of equations. Weight is the resultant force of attraction on the mass of a body due to a gravitational field—an astronaut weighs less in space than on earth, but his mass is the same. On earth, gravity tries to accelerate a body as long as it is in free fall. And the accepted standard is that units of weight are based upon an acceleration by gravity of 32.1740 ft/sec² or 386 in./sec². Hence, if weight is w in pounds and m is the mass, then:

$$m = \frac{w}{386} = 0.0026 \, w \tag{1.2}$$

Moment of inertia

Inertia is the inherent property of bodies to resist any change in their state. A measure of this property is the moment of inertia. Since the two are so closely related, the terms "moment of inertia" and "inertia" are often used interchangeably.

Equation 1.1 describes the linear acceleration of a mass. For a mass that is being rotated, an analogous equation applies, which says that:

$$T = J\alpha \tag{1.3}$$

where T is the torque, which is the distance from the axis of rotation to the point where the force is applied multiplied by the force exerted at this point; J is the moment of inertia, which is the sum of the products of every mass particle of a body multiplied by the square of its respective distance from the axis of rotation; and α is the angular acceleration in rad/sec², where the concept of radians will be discussed later. If J is expressed in lb-in.-sec², then T is in lb-in.

Friction and Damping

A body in motion may encounter two types of friction: Coulomb friction and viscous friction. Static friction, discussed above, resists the onset of motion, not motion once it takes place. Hence, it is not considered in this discussion on dynamics.

Coulomb friction is also called dry friction or sliding friction. It is caused by minute irregularities of one surface engaging those of the surface over which it slides. It is diminished by lubrication.

Coulomb friction depends upon the force with which the two surfaces are pressed together, but is nearly independent of the area of the surfaces in contact and of their relative speed.

Viscous friction, on the other hand, is due to the viscosity of a fluid or lubricant. Through lubrication, Coulomb friction becomes mostly viscous friction and is considered as such. Viscous friction is generally considered a force that is proportional to velocity, even though at higher speeds the force may increase with the square or some higher power of speed.

In dynamic systems, damping (i.e., checking of vibration or oscillation) is generally associated with viscous friction and, hence, is proportional to speed. Damping is essential to prevent a system from instability that could otherwise produce excessive oscillations of the end effector of a robot.

Stiffness and Compliance

Compliance is the property of a body to yield within its elastic limits under the influence of a force. When the force is removed, the body will return to its original shape. The inverse of compliance is stiffness. The stiffer a body is, the less compliant it is.

Consider a coil spring subjected to a stretching force. The spring will expand a distance x under the influence of the force F:

$$F = kx \qquad (1.4)$$

where k is the stiffness coefficient or spring constant.

The spring in being expanded stores potential energy. If the spring is part of a mechanical system, it restores this potential energy to the system when the force is removed.

In automatic control of systems the term "compliance" is often preferred to "stiffness."

Compliance appears in many forms. For example, compliance is the ability of shafts and similar structures to yield within their elastic limits and then to spring back when released. This can readily occur with actuator shafts of robot joints. Compliance is also the deflection of the end effector of a robot when a load is applied. Wherever energy is absorbed and then released again, there is compliance.

In hydraulic actuators, valves, and connecting lines, compliance is due to the slight compressibility of the hydraulic fluid and possibly to the elastic expansion of connecting hoses. Though this compliance is minute, it may become a factor in systems that work with high pressures and/or extended connecting lines and where a high response speed is required.

In pneumatically actuated robots, because of the compressibility of air, compliance obviously becomes an even more critical factor.

Time Response

Response of a robot joint (or any other controlled system) is not instantaneous; delays of different kinds occur between input and output. The time response is a specific characteristic of any system.

The reason for a time delay or lag is found in Newton's second law, $F = ma$, as expressed in Eq. 1.1. This equation contains a, the acceleration, which is a *gradual* increase in velocity. Any mass to which a force is applied is gradually accelerated and does not instantaneously assume a velocity. (It is illustrated in everyday life by one's stepping on a car's gas pedal. The car—fortunately—does not assume its new speed immediately, but accelerates gradually toward it.)

The shape of the acceleration or response curve depends on the mass and on the force applied to move it. Since the mass of the moving parts of a robot is not a monolithic block but many interconnected components, there are other factors involved, such as the above-mentioned frictions and compliances.

In order to compare time responses of different systems, a standard pattern of input is required. Several such patterns exist, one of the most frequently used being the step input. It is a practically instantaneous change of the input signal at a time that for the sake of convenience is considered the zero base for the subsequent response of the system.

Single-Capacity Systems

Damping is introduced by any device or characteristic that produces a counterforce (or, in the case of electrical networks, a countervoltage) that increases as a function of the rate of change of the input.

There are many mechanical systems that can be considered a combination of a damper and a spring (or compliance), as shown in Fig. 1–7a. The mass involved here is obviously small enough to be neglected. The initial force exerted by the spring is assumed to be zero. Let a displacement x occur at a given time $t = 0$. This is a step input, fast enough to be considered instantaneous. The effect is a compression of the spring (Fig. 1–7b), which in turn starts the motion of the plunger. The plunger of the dashpot would move with a velocity that, for a given damping fluid and orifice size, is a

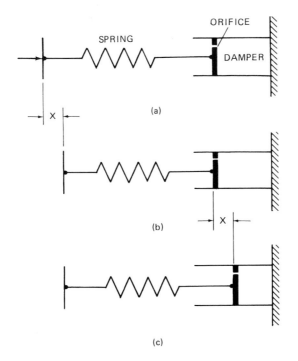

Fig. 1-7. Single-capacity dashpot-and-spring system.

function of the force exerted by the spring. This force diminishes gradually until it reaches its initial zero condition with the plunger in a new position (Fig. 1-7c). The system is now again in equilibrium.

The situation is the equivalent of an electric circuit, as shown in Fig. 1-8. The combination of resistance R and capacitance C has the same effect in the electrical network as the spring and damper in the mechanical system of Fig. 1-7. When a step change of voltage V_1 is applied, the resultant current change is absorbed by the capacitance in the same manner as the step input x was absorbed by the compression spring. Consequently, at time

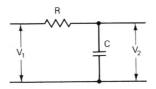

Fig. 1-8. Single-capacity electric network.

0 there will be no voltage output V_2. As the capacitor charge rises, voltage V_2 will gradually increase, in the same manner as was described for the plunger motion.

Plotting the increase of V_2 vs. time, as is done in Fig. 1–9, yields a curve of the same shape as would be produced by the plunger motion. Here, the output (voltage V_2) rises gradually to the level of the step input (voltage V_1), which is assumed to be 2 V. In the case of the dashpot–spring system, it would have been the gradual approach of the displacement x_2 to the step input x_1.

A straight line or tangent may be drawn at any point of the curve. This tangent has the same direction that the curve has at this point. It indicates the rate of change that takes place at that point. Thus, at point A the tangent is a line that leads from A to B.

Constructing a rectangular triangle ABC shows that the distance AC in this particular case is equal to 4 sec. Repeating the same at point A' (i.e., drawing the tangent $A'B'$ and the rectangular triangle $A'B'C'$), shows that the distance $A'C'$ is again 4 sec. In fact, no matter what point on the graph is chosen for this procedure, the horizontal of the triangle will always have the length of the same 4 sec. This is called the *time constant* of this RC network. The time constant depends solely on the magnitudes of R and C. In the case of the dashpot–spring system, it would be a function of the spring constant and the damping constant.

It can also be seen (though only very approximately on a simple graph) that V_2, the output voltage, reaches 1.264 V, or 63.2% of its final value, 4 sec after the step input V_1 is applied.

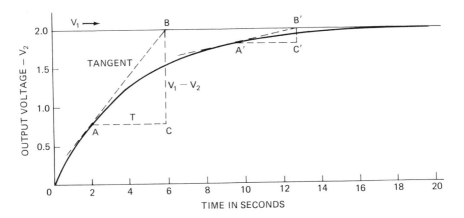

Fig. 1–9. Response of single-capacity electric network.

Hence, the time constant can be defined as the time required in a single-capacity system to complete 63.2% of the total change of output after a step input is applied.

Since it is difficult to determine a time constant and its corresponding output value accurately on a graph, mathematical methods are generally used, as will be described in the following.

The slope of the tangents drawn in Fig. 1-9 reads in mathematical terms dV_2/dt, a differential expression that indicates the change of V_2 per unit time, assuming that only an extremely small change at a given point is involved. Since the tangent at point A is only concerned with the slope at this very point, a very small change indeed is involved, and the expression dV_2/dt represents this change relative to time very well.

The standard expression for the current i that flows into an electric capacitor contains the same differential:

$$i = C \frac{dV_2}{dt} \tag{1.5}$$

where C is the capacitance.

Current that flows through the resistance is given by Ohm's law, which says that:

$$i = \frac{V_1 - V_2}{R} \tag{1.6}$$

Substituting Eq. 1.6 in Eq. 1.5 gives:

$$C \frac{dV_2}{dt} = \frac{V_1 - V_2}{R} \tag{1.7}$$

which can be written:

$$\frac{dV_2}{dt} = \frac{V_1 - V_2}{RC} \tag{1.8}$$

This equation says that the rate of change of the output voltage at any particular point is proportional to the difference between input and output voltages.

It was stated above that the slope of the tangent in Fig. 1-9 is dV_2/dt. It is also the ratio of the vertical BC in Fig. 1-9, which is $(V_1 - V_2)$ at that particular time, to the horizontal AC, which is the corresponding time constant T. This may be expressed by:

$$\frac{dV_2}{dt} = \frac{V_1 - V_2}{T} \qquad (1.9)$$

Comparison between Eqs. 1.9 and 1.8 shows that:

$$T = RC \qquad (1.10)$$

This means that the time constant is equal to the product of resistance and capacitance. If the resistance is expressed in megohms ($M\Omega$) and the capacitance in microfarads (μF), then the time constant is in seconds.

It is beyond the scope of this book to go through the steps necessary to convert Eq. 1.9 into the following:

$$V_2 = V_1 (1 - e^{-t/T}) \qquad (1.11)$$

This is, however, directly derived from Eq. 1.9. Here, e is the basis of natural logarithms. Most, if not all, engineering pocket calculators include the key e^x. It is only necessary to enter the value $x = -t/T$ in order to read off the value $e^{-t/T}$.

Consider a network with the time constant $T = 4$ sec and a step input of $V_1 = 2$ V. These are the values used in Fig. 1-9. The term $e^{-t/T}$ in Eq. 1.11 now becomes $e^{-t/4}$. On this basis and with the help of a pocket calculator, the values of Table 1-1 are obtained.

Comparison with Fig. 1-9 shows that within the approximations of a graph, the values for t and V_2 of Table 1-1 coincide with corresponding points of the curve. In fact, the curve was constructed by calculating a sufficient number of points by the above method.

It is in the nature of Eq. 1.11 that no matter how long the time t is, $e^{-t/T}$ will approach but never equal zero, and hence V_2 will never be equal to V_1. Even the solution of e^{-100} is 3.72×10^{-44}. But this, of course, is such an

Table 1-1. Solutions of Eq. 1.11
for $V_1 = 2$ V and $T = 4$ sec.

t	$e^{-t/4}$	$1 - e^{-t/4}$	V_2
2	0.607	0.393	0.796
6	0.223	0.777	1.554
10	0.082	0.918	1.836

extremely small number that for all practical purposes V_2 has become V_1 long before this. The fact is mentioned here because this "asymptotic" approach prevents a precise statement for the time it takes V_2 to reach V_1. A more or less artificial basis for comparison of response time of different single-capacity systems is therefore required—and this basis is the time constant.

Series-Connected Lags

The single-capacity system, which is described above, contains one single element that can absorb energy from a system and return it to the system. This is the case with the spring (or compliance) in Fig. 1–7, as well as with the capacitor in Fig. 1–8. These elements, together with the resistor in one case and a dashpot in the other, produce a lag that is expressed in the time constant. In either case, a single capacity is involved. This is actually rarely the case, even though a simplification of this sort may often be permissible.

Consider a robot with all its joints. Each one represents a certain compliance and hence a certain lag. One can consider this robot a chain of lags connected in series. The electrical equivalent is shown in Fig. 1–10. To separate one lag clearly from the next, an isolation amplifier is introduced. If additional series-connected lags were involved, the network could be expanded correspondingly.

There are now two time constants involved: $T_1 = R_1C_1$ and $T_2 = R_2C_2$. They lead to two variations of Eq. 1.9, which are:

$$\frac{dV_2}{dt} = \frac{V_1 - V_2}{T_1} \tag{1.12}$$

and:

$$\frac{dV_3}{dt} = \frac{V_2 - V_3}{T_2} \tag{1.13}$$

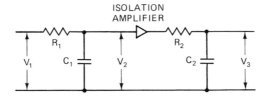

Fig. 1–10. Series-connected lags.

and also to the corresponding expressions:

$$V_2 = V_1 (1 - e^{-t/T_1}) \tag{1.14}$$

and:

$$V_3 = V_2 (1 - e^{-t/T_2}) \tag{1.15}$$

The latter may be written:

$$V_2 = \frac{V_3}{1 - e^{-t/T_2}} \tag{1.16}$$

and substitution of Eq. 1.16 into Eq. 1.14 leads, after rearranging, to:

$$V_3 = V_1 (1 - e^{-t/T_1}) (1 - e^{-t/T_2}) \tag{1.17}$$

Figure 1-11 plots response curves of V_3 to a step input V_1 on the basis of Eq. 1.17. It also shows the corresponding response of a single-capacity system for comparison. Thus, a single-capacity system with $T = 1$ sec is compared with a series-connected system with $T_1 = 1$ sec and $T_2 = 1$ sec. Similarly, a single-capacity system with $T = 4$ sec is compared with a series-connected system with $T_1 = 1$ sec and $T_2 = 4$ sec.

It can be seen that the time for the output to reach the value of the input is not too much influenced by the second capacity. It is important, however, that the maximum rate of rise is no longer at the beginning of the slope but develops gradually. The larger the difference between T_1 and T_2 is, the slower the initial response. It follows that with multiple lags in series, as occurs with robots, the initial response can be undesirably slow, unless countermeasures, as will be described, are used.

Dead Time

Dead time is a delay before the output even begins to respond. It can be caused by backlash due to loosely meshed gears or poorly fitted linkages, overlaps in hydraulic valves, and so on. These causes are equivalent to the previously mentioned dead zones, and may be considered static rather than dynamic phenomena. However, they show up in the dynamic behavior of the system.

If the motion of a robot joint lags behind the input, then the feedback signal is likewise delayed. Thus the joint has already moved beyond the

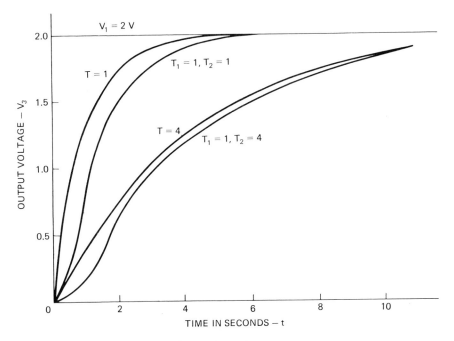

Fig. 1–11. Comparison of response curves.

intended position before the feedback signal zeroes out the input signal (see Fig. 1–6).

Because of the dead time, however, the joint continues to move until it receives a signal to stop. At this point, it has already overshot the desired position. The feedback signals this situation, and the error signal reverses to correct the position, but unfortunately the dead time persists, and this time it produces an undershooting of the desired position. Thus, the system will oscillate around the desired position because of dead time.

Systems with Sinusoidal Response

In the systems with single-capacity and series-connected lags, considered above, the influence of mass was neglected. Adding mass leads to the problem of instability. Consider the system depicted in Fig. 1–12. Here, a mass is added to the dashpot-and-spring system. Energy is now being stored in two elements: the spring and the mass. The following forces are involved:

- The spring exerts a force, which is the product of spring constant k and displacement x.

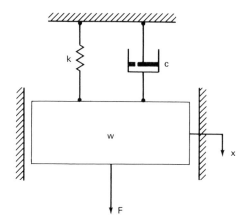

Fig. 1-12. System with mass, spring, and damper.

- The dashpot exerts a force, which is equal to the product of the damping constant c and the velocity x/t.
- The mass exerts a force, which, in accordance with Eq. 1.1, is equal to the product of the mass and its acceleration.

If the spring constant is small enough to be neglected, the system behaves like a single-capacity system. If the damping constant is small or negligible, the system will move up and down at a frequency that is described by a sinusoidal wave pattern. The visualization of such sine-wave motion requires a short detour:

A circle describes an angle of 360°. Instead of expressing the angle in degrees, it may be expressed in radians. In this case, 360° equals 2π radians, so one radian equals 57.3°. In the mathematics of automatic control, it is quite customary to use radians rather than degrees.

The sine value of any angle, whether in degrees or radians, can be read from practically any engineering pocket calculator. The angles are entered on the horizontal axis of Fig. 1-13, and their corresponding sine values on the vertical axis. By connecting the points of a number of such values, a sine wave can be plotted.

There are many motions that follow sinusoidal wave-shapes. Examples are sound waves, light waves, pendulums, piston rods, underdamped robot arms, and the system depicted in Fig. 1-12.

Any periodic motion consists of a repetition of sine curves. Figure 1-13 shows the part of the motion that is being repeated: a sine wave of 360°. It is the smallest unit of the periodic motion, and is called a cycle. If 20 such cycles are repeated in 1 sec, the frequency of the sinusoidal motion

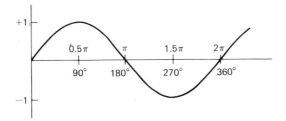

Fig. 1–13. Sine curve.

is 20 cycles/sec or 20 Hertz (Hz). Expressed in radians, it would be $20/2\pi = 3.18$ rad/sec. The duration of each cycle is its period. The greatest absolute value of the sine wave excursion is its amplitude.

INSTABILITY OF DYNAMIC SYSTEMS

Dynamic systems tend toward instability. For the practicing engineer, this instability is often enough "the curse of the dynamic system" in general and of the robot in particular. It can be prevented by damping as already stated on page 13. But because instability, or rather its inverse—stability— is of such importance, it is treated in greater detail in the following.

Damping Ratios and Oscillations

The effect of damping is that it gradually reduces the amplitude of the sinusoidal motion. There are three conditions that can occur in a dynamic system:

1. The system is overdamped; that is, it will not oscillate. This condition occurs when the damping action is greater than the tendency of the spring to sustain oscillations. It corresponds to a damping ratio, ζ, of more than one ($\zeta > 1$).
2. The system is critically damped. This is the limiting condition before oscillation sets in. It occurs when the above mentioned forces are equal, and corresponds to a damping ratio that equals unity ($\zeta = 1$).
3. The system is underdamped; that is, it will oscillate with decreasing amplitude. This condition occurs when the damping is less than the tendency of the spring to sustain oscillations. It corresponds to a damping ratio of less than unity ($\zeta < 1$).

The damping ratio is the ratio of the amount of damping to the damping that would be required to produce critical damping. Figure 1–14 shows the responses to various degrees of damping. In most but not all cases, including a robot arm, it is desirable to move at maximum speed without overshoot. This is the critically damped condition where the damping ratio is unity.

Oscillations without damping occur at the natural frequency of the system. The natural frequency ω_n is given by the equation:

$$\omega_n = \sqrt{\frac{k}{m}} \tag{1.18}$$

By applying Eq. 1.2, this can be written $\omega_n = 19.6\sqrt{k/w}$, where ω_n is the natural frequency in rad/sec, k is the coefficient of stiffness (the inverse of compliance) in lb/in., and w is the weight of the body in Fig. 1–12 in pounds.

Where damping is present, the frequency of oscillations changes slightly. It is then given by:

$$\omega = \omega_n \sqrt{1 - \zeta^2} \tag{1.19}$$

where ζ is the damping ratio.

It is convenient to plot response curves of such systems in dimensionless numbers, as done in Fig. 1–14. There, P_o/P_i is the ratio of output to input. It does not matter what dimensions are used for either, as long as they are the same for both. The same goes for $\omega_n t$ used for the horizontal axis of the graph, where t is the time for which the value is calculated. The quantity

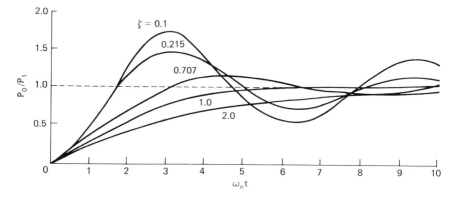

Fig. 1–14. Response curves of dynamic systems with different damping ratios.

$\omega_n t$ is dimensionless, because ω_n is expressed in rad/sec and t in seconds. The seconds cancel, and the radians are not considered a dimension.

To calculate points on the curves of Fig. 1-14, the equation

$$\frac{P_o}{P_i} = \left(\frac{e^{-\zeta \omega_n t}}{1 - \zeta^2} \right) \cos \left(\omega_n t \sqrt{1 - \zeta^2} + \cos^{-1} \sqrt{1 - \zeta^2} \right) \quad (1.20)$$

is used. Note that cos and $\cos^{-1}$ in this equation refer to angles expressed in radians.

Consider, for example, a system with a damping ratio of $\zeta = 0.1$ and a natural frequency of $\omega_n = 5$ rad/sec. To obtain P_o/P_i at time $t = 0.6$ sec, calculations may be arranged as follows:

$A = \omega_n t = 3$
$B = \zeta \omega_n t = 0.3$
$C = e^{-B} = 0.741$ (as read from a pocket calculator)
$D = \sqrt{1 - \zeta^2} = 0.995$
$E = C/D = 0.745$
$F = A \times D = 2.985$
$G = \cos^{-1} D = 0.1$ (as read from a pocket calculator)
$H = F + G = 3.085$
$J = \cos H = -0.998$ (as read from a pocket calculator)

and, thus:

$$P_o/P_i = 1 - (E \times J) = 1.744$$

Equation 1.20 is valid only for sinusoidal responses, which means for damping ratios of less than unity. For critical damping where $\zeta = 1$, the following equation applies:

$$\frac{P_o}{P_i} = 1 - e^{-\omega_n t} (1 + \omega_n t) \quad (1.21)$$

Critical damping is, as already stated, generally the most desirable, since it gives the fastest response without oscillation. To increase the response of a given system and yet stay within these conditions, it is necessary to increase its natural frequency, as is expressed by Eq. 1.21.

For example, at a given $\omega_n t = 2$, the ratio of output over input, P_o/P_i, is 0.6. If the natural frequency is doubled, then for the same time t the function $\omega_n t = 4$, and during this time the output has already risen to 0.91.

This means at the same point in time it is only 9% away from the final value instead of 40%.

Thus, the goal of robot design is to maximize the natural frequency of the arm. As has been pointed out, response—and this means natural frequency—slows considerably when one proceeds from single robot joints to the multiple joints of robot arms. In fact, Engelberger[4] has pointed out that if each joint of a robot arm has a natural frequency of 8 rad/sec, then the overall natural frequency of an arm with six joints in series is only 2.7 rad/sec.

The requirement of critical damping should not be a dogma. There are cases, including robot arms, where some over-shoot may be tolerated, since the benefit of faster response outweighs the occurrence of some transient oscillations.

Sometimes a band of $\pm 2\%$ around the final steady-state output is stipulated. The time required to reach this band is then referred to as settling time.

There are also systems that are inherently unstable but can be made stable by feedback. This condition is described in the following paragraphs.

Gain and Velocity Feedback

Work is the product of displacement and the component of the force in the direction of the displacement. If the force is constant and in the same direction as the displacement, then:

$$W = Fx \tag{1.22}$$

where W is the work, F the force, and x the displacement.

Power is work per unit time, or:

$$P = \frac{W}{t} = \frac{Fx}{t} \tag{1.23}$$

Since x/t is the velocity of the displacement, then—for a constant velocity, v—it is also true that:

$$P_m = Fv \tag{1.24}$$

The subscript m is used to indicate *mechanical* power.

The input to an electrical actuator that drives the joint of a robot is an electrical signal. Electrical power, P_e, is expressed by the product of voltage E and current I, or:

$$P_e = EI \qquad (1.25)$$

The output power of a robot joint is the product of force times velocity, which equals mechanical power, P_m. The input to the actuator that produces this output is usually an electric signal of a certain current and voltage, the product of which is electrical power, P_e. Thus, we have a relation of output to input that can be expressed by P_m/P_e.

If the joint has to supply a given force for the handling of a load by the robot, its velocity can be increased by increasing the current or voltage of the input (which one depends on the design of the specific actuator). The larger P_e, the larger is the velocity and, hence, P_m. The ratio of P_m/P_e is the gain of the system. As the gain increases, the tendency to instability also increases.

There are means of counteracting this. One way is by velocity feedback, to produce what is called derivative or rate action. It produces a corrective action superimposed over the usual position feedback. Its effect is the maintaining of the stability of the system even when the gain is increased. The block diagram of Fig. 1–15 shows schematically how derivative action is obtained by a secondary feedback loop. The signal that is transmitted through this loop is proportional to the *velocity* of the joint motion. Since it reduces the error signal as a function of output velocity, it has the same effect as viscous damping.

Velocity feedback can be accomplished by a tachometer attached to the axis of the joint. The signal is subtracted in a summing point from the error signal so that the signal to the joint actuator is reduced—or damped—as a function of the velocity of the output.

A system that would otherwise be unstable because of its high gain can thus be stabilized by the addition of derivative action.

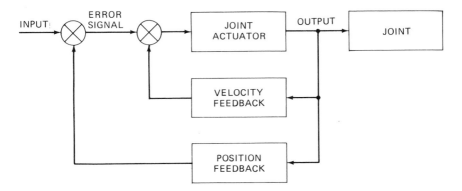

Fig. 1–15. Closed-loop actuator with derivative action.

Reducing Weight and Compliance

Another way to improve the dynamic behavior of a robot is, of course, to reduce the weight of all moving parts and provide maximum stiffness. Composite materials seem to offer a viable way to do this. An example is the graphite/epoxy robot arm made by Advanced Composite Products, Inc. for the Apprentice robot of Unimation Inc. The robot, shown in Fig. 1-16, is designed for arc welding. The arm, which carries the wrist and welding torch on its lower end, is 53.5 in. long. It used to be made of a steel tube; now a composite material is used that consists of pitch graphite fibers, bound by epoxy and wound over a mandrel and cured. The fibers run in several orientations to provide bending stiffness and hoop strength. Ad-

Fig. 1-16. "Apprentice" welding robot. (*Courtesy of Unimation Inc.*)

ditional glass fibers wound over the outside protect against corrosion between graphite and mating aluminum hardware (thus eliminating chrome plating).

The composite arm weighs 65% less than the steel arm, but the composite structure increased the stiffness by 48%. This change made it possible to increase the transfer speed of the arm from 2 in./sec to 3 in./sec—an improvement of 50%—without affecting the system's stability.

Another significant weight reduction is possible by using rare-earth magnets, such as samarium-cobalt, in the dc servomotors of robots, thus making high-speed, low-weight actuators possible. A typical application is described below.

Acceleration and Deceleration

Acceleration and deceleration are important damping factors for any motion. However, and particularly where short motions are involved, they can become an essential part of the total time elapsed in moving from one position to another. Consequently, acceleration and deceleration must take as little time as possible. Two examples of how this has been accomplished in actual models are given below.

ASEA. The IRB 1000 pendulum robot from ASEA is suspended on gimbals and swings around its center of gravity as shown in Fig. 1–17. The company reports that this design contributes to an acceleration that is said to be 50% higher than that of conventional jointed-arm robots. The six-axis robot can be mounted on an elevated track for lateral travel along a seventh axis, as shown in the illustration.

Designed specifically for assembly, the gimbal-type mechanism provides three rotary motions plus a vertical stroke. Axes A and B are pendulum motions at 90° to each other, each through a swing of $\pm 30°$ and a maximum speed of 86°/sec. Up-and-down motion along the Z-axis is over a working range of 13.8 in. at a maximum speed of 39 in./sec. The arm rotates $\pm 185°$ through the D axis. The wrist bends and swivels through $\pm 120°$ and $\pm 355°$, respectively. The IRB_{1000} can handle a payload of 6.6 lb: Repeatability is ± 0.004 in.

Digital Automation Corporation. The other example is MACH 1, a robot produced by Digital Automation Corporation, which is designed primarily for semiconductor manufacturing. Consequently movements are short, and special attention must be given to obtaining fast accelerations and decelerations. The robot is small and compact, with its load capacity of 2 lb ample for the work considered. It moves within a limited work en-

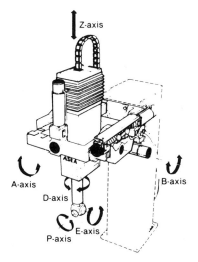

Fig. 1–17. IRB 1000 pendulum robot. (*Courtesy of ASEA Robotics*)

velope that measures $3 \times 7.5 \times 10$ in. Three degrees of freedom are provided: 2 in. in the X-axis, 3.3 in. in the Y-axis, and a 360° rotation about the Z-axis. The precision is high with a resolution of 0.0001 in. and a repeatability of 0.00002 in.

Maximum attention has been given to reducing acceleration and deceleration time for the short motion involved. This has been done in part by using the above-mentioned samarium-cobalt dc servomotor, which reaches its speed of 6000 rpm in less than one full turn. Its torque is low, but by using speed-reducing methods, the speed is reduced and the torque increased, yet acceleration and deceleration remain high. Efficiency of the drive train is 90%, which means friction losses are extremely low, a condition that further contributes to improved dynamic behavior.

Another means for improved dynamic characteristics of the MACH 1 is its "adaptive dynamics control" (ADC), which controls settling time and other dynamic phenomena by means of microcomputers. As a result of ADC, the robot has a settling time of less than 5 msec, while more conservative systems may have settling times of 100 to 200 msec.

In addition to the extra time required for deceleration, a "fine" motion frequently must be added. For example, with assembly robots fine motion takes place when the parts to be assembled are already in contact, particularly when remote-center compliance (see Chapter 3) is used, which provides final adjustments to slight offsets between parts to be assembled. The time required for this fine motion may be minute, yet may affect the overall assembly time.

Further Developments in Dynamic Analysis

The mathematics used in this chapter have been those of classic control theory. They have developed into computer-oriented procedures for robot dynamics. Such procedures are described, for example, by R. L. Huston and F. A. Kelly[5] of the University of Cincinnati, who proposed the use of Kane's equation. In 1962, T. R. Kane[6] had introduced a procedure for obtaining the governing dynamic equations of mechanical systems. Using these equations, Huston and Kelly included link flexibility and joint compliance as well as the accommodation of constraint equations and the solution of closed-loop problems with multi-arm robots. The mathematics of these procedures are too specialized to be included here. The interested reader may refer to the sources listed.

Steven Dubowsky[7] of the Massachusetts Institute of Technology investigated what problems have to be solved to improve high-speed performance of robots. According to Dubowsky, these problems are:

- Nonlinear dynamic effects that result from inertia effects of manipulated objects.
- Nonlinear effects caused by change of manipulator configuration as it moves through the work envelope.
- Structural flexibility.
- Discrete time effects resulting from interactions among microcomputer, resonances of system structure, and nonlinear dynamic effects. They can influence position accuracy and system stability.
- Suitable computer methods for solving the dynamic behavior of the robot.

Dubowsky points out that, although mathematical methods are available to describe the dynamic behavior of the three-dimensional rigid robot structure, they result in equations that so far have no closed-form analytical solution.

He also believes that the flexibility of robots is a problem that requires particular attention. At first glance, it appears best to build robots as stiff as possible. However, use of such structures may well increase the mass in motion and worsen the situation instead of help it. The possibility of using composite materials that are light and stiff is one alternative. Another one is to take advantage of the flexibility, just as in a fishing rod flexibility is used to increase the casting reach. However, this concept may have a long way to go before its practical implementation can be achieved.

However, the concept of a compliant robotic manipulator was investigated at Carnegie-Mellon University. J. J. Mendelson and J. R. Rinderle,[8] who conducted the research, observed that "the current practice of designing manipulators using stiff and heavy structural links has resulted in robots which are heavy, slow, and difficult to control."

The investigation had two separate goals: (1) to achieve fast, accurate positioning; and (2) to isolate the positioning of the arm from dynamic disturbances at the end effector, so that when the robot was used for such tasks as drilling or deburring, the tool forces would not excite vibrations in the arm.

To obtain fast, accurate positioning, a damped structure consisting of an aluminum beam was coated with a viscoelastic material as a means of improving the dynamic response. The aluminum has a high mechanical impedance, storing over 95% of the strain energy during vibration, and dissipating only a small fraction of it; whereas the viscoelastic material can dissipate a large fraction of the energy it stores. Overall energy dissipation was inhibited by this combination, which prevented transfer of energy from the aluminum to the viscoelastic material. The conclusion was that composite materials may have both sufficient strength and damping; however, the facilities needed to fabricate and test composite structures were not available.

To isolate dynamic disturbances, the end effector was suspended between springs and thus isolated from the wrist and the rest of the robot. The spring-damper suspension could be tuned to obtain optimum tool performance, and thus interaction with the arm and arm controller could be minimized.

Measurement of Dynamic Errors

Robots can develop deterioration of their dynamic performance, due to loosening of mechanical components, errors in their electronic or other controls, and so on. Periodic checking of performance to identify and correct such causes of error is desirable. One approach is the dynamic performance calibration test[9] developed at the National Bureau of Standards (NBS). The purpose of the test is to detect defects by testing the control and drive system of individual robot joints.

NBS used a Puma 600 robot from Unimation Inc. The diagram of Fig. 1–18 shows the robot together with the instrumentation used for the test. The arm was extended horizontally. A noise generator in which the output is a random function of time is connected to a low-pass filter, which cuts off frequencies above 50 Hz. The remaining noise signal was injected into

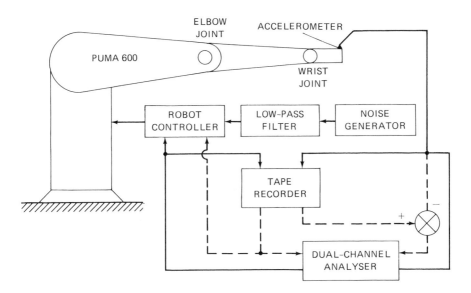

Fig. 1–18. Puma 600 robot with instrumentation for NBS test.

the robot controller to make the wrist joint oscillate at a small amplitude. The oscillations were picked up by the accelerometer.

The oscillations of the wrist were compared with the noise signal frequencies. Both were recorded by the tape recorder and correlated by a dual-channel analyzer.

Two different tests were performed, a baseline test and a performance test. The difference between the two is that during baseline tests, the oscillations are produced by the noise generator output after it passes through the low-pass filter. The solid lines in Fig. 1-18 represent the connections for the baseline test. During the performance test, the excitation signal is produced by the tape recorder from the prerecorded baseline test signal. Here, the connections are represented by the dashed lines. By subtracting the response of the performance test from the baseline test response, the response error is determined.

The test was applied to two different types of simulated defects. One was a defect in the operation of the wrist-joint controller; the other, a loosening of the end effector. In both cases, the system was capable of detecting the defects.

SUMMARY

The most important equations of this chapter can be summarized as follows:

$$F = ma$$
$$m = 0.0026 \, w$$
$$T = J\alpha$$
$$F = kx$$
$$\frac{dV_2}{dt} = \frac{V_1 - V_2}{RC}$$
$$T = RC$$
$$V_2 = V_1 \left(1 - e^{-t/T}\right)$$
$$V_3 = V_1 \left(1 - e^{-t/T_1}\right)\left(1 - e^{-t/T_2}\right)$$
$$\omega = \omega_n \sqrt{1 - \zeta^2}$$
$$\frac{P_o}{P_i} = \frac{e^{-\zeta\omega_n t}}{1 - \zeta^2} \cos\left(\omega_n t \sqrt{1 - \zeta^2} + \cos^{-1}\sqrt{1 - \zeta^2}\right)$$
$$\frac{P_o}{P_i} = 1 - e^{-\omega_n t}\left(1 + \omega_n t\right) \quad \text{(when } \zeta = 1\text{)}$$
$$P_m = Fv \quad \text{(mechanical power)}$$
$$P_e = EI \quad \text{(electrical power)}$$

where:

a = Acceleration, in./sec^2
C = Electrical capacitance, μF
dV/dt = Differential, expressing change of voltage per unit time, V/sec
E = Voltage, V
e = Base of natural logarithm
F = Force, lb
I = Current, A
J = Moment of inertia, lb in. sec^2
k = Spring constant, lb/in.
m = Mass, lb sec^2/in.
P_o/P_i = Ratio of power "out" to power "in"
R = Electrical resistance, ohms (Ω)
T = Torque, lb in. or time constant, sec
t = Time, sec
v = Velocity, in./sec
w = Weight, lb
x = Linear displacement, in.
α = Angular acceleration, rad/sec^2
ζ = Damping ratio
ω = Frequency, rad/sec
ω_n = Natural frequency, rad/sec

REFERENCES

1. Ranky, P. G., and Ho, C. Y. *Robot Modeling, Control and Applications with Software.* Kempston, Bedford, United Kingdom: IFS (Publications), 1985.
2. Lumelsky, V. J. Control of robot motion. *Proc. SPIE Convention on Robotics and Robot Systems*, San Diego, CA., August, 1983, pp. 84–96.
3. Paul, R. P. *Robot Manipulators.* Cambridge, MA: MIT Press, 1981.
4. Engelberger, Joseph F. *Robotics in Practice.* New York: AMACOM, A Division of American Management Associations. 1980.
5. Huston, R. L., and Kelly, F. A. New approaches in robot dynamics and control. ASME Second International Computer Engineering Conference, San Diego, CA, 1982.
6. Kane, T. R. Dynamics of nonholonomic systems. *Journal of Applied Mechanics*, Vol. 28, 1961, pp. 574–578.
7. Dubowsky, S. Dynamics for manipulation: Areas of future research. Workshop on Research Needed to Advance the State of Knowledge in Robotics, Newport, RI, 1980. National Science Foundation, Report 80 02 23.
8. Mendelson, J. J., and Rinderle, J. R. Design of a compliant robotic manipulator. Conference on Robotics Research: The Next Five Years and Beyond. Paper MS84–495. Dearborn, MI: Society of Manufacturing Engineers, 1984.
9. Dagalakis, N. G. Analysis of robot performance operation. *Proc. 13th International Symposium on Industrial Robots*, Vol. 1, Chicago, 1983.

Chapter 2
CONFIGURATIONS

DEGREES OF FREEDOM

Outwardly, the moving parts of a robot consist of a series of segments, which either are jointed or slide relative to one another for the purpose of reaching, grasping, and moving objects. Consider the $T^3$566 robot of Cincinnati Milacron, illustrated in Fig. 2–1. Its construction is such that it tries

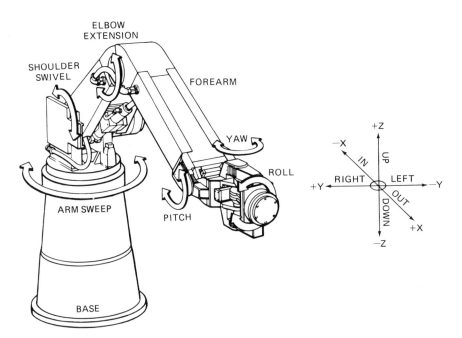

Fig. 2–1. Degrees of freedom of T^3robot. (*Courtesy of Cincinnati Milacron*)

to simulate the movements of a human arm. There are the shoulder, the elbow, and the wrist. The end is a flat plate, the so-called tool mounting plate (also known as the wrist mounting surface, faceplate, socket, etc.), to which the end effector can be attached. End effectors are described in Chapter 3.

In spite of the fact that arm sweep, shoulder swivel, and elbow extension are produced by rotary motions, the change of position they produce in the position of the wrist is defined by translational motions along three mutually perpendicular axes, X, Y, and Z. In addition, the wrist itself will rotate through three axes, exhibiting the following motions: pitch, which is rotation about a horizontal axis; yaw, which is rotation about a vertical axis; and roll, which is rotation about a third axis that is perpendicular to both the pitch and yaw axes. The six motions that can thus be produced, three translational and three rotational, represent the six degrees of freedom of this robot.

It is a principle of physics[1] that "a rigid body moving freely through space has six degrees of freedom, three translations and three rotations." This is true of the center of mass of a molecule as well as of the end effector of a robot, as long as it is provided with the necessary joints.

A degree of freedom has been defined[2] as "one of a limited number of ways in which a point or a body may move or in which a dynamic system may change, each way being expressed by an independent variable and all required to be specified if the physical state of the body or system is to be completely defined."

In programming a robot, it is the tool center point (TCP) whose position is usually of interest. This is a point on the end effector such as the center of the gripper jaws or the end of the welding gun, or a point in the center of a paint spray as it emerges from the nozzle. The position of the TCP in space can be described by the degrees of freedom of the robot. The data defining the TCP are entered by the programmer and thus become the reference point for all motions of the robot.

Basically each robot has three *main* axes or degrees of freedom, and they determine its maximum reach, as expressed by the robot's work envelope. In the particular case of the robot shown in Fig. 2–1, these three degrees of freedom are given by the following motions:

1. Arm sweep, a horizontal rotation of 240°.
2. Shoulder swivel to lift the arm upward. Together with the elbow extension, it determines the vertical reach of the robot, which stretches from the floor level to a height of 154 in.
3. Elbow extension, which, together with the shoulder swivel, determines the horizontal reach of the robot, which is 97 in.

The number of main axes can be extended, for example, by adding rails on which the robot can move. However, the limiting factor in such additions is usually the number of axes the control electronics of the robot can accommodate.

In addition to the main axes, the robot has *wrist* axes that determine the mobility of the end effector over and beyond the capability of the main axes of the robot arm to obtain the target position. Generally, robots have up to three wrist axes. To these can be added further degrees of freedom located in the end effector; but here, too, there is the limitation of programmability, which may determine the extent of additional axes. In Fig. 2-1, these axes are the following:

1. Yaw, as the angular displacement between left and right and vice versa. The maximum displacement is 172°.
2. Pitch, as the angular displacement up or down. The maximum displacement is 180°.
3. Roll, as the rotation of the faceplate to which the end effector is attached. The maximum roll is 270°.

Thus, the total number of degrees of freedom of the T^3566 robot is six: three main axes and three wrist axes.

While six degrees of freedom are needed to express the position and orientation of a rigid body moving freely through space, this does not necessarily mean that these six degrees of freedom are needed to *get* to a given set of positions and/or orientations. Sometimes fewer degrees of freedom are needed, sometimes more.

It is apparent that if a human arm with shoulder, elbow, and wrist were limited to six degrees of freedom, it would be quite lacking in its customary flexibility and flow of motion. The robot is clumsy by comparison to the human arm, but its clumsiness is no detriment to its operation in many industrial applications where it is programmed to move or manipulate materials, parts, tools, or specialized devices and perform a variety of tasks.

The T^3566 robot uses only rotary motions. However, motions are different with different robots; there may be sliding motions, produced by rack and pinion, ball-bearing screws, hydraulic cylinders, and so on, as well as rotary motions. In referring to specific designs, the abbreviation R for rotary and S for sliding may be used. For example, an S2R1 arm would be one with two sliding motions and one rotary motion.

Other forms of joints are possible. Thus, the robot built by Spine Robotics AB in Sweden, shown in Fig. 2-2, is made up of upper and lower sections.[3] Each section is comprised of a series of disks held together by

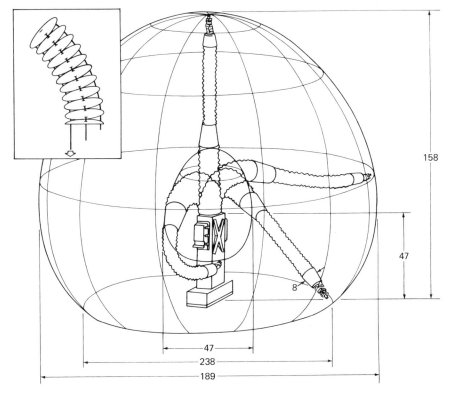

Fig. 2-2. Spine robot. (*Courtesy of Spine Robotics AB*)

two pairs of hydraulically actuated cables in tension, as shown in the insert of the illustration. Hence, there are four cables in each section, for a total of eight. Movement of the arm is effected by pulling these cables by means of hydraulic cylinders located in the base of the unit. Computerized coordination of tension in all eight cables provides four degrees of freedom. Wrist action provides three more degrees of freedom. An independent set of low-tension cables is used for position feedback.

Repeatability is about 0.08 in. Load capacity is 6.6 lb. The arm can work to a height of 13 ft and stretch laterally in any direction up to 8 ft. This means it has a larger work envelope than most other robots.

Many robots have fewer than six degrees of freedom. The term "limited-degrees-of-freedom robot" is frequently used for such robots, although the word "limited" is somewhat misleading. An assembly robot, for example, may not require more than three or four axes, and there is nothing limited about it within the work for which it is intended. Adding unnecessary degrees of freedom can be very uneconomical.

WORK ENVELOPE, PAYLOAD, AND COORDINATES

The work envelope is the operational space of the robot; it determines its reach in all directions. It can be defined as all the points in space that can be touched by the end of the robot's wrist; that is, by the faceplate to which the end effector is to be attached. It should be noted here that the robot will actually be able to reach beyond its work envelope once the end effector is attached. This extra reach must be taken into account when placement of equipment and safety of personnel are considered.

The situation is somewhat similar with the payload of a robot. Payload, also called load or handling capacity, is the maximum weight that can be safely handled by the robot in normal and continuous operation. However, payload includes the weight of the end effector. Thus, the weight of the object that the robot is actually capable of carrying is the load capacity minus the weight of the end effector. A rule of thumb that is sometimes used is that the weight of the end effector is equal to 60% of the load it can carry. Thus, if the payload of a robot is 100 lb, only 40 lb would be left for the load that can be lifted, since 60 lb may go for the end effector.

Main Axes

Basically, the robot operates with three and sometimes four *main* axes that define its work envelope. In the example of Fig. 2-1, these are the arm sweep, the shoulder swivel, and the elbow extension. The axes that represent the wrist action (pitch, yaw, and roll) do not affect the work envelope.

The purpose of the robot motion is to reach a point in space. To determine the position of such a point, a system of coordinates is used. Systems that are of particular interest for robotic technology are shown in Fig. 2-3. They are:

- Cartesian (rectangular or rectilinear) coordinates
- Cylindrical (or polar) coordinates
- Spherical coordinates

Robots may be categorized according to the coordinate system that best represents their motions. However, very few actual robots describe in their work envelope an exact rectangle or cylinder or sphere. Hence, a crude approximation is required in order to ascribe a robot to one or the other of these categories.

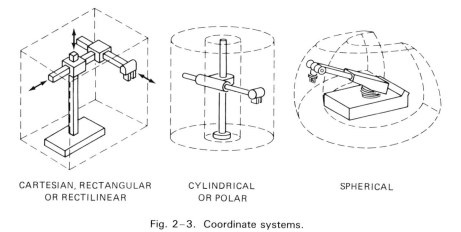

CARTESIAN, RECTANGULAR   CYLINDRICAL   SPHERICAL
OR RECTILINEAR   OR POLAR

Fig. 2–3. Coordinate systems.

Coordinate Systems for Work Envelope and Programming

Another factor to consider in categorizing robot motion is that a robot may operate, for instance, within a cylindrical envelope but be programmed in terms of rectangular coordinates. Conversion takes place in the electronics of the robot. Some robots permit programming in two or three different systems, since for the user it may be simpler to think in linear dimensions instead of angles.

For the computer, it is necessary to use some sort of computational algorithm with a series of equations to transform one sort of coordinates into another and to solve the problems in a finite number of steps. Thus, a six-axis robot controller could be required to perform a number of almost simultaneous transformation computations including 30 multiplications and divisions, 15 additions and subtractions, three square root extractions, seven squaring operations, six cosine and sine calculations, and six arctangent computations.[4]

In the Cincinnati Milacron robotic controller, the kinematic equations of motion are solved in the computer. The joint angles for each axis are determined, and are used as reference signals to each of the corresponding servo controllers. The computations are performed every 15 msec, and the computer outputs update the information to the servo actuators every 3 msec.

To have the flexibility to program robots in various coordinate systems, we must, when talking about coordinate systems, distinguish between work envelope and programming coordinates. The following discussion is primarily concerned with the work envelope.

Cartesian Coordinates

With Cartesian coordinates, space is considered a rectangular solid, as shown in Fig. 2-4. (In Figs. 2-4, 2-6, and 2-9, the three-dimensional geometrical solids may be considered to be clear plastic in which points and their position can be located and clearly seen.)

There are three axes in Fig. 2-4: X, Y, and Z. A coordinate system of two axes would have only X and Y, with X the horizontal axis and Y the vertical axis. In a three-dimensional system, the vertical axis is Z, as shown in the diagram.

Axes X and Y are horizontal, but at right angles to each other; the third axis, Z, is vertical and at right angles to both X and Y. All three axes are joined in a common point of origin O, starting from which they may extend to any length. They may also extend in opposite (negative) directions from the point of origin O. Dimensions are then marked with a minus sign.

Consider a point P that is located somewhere within the rectangular solid. To locate it, one could say: Move 4 in. in the direction of axis X, then 3 in. in the direction of axis Z, and then 3 in. in the direction of axis Y. In Cartesian coordinates, this means $x = 4$ in., $y = 3$ in., and $z = 3$ in.

While the axes as such are designated by capital letters, the length of any movement parallel to these axes is given in small letters. Any point within the rectangular solid is defined by the magnitude of the linear dimensions x, y, and z.

An example of a robot with a work envelope readily described by Cartesian coordinates is the robot shown in Fig. 2-5. It is a small robot intended

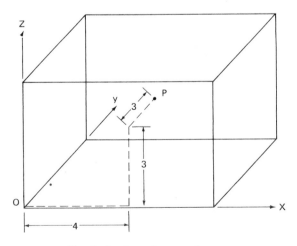

Fig. 2-4. Cartesian coordinates.

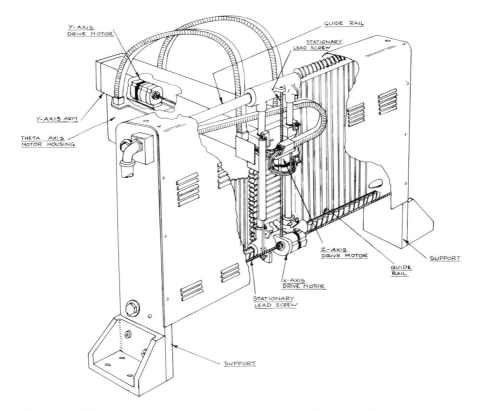

Fig. 2-5. Robot operating within Cartesian coordinates. (*Courtesy of Control Automation Inc.*)

to be mounted on a table for electronic assembly operations. Its producer is Control Automation, Inc.

The robot has four degrees of freedom: X, Y, and Z plus the rotary motion of the wrist, which is produced in the "4th-axis" motor housing of the illustration. As pointed out before, the work envelope and coordinates are solely determined by the main axes, not by those of the wrist. The work envelope is given by the 22 in. motion in the X-axis, the 14 in. motion in the Y-axis, and the 4.5 in. motion in the Z-axis.

Each one of these axes is equipped with its own servomotor. The robot can carry a load of 5 lb. To obtain a positioning speed of 50 in./sec and a positioning repeatability of ± 0.001 in., it was necessary to reduce mass and moment of inertia as much as possible.

Thus, to produce the motion, a Teflon-coated lead screw was used in a somewhat unconventional arrangement. Generally, the nut of the lead screw is kept stationary, and the lead screw rotated, so that the latter advances while moving through the screw. This was not considered practical in this case, because of the long travel, high slew speed, and minimal lost motion that were required. The large mass and high moment of inertia of the lead screws and their drive trains would have made it difficult to meet these requirements in an economical way.

The solution was to reverse the role of nut and lead screw and to move the nut while keeping the lead screw stationary. The nut is driven by the servomotor, which is concentric with the lead screw without touching it and moves in a linear direction with the nut. This arrangement is used for all three main axes. Thus, mass and moment of inertia are considerably reduced, and dynamic as well as cost requirements can be met.

Cylindrical Coordinates

With cylindrical coordinates, concepts change somewhat. Space is no longer considered rectangular but is a cylindrical solid, as shown in Fig. 2–6. Two of the coordinates, X and Z, are left as before. The Z-axis rises through the center of the cylinder. The Y-axis although not necessarily involved, can be imagined to be at right angles to X as well as Z.

The first dimension used to locate a point P within the cylinder is along the Z-axis, to determine point m which corresponds to the height of the point over the base. The distance from the base to point m is z, which in

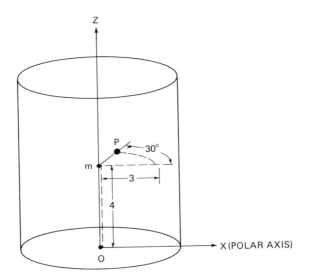

Fig. 2–6. Cylindrical coordinates.

this case equals 4 in. With a compass, a circle is drawn around m, with a radius that equals the distance to P. The circle is drawn to the X-axis, showing in this case a radius length of $x = 3$ in. The angle between the radius from m to P and the X-axis is θ, which in this case is 30°. Thus, it is possible to define any point in the cylinder by the height along axis Z, the radius along the axis X, and the value of the angle θ.

Axis X in this case must be considered an arbitrary line along the bottom of the cylinder, say at the 3 o'clock position, that serves for reference purposes. It is therefore frequently referred to as the polar axis rather than the X-axis.

It would be quite possible also to use Cartesian coordinates to determine a point in a cylindrical body. In this case, x would be the sine of the angle times the radius, and y would be the cosine of the angle times the radius. Thus, the Cartesian coordinates for the above example would be $x = 2.6$ in., $y = 1.5$ in., and $z = 4$ in.

Model M-3 of GMF Robotics Corporation is an example of a cylindrical coordinate robot. Its concept and work envelope are shown in Fig. 2–7. The M-3 is a heavy-duty robot with three main axes. What has been called X-axis in this text is designated as R-axis by GMF. Thus, it has an R-axis, a Z-axis, and the θ-axis. Two wrist axes, α and β, may be added. The pay load of this robot is rated at 264 lb. Its primary usage is heavy part handling and loading, automatic palletizing, and applications in forging and die casting operations.

The arm can move up and down along the Z-axis with a travel range of 47.2 in. and a speed of 20 in./sec. It can move in and out along the R-axis from a minimum of 43.5 in. to a maximum of 90.7 in. This means a reach of 47.2 in. The speed this motion can attain is 40 in./sec. Positions within the Z–R–plane, which forms a square of 47.2 × 47.2 in., are obtained by the rotary motion, θ, of the base. The maximum speed of the rotation is 60°/sec.

As seen in the illustration, the cylinder of the work envelope is not complete. There is (a) a hollow core with a radius of 43.5 in. and (b) a sector of 30° which is not covered by the base sweep.

The repeatability in returning to any point that has been programmed into the robot is ±0.039 in. The robot, when programmed, will calculate and execute automatically the most efficient path between two points through the three axes.

Scara Robots.   Scara stands for the Selective Compliance Assembly Robot Arm. It was developed in a joint project by several Japanese companies with the intent of obtaining an uncomplicated device operating with three or four degrees of freedom. It is primarily targeted for assembly work re-

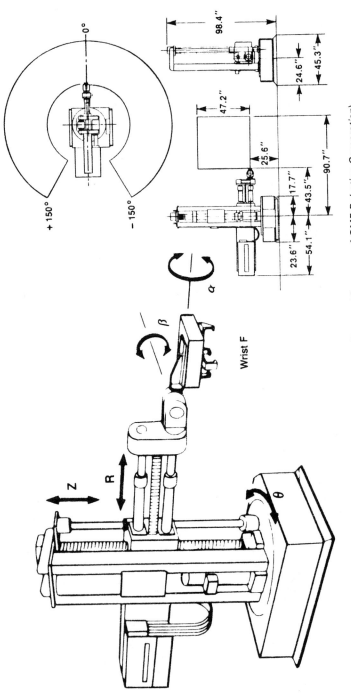

Fig. 2-7. Model M-3 robot with cylindrical coordinates. (*Courtesy of GMF Robotics Corporation*)

quiring vertical insertion of parts, and it has been claimed that scara units can be applied to more than 80% of all types of assembly operations. The concept has already been adopted by a great many producers of assembly robots.

Robots that rotate their shoulder and swivel about a horizontal axis, as in Fig. 2–1, are occasionally considered to have a problem of motion control. "Sometimes the robot must handle a large inertia on top of the long arm, and it causes vibration. The problem is severer in case of assembly robots because the cycle time must be minimum."[5] By making the joints swing about vertical instead of horizontal axes, as shown in Fig. 2–8, this problem is largely eliminated. The design emphasizes stiffness in the horizontal direction, so that changes in payload have relatively little effect on the motion properties of the robot. A third axis provides up-and-down motion. Thus, the work envelope becomes cylindrical, using two rotary motions and one sliding motion.

A particular feature of scara-type robots is their compliance. When an assembly robot has to insert a peg in a hole, close tolerances often exceed the capabilities of the robot to hit the exact position. Rather, it is necessary to "slide" the peg into the hole. The scara design simplifies this process by virtue of the two vertical rotary axes and their servomotors. It provides full rigidity in the Z-axis, but because of some "give" in the servomotor, a minor lateral motion can be provided that supports the peg in sliding into the hole.

One other virtue of the scara is its relatively simple design, which reduces manufacturing costs and maintenance.

The scara-type robot in Fig. 2–8 is the A3 assembly robot of General Electric Co., which is the U.S. equivalent of the Japanese Hitachi robot. The horizontal work envelope covers a rotation of 170° with a reach of 27.6 in. Motion in the Z-axis is 7.9 in. A fourth axis, which is a 360° rotation of the wrist, is available as an option.

High speed and accuracy are essential features in assembly operation. The A3 specifications are particularly impressive in this respect: 59 in./sec in the horizontal plane and 10 in./sec in the vertical, with a positioning repeatability of ±0.002 in. However, as pointed out before, it is necessary to take acceleration and deceleration—and not only speed—into account when calculating the productivity of this or any other robot. While the robot is primarily suited for vertical assembly work, there are also models available for horizontal work.

The A3 can be programmed for simultaneous three-axis continuous path control. It will perform vertical parts insertion, palletizing, and parts loading and unloading, as well as efficiently manipulate detachable assembly tools.

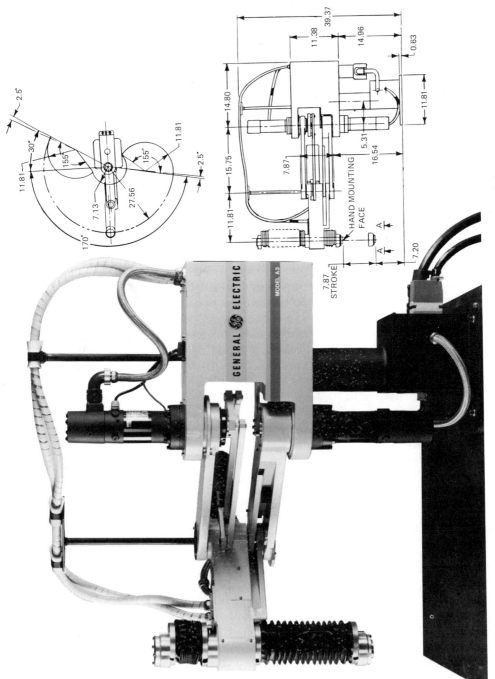

Fig. 2–8. Scara-type robot, Type A3. (*Courtesy of General Electric Co.*)

48

A portable teach box enables the operator to enter almost all program data. The control system is equipped with a 9-in. video screen to aid in programming and diagnostics. A magnetic bubble memory with a 16-bit microprocessor permits storage and execution of 508 program steps. An additional 512 steps are available as an option.

Furthermore, the robot can be equipped with visual and/or tactile sensors. This great flexibility permits the execution of a multitude of tasks with the only requirement being that the parts handled including the end effector not weigh more than 4.4 lb.

Spherical Coordinates

With spherical coordinates, spatial relationships are more difficult to visualize than with Cartesian and cylindrical coordinates. Space here is considered a sphere, as shown in Fig. 2-9.

A horizontal plane may be imagined to cut through the sphere. The center of the sphere lies on this plane, and axes X and Y also lie on it. Axis Z is vertical to the plane. The point of origin of all three axes coincides with the center of the sphere. The radius, R, pivots around the center of the sphere in any desired direction.

To locate a point within the sphere, one could say: Pivot radius R through

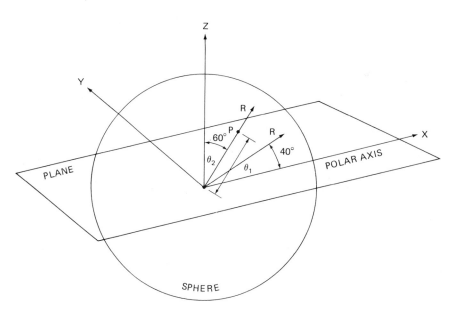

Fig. 2-9. Spherical coordinates.

an angle of 40° away from the polar axis in the plane as shown. Lift the radius vertically to the plane until it forms an angle of 60° with the Z-axis. Measure 4 in. along the radius, and the desired point P is located. In terms of spherical coordinates this means $\theta_1 = 40°$, $\theta_2 = 60°$, and $r = 4$ in.

The UNIMATE Series 4000, as produced by Unimation Inc., a Westinghouse company, is shown in Fig. 2–10. Its work envelope was shown in Fig. 2–3, as an example of spherical coordinates.

The UNIMATE Series 4000 is a heavy-duty robot suited for payloads of up to 450 lb. The vertical up-and-down rotation about the Y-axis covers an angle of 230° below the horizontal, and 28° above the horizontal. The swing around the Z-axis extends over 200°.

It is interesting to observe how programming methods can change trajectory, as is illustrated in Figs. 2–11 and 2–12. The point-to-point moves can be programmed in either a joint interpolated mode or a straight-line interpolated mode. Straight-line moves can be taught between any two points using "world" or "tool" teaching modes. World coordinates are referenced to the earth or, in more practical terms, to the shop floor. Here,

Fig. 2–10. UNIMATE Series 4000 robot with spherical coordinates. (*Courtesy of Unimation Inc.*)

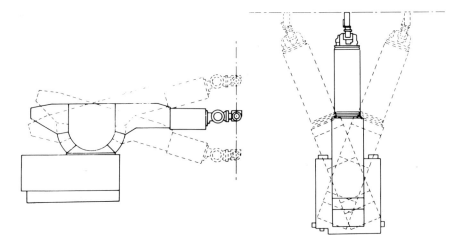

Fig. 2–11. World coordinates. (*Courtesy of Unimation Inc.*)

the arm will move along the programmed trajectory with the wrist moving simultaneously to keep the tool vertical to the trajectory. In tool coordinates, it is the tool motion and trajectory that is being programmed, and the arm will automatically and simultaneously follow the requirements set by the tool.

The IRB 90S/2 robot from ASEA Robotics Inc. is an articulated robot that offers six degrees of freedom and a payload capacity of 198 lb. It is designed for spot welding and material-handling applications that require

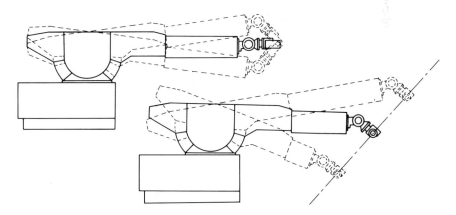

Fig. 2–12. Tool coordinates. (*Courtesy of Unimation Inc.*)

great flexibility to reach spots that are difficult to access. The robot can be programmed in three different coordinate systems. Its approximately cylindrical work envelope is shown in Fig. 2–13. The robot base can rotate through an angle of ±135°, and the arm can move from a minimum radius of about 46 in. to a maximum, without gripper, of approximately 98 in. The vertical motion is 25° down and 65° up.

Figure 2–14 represents the work envelope of another articulated robot, the M50 material-handling robot manufactured by International Robomation/Intelligence. The upper diagram shows the horizontal view of the work envelope; the lower diagram, the vertical view. In the horizontal view, the horizontal reach of the robot arm is 79.8 in. However, there is a circle in the middle that has a radius of 8.8 in. Thus the horizontal stroke is only 71 in. and differs in this respect from the horizontal reach. Note also that the circle that the gripper is capable of describing extends beyond the work envelope. In the vertical view, the same circles of 79.8 in. for the vertical reach and 71 in. for the vertical stroke apply. However, because of the column on which the robot is mounted, the circle is interrupted; and because the shoulder motion is limited to ±90° from the horizontal, after the arm has reached its vertical position the circle radius diminishes rapidly.

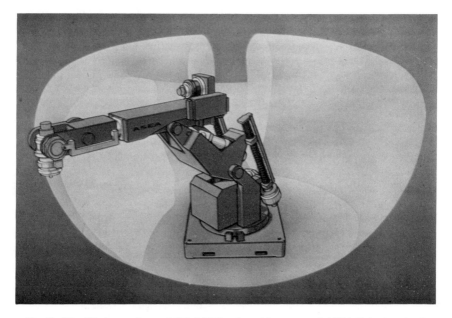

Fig. 2–13. Work envelope of IRB 90S/2 robot. (*Courtesy of ASEA Robotics, Inc.*)

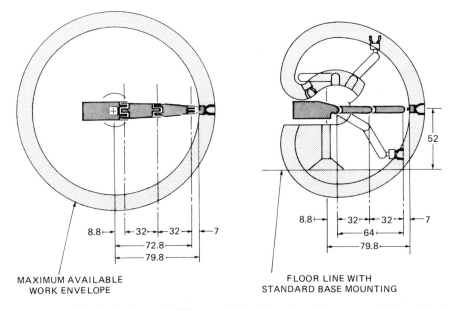

MAXIMUM AVAILABLE
WORK ENVELOPE

FLOOR LINE WITH
STANDARD BASE MOUNTING

Fig. 2-14. Work envelope of M50 material handling robot. (*Courtesy of International Robomation/Intelligence*)

BALANCED ARMS

Robot arms tend to be heavy. The actuators often have to be capable of lifting not only the payload, but also large parts of the structure of the robot. To reduce the necessary force, counterbalance systems are often used. An example is the articulated robot that was developed by Citizens Robotics Corp., shown in Fig. 2-15. Five of the six axes are belt-driven by motors that comprise the motor package mounted on the frame, which pivots about a shaft at the rear of the arm. This shaft is located behind the arm's shoulder pivot.

The weight of the motor package is used to counterbalance the weight of the arm. Raising or lowering the front end of the arm causes the motor package to swing to a new position that maintains the balanced condition. The motions are transmitted to the various joints via timing belts.

The counterbalancing gives smoother operation by eliminating changes in arm position that otherwise will occur when an external load is applied unless a very rigid, and consequently costly, structure is used. The only power the motor has to develop for this robot is the power required actually to lift the load. A model that has six degrees of freedom weighs about 125

Fig. 2–15. Counterbalance robot. (*Courtesy of Citizens Robotics Corp.*)

lb, yet is capable of lifting a load of 25 lb by using a drive motor of only 0.4 hp.

This is a low-cost robot that is ideally suited for small shops and experimental work, so dynamic response is not of primary interest. It is necessary to realize this because the dynamic response of such a counterbalanced robot is somewhat handicapped, since the counterweights increase the mass to be moved. Thus the repeatability is limited to about ±0.02 in., and the speed of each axis to 18 in./sec.

Another example of balanced arms is given by the IRB 1000 pendulum robot from ASEA, described in Chapter 1.

PARALLELOGRAM LINKAGES

The principle of the parallelogram linkage is illustrated in Fig. 2–16. Two servomotors are provided: *A* and *B*. They are located in the same horizontal line, but one drives link *E*, the other link *C*. When motor *A* is activated, it rotates link *E*, thereby pivoting link *F* about point *G*. This gives the up-and-down motion of the robot arm. On the other hand, when motor *B* is activated, it rotates link *C* about pivot *H*, thereby providing the back-and-forth movement of the robot arm. In General Electric's P50 robot, for example, servomotor *A* moves the forearm through +25° and −45°, while servomotor *B* moves it through +50° and −45°. Since link *C* has a length of 24 in., this means a horizontal reach of about 32 in.

Several robot manufacturers today use a cylindrical robot with parallelogram linkages. An example is the Model P50 of General Electric, illustrated in Fig. 2–17. Two servomotors, *A* and *B*, separately drive the *C* and *D* links of the parallelogram, respectively. Two other servomotors, one of

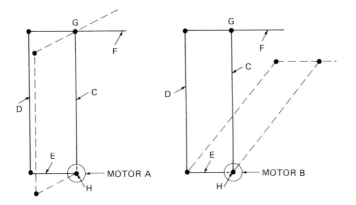

Fig. 2–16. Principle of parallelogram linkage for robots.

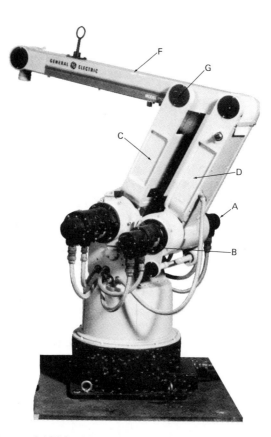

Fig. 2-17. Robot model P50 with parallelogram link mechanism. (*Courtesy of General Electric Co.*)

which is not visible in the picture, are for the bend (pitch) and twist (roll) motions of the wrist. These motions are transmitted to the wrist by roller chains. The four-joint parallelogram link mechanism for forearm and upper-arm motions permits locating the servomotors for all motions at a point where the effect of their moment of inertia on the link mechanism is zero.

Parallelograms have the advantage that they are very compact mechanisms that provide large working envelopes, so that the robot takes up a relatively small portion of the work cell. Furthermore, because of their stiffness, their dynamic characteristics are such that they permit fast and accurate control in positioning the end effector.

PANTOGRAPH LINKAGES

The pantograph linkage is similar to the parallelogram linkage. Its principle is illustrated in Fig. 2–18.

If point Q in Fig. 2–18a is fixed and point R is moved, then point P follows R. However, the contour that P follows is larger than that of R by a factor that is equal to the ratio n/m as shown for the corresponding segments in the illustration. If $m = 1$, then the magnification factor is equal to n.

On the other hand, if point R is fixed as shown in Fig. 2–18b, then motion of point Q is magnified at P by the factor $(m + n)$. And if $m = 1$, then the magnification factor is equal to $1 + n$.

This means that motion of point P can be controlled by Cartesian coordinates in the Z–X-plane when linear actuators are installed at Q or R. This is illustrated in Fig. 2–18c. The third axis would have to be supplied by a rotating base.

Hirose and Umetani[6] of the Tokyo Institute of Technology proposed a configuration that extends the pantograph principle to a robot with rectangular coordinates for all three axes. Their mechanism, which they call a Pantomec, is illustrated in Fig. 2–19. Points Q and R are free to rotate about axis Z. By moving R linearly along the Z-axis, a scaled-up vertical motion for point P is obtained. By moving point Q along the Y- or X-axis, the corresponding scaled-up motion of point P is obtained. Thus a Cartesian-coordinate robot results with the simplified programming characteristic of such configurations.

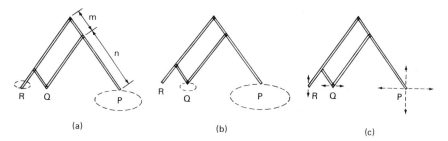

Fig. 2–18. Principle of pantograph. (*From Hirose and Umetami, Ref. 6*)

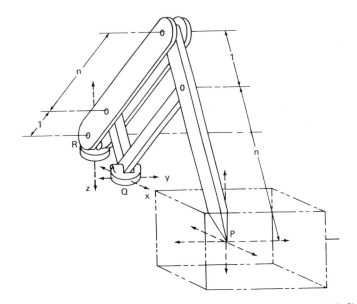

Fig. 2-19. Pantomec mechanism. (*From Hirose and Umetami, Ref. 6*)

WRISTS

Design Concepts

Wrists have up to three degrees of freedom. They give the wrist, and with it the end effector, different angular orientations. Each degree of freedom requires its own actuator. With direct drives, the actuator is mounted directly to the corresponding joint on the wrist. With remote drives, the actuation is transmitted by roller chains or other means. As will be shown further below, not all wrist configurations are easily actuated by remote drive.

Wrist compactness, weight, mechanical stiffness, dexterity, speed, and spatial orientation volume of payload (SOV) are important characteristics of a wrist. Requirements such as cost-effectiveness, ruggedness, and reliability are implicit.

Compactness is needed because interference with other parts must be minimized. Weight is important because it limits the load capacity of the robot. Also, by being at the outer end of the arm, the moment of inertia of the weight is of particular significance, since it influences the dynamic response of the robot—as, of course, its mechanical stiffness does also.

Dexterity or orientation flexibility is needed. However, for economical reasons, it should not exceed the particular usage of the robot. Applications such as arc welding and spray coating require high dexterity to permit the intricate motions that accompany this type of work.

The speed of wrist motion is critical for the productivity of the robot. If, for example, the wrist has three degrees of freedom, it makes a significant difference whether the motions occur successively or simultaneously. It affects not only the speed of the wrist motion but also design and complexity of its control system. Precision is equally important; it is obvious that no robot can be better than the precision of its wrist.

The end effector attached to the wrist, as mentioned before, is part of the payload. The SOV (spatial orientation volume) is a measure of the volume of space that the payload can sweep through while the arm remains in a fixed position. A small SOV implies wrist compactness. It can best be obtained by using remote drives and intersecting axes of the wrist action.

Kinematics differentiate between roll axes and bend axes. Two members that are linked by a common shaft and can rotate relative to each other without mutual obstruction for any angular displacement, whether it be for less or more than 360°, are considered roll axes. On the other hand, if the two members cannot pass each other, then the members are said to rotate about a bend axis. Wrist designs use roll as well as bend axes.

The degrees of freedom of the wrist correspond to the orientations that can be obtained by the payload. It may be a roll, pitch, or yaw orientation. The roll orientation always requires a roll axis. Pitch and yaw orientation may be obtained by either roll or bend axes.

Roll is rotation of the payload around a longitudinal axis, pitch moves the payload up and down about an axis at right angles to the longitudinal axis, and yaw is the rotation from left to right about an axis that is vertical to both pitch and roll.

Ted Stackhouse[7] has pointed out that each wrist axis may be considered a building block, and that the sequence in which these building blocks are arranged gives the wrist different capabilities. For example, if a wrist has two degrees of freedom and a roll axis precedes a bend axis, then the payload is limited to pitching and yawing. However, if a bend axis precedes a roll axis, then the payload may either pitch or yaw as well as roll.

Stackhouse arranged the most commonly used two- and three-axis configurations into six basic groups, as shown in Table 2-1.

These six groups of wrist configurations are shown in Fig. 2–20. In these illustrations *B* stands for bend, *P* for pitch, *R* for roll, and *Y* for yaw. The description of these groups that follows is based on evaluations by Stackhouse.

Groups 1 and 2 are wrists with two degrees of freedom. The tasks they can handle are obviously more limited than those that are possible with wrists with three degrees of freedom, but they can have a very small SOV

Table 2–1. Common wrist configurations.

GROUP	WRIST AXES	WRIST ORIENTATION OF PAYLOAD
1	Roll–roll	Pitch and yaw or pitch and roll
2	Bend–roll	Pitch and roll
3	Bend–bend–roll	Pitch, yaw, and roll
4	Bend–roll–roll	Pitch, yaw, and roll
5	Roll–bend–roll	Pitch, yaw, and roll
6	Roll–roll–roll	Pitch, yaw, and roll

of the payload when remote drives are used. The cost of a wrist with two degrees of freedom is obviously less than that of one with three degrees of freedom.

Group 3 shows two commonly used configurations that make use of two succeeding bend axes followed by one roll axis (BBR). In Group 3A, the first bend axis provides pitching orientation, the second yawing orientation. This design is difficult to drive remotely. However, if direct drives are used, it may yield the most cost-effective design for a wrist with three degrees of

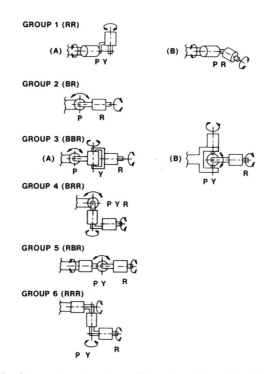

Fig. 2–20. Schematics of wrist configurations. (*From Stackhouse, Ref. 7*)

freedom. Considerations in chosing this design are the lack of compactness and the comparatively large SOV of the payload. The design is well suited to medium to large payload capacities and is capable of giving large angular orientations.

The alternate arrangement, shown in Group 3B, uses two intersecting bend axes for pitching and yawing motions in a configuration that resembles a universal joint. This is the most restrictive of the three-axis designs. It is hard to package compactly and effectively with remote drives. Even though all three axes intersect, the overall wrist size and length needed for the roll axis result in a fairly large SOV of the payload.

Group 4 differs from 3, since one bend axis is used for pitching, and mounted perpendicularly on it is a roll axis for the yawing motion. A further roll axis is mounted perpendicularly on the end of the first roll axis and provides the roll orientation of the payload. As with group 3, the wrist is difficult to construct in a compact design. However, it is suitable for a medium to large payload and has a very good orientation capability.

Group 5 uses a roll axis for pitch and a bend axis for yaw, plus another roll axis for roll orientation. All three axes intersect and can be compactly packaged for remote drives. At the same time, the SOV of the payload for this design can be smaller than for any other wrist with three degrees of freedom.

Finally, in group 6, all three wrist orientations are obtained by way of roll axes. This combination yields more orientation flexibility for the payload than any other wrist configuration, since all axes allow continuous rotation. Remote drives are easily applied. Otherwise, the wrists of groups 5 and 6 are equally compact and well suited for small and medium payload handling.

The Three-Roll Wrist

This concept, which was represented by group 6 in Fig. 2–20, has been used by Cincinnati Milacron in their T^3726, T^3746 and T^3776 robots, which have load carrying capacities of 14, 70, and 150 lb, respectively. The actual appearance of the wrist is shown in Fig. 2–21. It depicts the T^3726 with the three-roll wrist carrying a twin pair of grippers.

Figure 2–22 illustrates the three roll axes: $R1$, $R2$, and $R3$. They all intersect at point A. By coordinating the rotation of axes $R1$ and $R2$, it is possible to rotate $R3$ about point A without changing the plane in which it moves—a desirable feature in many application.

Suppose that point B on axis $R3$ is to be moved to B' but must stay within the plane of the paper at all times. Rotating only axis $R2$ would move B closer to B', but it would travel out of the plane of the paper. However,

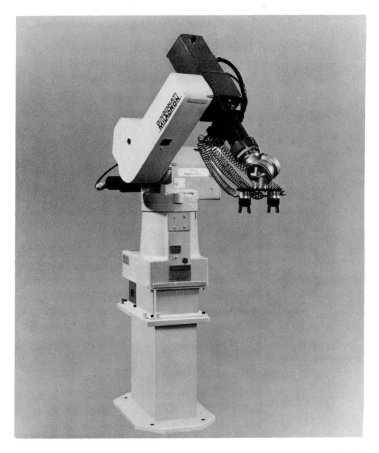

Fig. 2-21. T³726 robot with three-roll wrist. (*Courtesy of Cincinnati Milacron*)

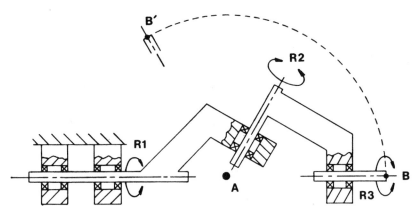

Fig. 2-22. Three-roll wrist linkage. (*From Stackhouse, Ref. 7*)

using axis $R1$ simultaneously corrects this tendency. By proper programming of the three motions, the motion of point B to B' becomes an arc, as shown by the dotted line. Furthermore, the arc BB' can be rotated around axis $R1$ to form a practically spherical surface about point A. Thus, the wrist is able to position point B to any point within this partial spherical surface. This by no means limits the orientation flexibility. Many other geometrical combinations can be constructed with the same wrist.

Figure 2–23 shows the dense packaging that can be obtained by rearranging the components of Fig. 2–22.

Electrohydraulic Wrist

Moog Inc. developed an electrohydraulic wrist of high velocity and precision without seal maintenance problems. It is shown in Fig. 2–24. The design is that of a yaw axis, followed by a pitch axis that is succeeded by a roll axis. Two types of wrist motion are available. In Type "A," the yaw actuator provides bend motion about an axis parallel to the base. In Type "B," the yaw actuator provides a rotary motion about an axis perpendicular to the base. A bend motion about the pitch axis and a rotary motion about the roll axis are common to both wrist types.

All three degrees of freedom are provided with their own single-vane rotary actuator with integral servovalve. The structural brackets house the feedback transducers for closed-loop control of all motions. There are 270° of motion in the roll axis and 220° of motion in the yaw as well as in the pitch axis. Depending on load, repeatability can be better than 0.005 in. Unloaded rotational speed is greater than 1000°/sec at the rated flow of 1 gpm.

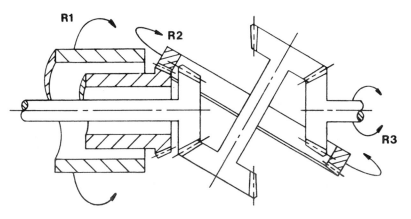

Fig. 2–23. Three-roll wrist drive mechanism. (*From Stackhouse, Ref. 7*)

Fig. 2-24. Electrohydraulic wrist for payloads of up to 7 lb. (*Courtesy of Moog Inc.*)

These actuators have about half the weight of a direct drive electric motor of equal power. Weight and volume are further reduced by routing electric and fluid power lines through channels cast into the wrist elements. The function of the electrohydraulic servovalves is described on page 164.

Depending on type, the wrist weighs 17 to 35 lb, with rated payloads of 4 to 28 lb.

Where larger load capacity is required, the Hyd-ro-wrist may be used. It is a product of Bird-Johnson Co. Figure 2-25 shows this three-axis robotic module with one Hyd-ro-ac rotary vane actuator for each of the axes. The Hyd-ro-ac actuator is further described on page 176. Pitch and yaw axes can move through 180°, while the roll axis covers 280°. As in the preceding example, the Hyd-ro-wrist is designed to accommodate one servovalve and feedback transducer for each axis of motion. There are four different models available. They are for load capacities of 50, 120, 200, and 300 lb,

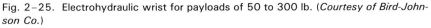

Fig. 2-25. Electrohydraulic wrist for payloads of 50 to 300 lb. (*Courtesy of Bird-Johnson Co.*)

respectively. The model for 120 lb, for example, has a maximum flow rating of 5 gpm. Since the actuator has a displacement of 0.021 in.³ per degree of travel, it has a maximum rotational speed of 917°/sec.

Wrist Mounting Surface

The wrist mounting surface, tool mounting plate, faceplate, or socket is the part of the wrist that comes into contact with the end effector, either directly or by way of a remote-center compliance device (see Chapter 3). It must be able not only to support the weight of the end effector but also to withstand inertial forces that result from rapid acceleration. If the end effector is a gripper, then the wrist socket also must support the weight of any object the gripper happens to carry.

End effectors are often highly specialized and not produced by the robot manufacturer. Therefore, standardization of the mounting surface is highly desirable. The following requirements for such standardization were laid down in a workshop on robot interfaces sponsored by the National Bureau of Standards and the Air Force:[8]

1. Mechanical fastening: The wrist mounting surface, to which the end effector is attached, must withstand rated static and dynamic loads.
2. Facilities must be provided for locating the end effector in the same attitude each time it is mounted.
3. Actuators, powered tools, and sensors on the end effector must be provided with standardized connections for electrical, electronic, hydraulic, and pneumatic lines.
4. Means must be provided to remove and replace end effectors. In some cases, the robot must be capable of changing tools itself.

Work on this standardization program has so far yielded no significant results within the industry.

It is desirable to provide wrist sockets with breakaway protection for the end effector. Excessive force on the end effector should cause the following two actions to occur:[9] (1) the mechanical connection should become compliant, and (2) sensor(s) in the wrist socket should signal the work station control computer that an unexpected condition has occurred. The computer should trigger immediate action to prevent further damage.

Among the different designs for this purpose are the following:

- Mechanical fuses, such as shear pins that break or thin-walled tubes that buckle under excessive stress.
- Detents that are held rigidly in position with respect to one another by spring-loaded detent mechanisms.
- Preloaded springs that hold in contact one or more pairs of structural elements. A force or torque acting in any direction on the end effector will tend to separate one or more of these pairs of elements in order to provide the breakaway action.

REFERENCES

1. Den Hartog, J. P. *Mechanical Vibrations*. New York: McGraw-Hill Book Co., 1956.
2. Smith, B. M., et al. A glossary of terms for robotics—revised. SME Paper MS83-914. Dearborn, MI: Society of Manufacturing Engineers, 1983.
3. Drozda, T. J. The spine robot . . , the verdict's yet to come. *Manufacturing Engineering*, September, 1984, pp. 110–112.
4. Morris, H. M. Robotic control systems; more than simply collections of servo loops. *Control Engineering*, May, 1984, pp. 74–79.

5. Makino, G., and Furuya, N. Motion control of a jointed arm robot utilizing a microcomputer. 11th International Symposium on Industrial Robots, Tokyo, Japan, 1981.
6. Hirose, S., and Umetani, Y. A Cartesian coordinates manipulator with articulated structure. 11th International Symposium on Industrial Robots, Tokyo, Japan, 1981.
7. Stackhouse, T. A new concept in robot wrist flexibility. Ninth International Symposium on Industrial Robots, Washington, DC, 1979.
8. Wheatley, T. E., et al. (eds.). *Proc. NBS/Air Force ICAM Workshop on Robot Interfaces*, Wright-Patterson Air Force Base, OH, 1980.
9. ICAM—Robotics Application Guide. Technical Report AFWAL-TR-80-4042, Vol. II. General Dynamics Corporation, Fort Worth, TX, 1980.

Chapter 3
MECHANICAL PARTS

END EFFECTORS

The purpose of the end effector is to perform the robot's operating functions, such as gripping, positioning, drilling, spray painting, welding, gluing, grinding, and so on. The end effector can be a gripper or other device to pick up and hold an object. It can also be a tool, a laser or welding gun, a paint sprayer, or other device that is attached to the wrist of the robot. The tool may be either held by a gripper or mounted directly on the wrist mounting surface. The end effector may also be a device for sensing, measuring, and so on. In other words, the end effector is the "business end" of the robot.

Figure 3-1 gives an example of the variety of grippers that are available. They represent models that are offered by Pentel Co., Ltd. In addition, a broad range of tools are offered that can be attached directly to the wrist. They include screwdrivers, soldering irons, wire-lead cutters, and so forth.

Grippers pick up and hold the object by (a) supporting action, (b) mechanical (clamping) action, (c) vacuum cups, or (d) electromagnets, which are discussed in that order in the following pages.

Support Grippers

Support grippers usually resemble a palmar prehension: a flat hand that reaches under the material or object and lifts it. In some other cases, however, they may use a suitable hook to lift an object. Figure 3-2 illustrates some support grippers made by Air Technical Industries. They are:

A. A shovel to pick up parts from a flat surface or from the floor.
B. A push–pull ram that is adaptable to the robot wrist to push, press, or retrieve workpieces, parts, or containers from hard-to-reach places, such as furnaces or otherwise unaccessible areas.

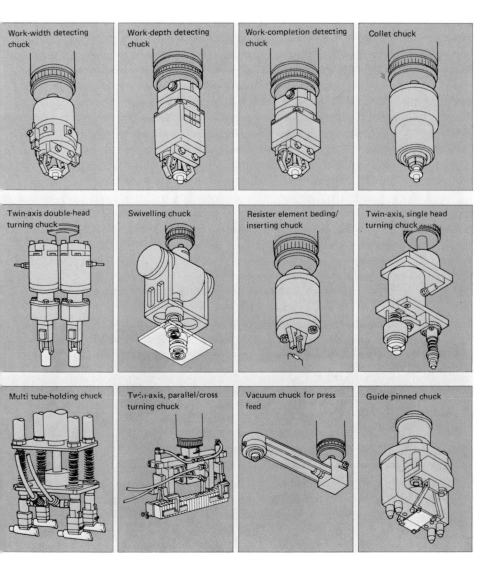

Fig. 3-1. Grippers. (*Courtesy of Pentel Co., Ltd.*)

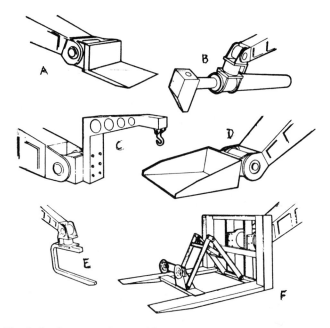

Fig. 3-2. Support grippers. (*Courtesy of Air Technical Industries*)

C. A short gooseneck boom attachment with a hook at the end to lift objects.
D. A scoop attachment for handling powdered material or granules.
E. A C-ram to lift coils or rolls.
F. A fork attachment for handling boxes, crates, or pallets that can load and unload itself by means of a push–pull attachment with a vacuum cup.

A specialized support gripper is illustrated in Fig. 3-3. This is a WMI palletizing robot, Model AC-1, from WMI Robot Systems, Inc. Its payload is 320 lb. It can be equipped with a number of palletizing end effectors. An example is shown in the illustration. The programmed operation is permanently recorded on a standard cassette. Subsequent palletizing of the same product can be accommodated by reading the tape into the robot's control system.

Mechanical Grippers

The two most frequent types of mechanical grippers are jaw types and finger types. In either case, they exert a clamping force on an object, either

Fig. 3-3. Palletizing robot. (*Courtesy of WMI Robot Systems, Inc.*)

by expanding within the inside of the object (sometimes by means of an inflatable bladder) or by closing on it from the outside. Limitation of clamping force can be desirable for safety reasons. On the other hand, there must be enough friction to prevent the picked-up object from accidentally pulling out of the gripper. The friction coefficient, softness, chemical resistance, and so on, must be suited to the application. Wearability is an important factor. Compliant padding is sometimes necessary to protect the finish of the workpiece.

Jaw-Type Grippers. Michael S. Konstantinov[1] of the Central Laboratory for Robots and Manipulators in Bulgaria has proposed a classification system for robot grippers. He tabulated 12 characteristic clamping grippers, as shown in Fig. 3-4. The upper three diagrams give basic types in which the first and second bring the fingers together at an angle, while the third one keeps the fingers parallel. These first three examples have the

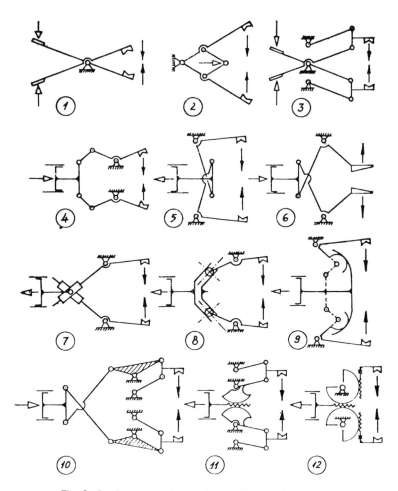

Fig. 3-4. Jaw-type grippers. (*From Konstantinov, Ref. 1*)

advantage of giving larger gripping forces than the other types, though the others offer lighter and often simpler construction than the first three. The first and third diagrams require that force be applied at two points, while the second and all the other diagrams require force application at one point only. The fourth through twelfth grippers are shown being actuated by pistons. Together, these examples illustrate the basic concepts of almost any jaw-type gripper in common practice. Other compilations of gripper configurations have been made by Lundstrom[2] and Engelberger.[3]

The majority of grippers are simple two- or three-finger open-and-close devices. Figure 3-5 shows a good example in grippers from Dixon Automatic Tool, Inc. These grippers are primarily intended for Dixon's AP-320

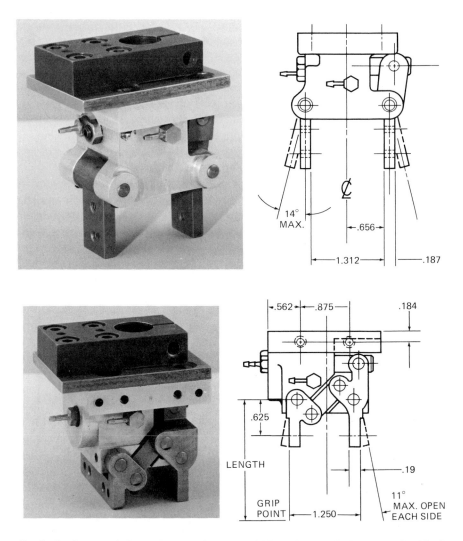

Fig. 3–5. Open-and-close grippers. (*Courtesy of Dixon Automatic Tool, Inc., Rockford, Illinois*)

Robotic Assembler. The AP-320 is designed for automated assembly. The piece part is picked up from a feed system and accurately placed with a placement accuracy within ±0.002 in.

The grippers come in two models: Model GR-400 has a repeatability of the jaw opening of ±0.003 in., while the repeatability for the precision version, Model GR-100, is ±0.001 in.

Either gripper is provided with jaw holders, so that jaws of any desired configuration can be fastened to it. Furthermore, a rotator module can be supplied. It can rotate the part through an angle of 180°.

The jaw holders of the GR-100 can open through an angle of 11° maximum, while the GR-400 can open through 14° maximum. Adjustments are provided to reduce these angles to any desired opening. To calculate the approximate clamping force, the formula $F = 0.34\, p/L$ applies. Here, F is the clamping force of the gripper at the end of the jaws, in pounds; p is the air pressure that actuates the gripper, in psi; and L is the jaw length, in inches, as shown in the illustration. Thus, if the air pressure is 100 psi and the jaw length 3 in., the clamping force is about 11.3 lb.

This does not imply that the gripper can carry a load of that much. In the first place, the maximum carrying capacity is specified at 5 lb. In the second place, in those cases where the load hangs vertically and the gripper is provided with friction pads to hold the load, the carrying capacity is given by the clamping force multiplied by the friction coefficient and divided by a safety factor. Thus, if the friction factor is 0.4 and the safety factor is 2, then the carrying capacity in the above case would be 11.3 × 0.4/2 = 2.26 lb. For other cases, such as loads that exert a lever moment on the gripper, Engelberger[3] gives some good examples.

Gripper Pads. Frequently, it is necessary to protect the workpieces from scratches and other damage that could be caused by the fingers of the end effector. In this case, the pinch surfaces of the fingers can be provided with padding. One material that is widely used for gripper pads is a 70 to 90 durometer polyurethane bonded to steel. The polyurethane can be frozen and, in this condition, easily shaped or machined to any configuration. Such a pad will last through thousands of hours of alternate compressions and releases. It has good wearability and a high coefficient of friction.

Barry Wright Corporation makes durable, nonslip elastomer gripper pads, as shown in Fig. 3-6. The elastomer is a proprietary compound by Barry Wright with a hardness of 60 ± 5 Shore D. It can operate over a temperature range of −20°F to 180°F and resists oils, most industrial cutting fluids, and corrosive elements. The elastomer may either be furnished as a pad or be vulcanized to a steel or aluminum backplate for easy attachment to the gripper. It can be cut with a band saw to the desired shape. Holes may be drilled to provide attachment.

V-Notch Grippers. If the robot has to handle round stock or cylindrical objects, friction force alone may not suffice to hold the object, and a V-notch gripper may be used. An example is shown in Fig. 3-7, depicting the IRI S1 gripper made by International Robomation/Intelligence. It can grasp workpieces of up to 2 in. diameter. However, in order to accommodate

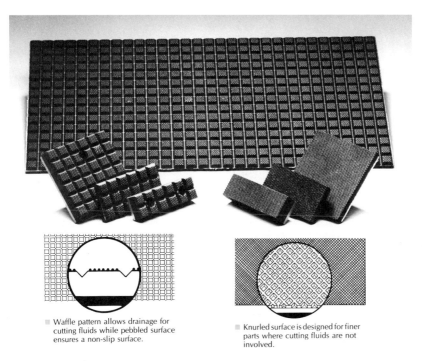

■ Waffle pattern allows drainage for cutting fluids while pebbled surface ensures a non-slip surface.

■ Knurled surface is designed for finer parts where cutting fluids are not involved.

Fig. 3-6. Gripper pads. (*Courtesy of Barry Wright Corp.*)

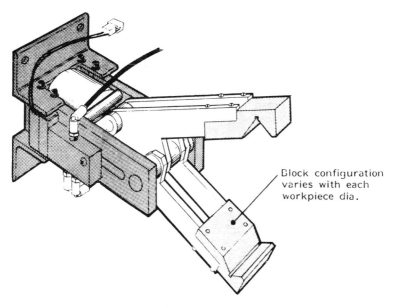

Block configuration varies with each workpiece dia.

Fig. 3-7. V-notch gripper. (*Courtesy of International Robomation/Intelligence*)

various diameters within this range, it is necessary to change the V-notch blocks, each of which is fastened by four bolts.

The gripper is operated by air pressure of 90 to 120 psi and can exert a grip force of 2000 lb when fully closed. At a pressure of 100 psi it can move its grip from open to close within 0.15 sec.

The disadvantage of this type of V-shaped gripper is the changing of blocks for different diameters. If frequent switching between gripper blocks is required, time loss can become a significant factor.

If the block is not changed for a different diameter, the gripper will clamp the workpiece in a position that differs from the normal concentric position, producing an undesirable offset. However, there are remedies for this.

One solution is the use of grippers with parallel fingers. However, parallel mechanisms are usually larger and heavier than the simple linkage gripper.

Huang Quingsen[4] of the Peking Institute of Aeronautics (China), working as a research fellow at the mechanical engineering department of Birmingham University (U.K.), developed a V-notch gripper for the concentric gripping of cylindrical components.

The gripper may be built either with two V-shaped fingers or with one V-shaped and one flat finger. The latter case is shown in Fig. 3-8. The design is such that during opening and closing of the gripper, the centers of inscribed circles enclosed by the V-shaped finger and the flat finger must remain coincident. Similarly, for the arrangement with two V-shaped fingers, the concept would require that when they open or close, the bisectors of the fingers always intersect at a fixed point.

These movements are accomplished by two sets of special cams. For an end effector with two V-shaped fingers, the cams would be identical. For

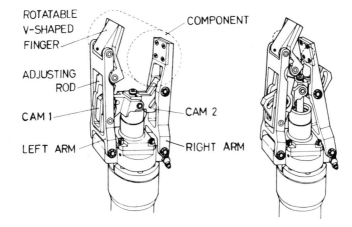

Fig. 3-8. Universal V-notch gripper. (*From Quingsen, Ref. 4*)

the design shown in Fig. 3–8, two different cams are needed because the motions in the two fingers are different from each other.

In tests, the same end effector was used for components ranging in diameter from 20 mm to 80 mm in steps of 10 mm. The gripper picked up each component and placed it in an operating machine. Each time, the particular position in which the gripper placed the component was measured. It was shown that the centers of the different components were almost at the same position. The largest difference was less than 0.2 mm. For an ordinary V-shaped gripper the corresponding offset value would have been 24.3 mm.

Dual Grippers.   Equipping a robot with dual grippers can substantially reduce the load/unload cycles of machine tools. Figure 3–9, for example, shows the dual gripper that is available with the RR625 robot from Reis Machines, Inc. The robot can lift payloads of up to 90 lb. As seen in the illustration, a five-gripper manifold is provided, so that grippers can be mounted in a variety of positions.

A typical working sequence of a dual gripper, consisting of grippers *A* and *B*, that exchanges workpieces in a machine tool chuck would be the following:

- Gripper *A* picks up workpiece No. 2.
- Robot moves arm to the machine tool chuck.
- Gripper *B* takes workpiece No. 1 out of the chuck.
- Gripper *A* inserts workpiece No. 2 into chuck.
- Robot moves arm to a programmed location, where gripper *B* puts workpiece No. 1 down.
- Robot arm moves to another programmed location, where gripper *A* picks up workpiece No. 3.
- The cycle is repeated.

Another application for dual grippers is the handling of workpieces of essentially different configuration or size. In such cases, the wrist would carry two different grippers, each one suitable for one class of workpieces.

Automatic Changers.   Like machining centers—which are capable of changing tools by moving to an automatic tool changer and depositing one tool and removing another in a programmed sequence—robots too can be equipped with an automatic change system, to exchange end effectors. A typical example of this method is the Automatic Handchanger (AHC) developed by GMF Robotics Corp. for its Model A-0, A-1, and M-1 robots.

When the robot is so equipped, it can perform several different operations without the need for manually changing end effectors from one op-

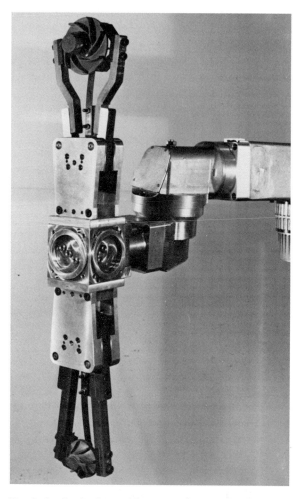

Fig. 3-9. Dual gripper. (*Courtesy of Reis Machines, Inc.*)

eration to another. The length of time necessary for the automatic exchange of robot end effectors, when utilizing an AHC, varies from application to application. Where such exchanges are frequently required, the elimination of robotic downtime during manual gripper changeover, as well as the elimination of manual labor, produces improvements in efficiency, machine uptime, and productivity.

Each different end effector is permanently attached to an adapter that fits a female socket in the gripper mechanism mounted on the faceplate of the wrist. A locating pin engages a hole in the adapter to maintain radial location.

Fig. 3-10. GMF Robotics automatic hand changer system. (*Courtesy of GMF Robotics Corp.*)

To exchange grippers, the robot positions the adapter's grooved flange in a U-slot in a storage rack, as shown in Fig. 3-10. Air pressure applied to the top of the built-in cylinder moves an air piston down. This lets the locking balls recede into the narrowed section of the piston rod, thereby releasing the adapter with its end effector. The robot then moves through the next programmed step, which is to move to another position on the storage rack and pick up the corresponding adapter and end effector combination.

To engage a new gripper, the robot goes through an overtravel motion that compresses a spring in the adapter plate to ensure full engagement of the AHC and gripper. The air cylinder is pressurized, forcing the locking balls to move outward and grip the inside of the new adapter. The spring in the bottom of the air cylinder provides gripping force even when the air supply is shut off. An air supply line is provided for tools that are held in the gripper.

In the event of a jam-up or overload condition, a safety joint is mounted between the robot's gripper mounting plate and the AHC to prevent damage to the robot or its gripper. The safety joint is designed to break at a predetermined force, allowing a microswitch to be actuated that results in an emergency stop of all robot motion. A pivoting safety plate holds the gripper to the robot even after the safety joint is broken. The AHC adds 4 to 20 in. in tooling length and 9 to 13 lb in weight at the end of the robot's arm.

Multi-position End Effectors. Another way to enhance the flexibility of end effectors is to combine several in one unit attached to the faceplate of the wrist. An example is the six-position multiple gripper from ASEA Robotics, shown in Fig. 3–11. Basically, it consists of a pneumatically driven disk or turret with a set of six grippers or tools. An internal air motor drives the disk to bring any one of the grippers or tools in position, and an air cylinder locks it firmly to the wrist.

The grippers can not only handle six parts simultaneously, but these can be different parts with different shapes. It is a particularly good approach to the assembly of small parts where short cycle times preclude picking up and moving parts individually.

However, another usage for a multi-position end effector is in the execution of a multiplicity of tasks, as shown in Fig. 3–12. Here, an ASEA IRB 60 robot is used for drilling and riveting airframe structures.[5] It is mounted on a 26-ft-long track to give it the necessary mobility. Its end effector was developed by United Technologies Research Center. It combines five different tools, which are used for the following purposes:

- The Renishaw probe serves to locate the robot at a work station.
- The hole sensor locates where drilling and riveting must take place.
- The drill is equipped with an air-driven motor that runs on shop air. It is provided with automatic feed and delivers a minimum thrust of 60 lb.
- The sealant applicator deposits a corrosion inhibitor on the hole perimeter before the rivet is inserted. The presence of the sealant is sensed after application by an infrared emitter and detector. The sealant absorbs the sensor-emitted signal. If the signal is not absorbed, the operation is stopped before the rivet is inserted.
- The rivet applicator can insert any one of five rivets required to fasten the skin, without changing jaws.

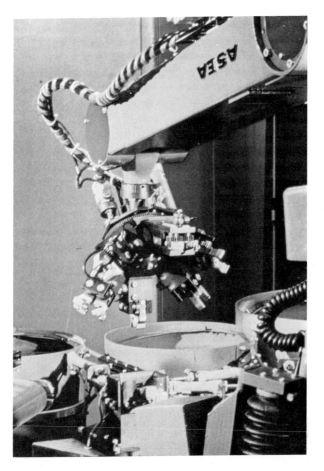

Fig. 3–11. Six-position gripper. (*Courtesy of ASEA Robotics Inc.*)

Articulated Hand.   The ideal gripper would be universal in the variety of its applications. It would be capable of six different basic operations:

- Spherical grasp, such as holding a ball or other round object.
- Cylindrical grasp, such as holding a hollow thin-walled tube.
- Tip grasp, such as picking up electronic components and other small objects.
- Hook grasp, such as picking up an object by its handle.
- Planar grasp, such as holding a pallet from underneath.
- Lateral grasp, such as grasping a workpiece from the side and moving it over to the next point of operation.

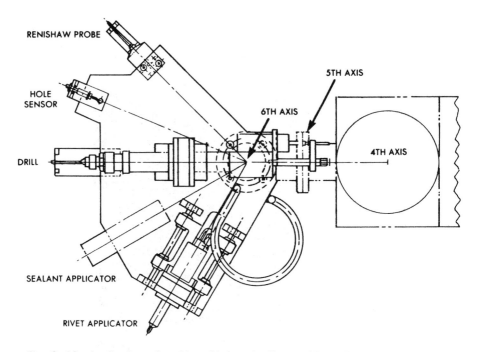

RENISHAW PROBE

HOLE
SENSOR

DRILL

5TH AXIS

6TH AXIS

4TH AXIS

SEALANT APPLICATOR

RIVET APPLICATOR

Fig. 3-12. Application of multi-position end effector. (*From Fitzpatrick and Barto, Ref. 8*)

Obviously, the human hand, even with its limited strength and endurance, can fulfill all these requirements, and thus is often used as the paradigm for a universal gripper. There is a practical problem in doing so, however, in that the human hand is composed of 48 actuators acting independently or in combination to perform its tasks.[6] Besides, the human hand operates with 22 degrees of freedom: 6 independent variables locate the wrist and 16 variables locate the fingers.

If only for economic reasons, it is essential for robotic end effectors to arrive at a much simpler solution than the complex arrangement of the human hand. But the provision of additional degrees of freedom in the end effector offers an important advantage that should not be overlooked, which, as J. Kenneth Salisbury has pointed out,[7] "stems from the advantages of placing additional degrees of freedom near the grasped object. This makes it possible to augment the resolution, sensitivity and bandwidth of the arm's joints. . . . A logical combination is the use of large links (arm) to achieve a big enough working volume in series with smaller, redundant links (hand) providing quick response."

Probably the first attempt to develop an articulated hand was made by Skinner[8] in 1974. The so-called Skinner hand, illustrated in Fig. 3–13, is a three-fingered design, like most others that followed it. Each finger is capable of grasping an object. A joint at the base of each finger allows it to twist about its long axis. This kind of hand could be used for holding an object by friction or physical constraint, and could fulfill a number of the conditions for a universal gripper as postulated above. It was never put into production.

A more recent development is the Stanford/JPL hand, shown in Fig. 3–14. It originated at the Artificial Intelligence Project of the Computer Science Department at Stanford University[9] under the sponsorship of NASA and NSF, and is now a product of Salisbury Robotics, Inc. The three-finger articulated gripping and manipulating device can exert controlled forces and moments upon grasped objects in arbitrary directions, and can impart arbitrary small translations and rotations of the grasped object. It permits complex assembly tasks to be performed without the need of specialized grippers.

The hand's weight and volume have been kept to a minimum by locating the actuators on the forearm. Actuation forces are conducted to the hand

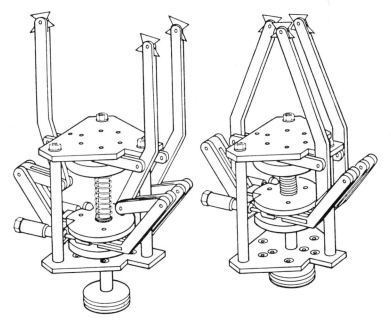

Fig. 3–13. Skinner hand. (*From Ref. 8*)

Fig. 3-14. Stanford/JPL hand. (*Courtesy of Salisbury Robotics, Inc.*)

by 12 Teflon-coated tension cables (tendons), of which Fig. 3-14 shows four. Normally, they are routed through flexible conduit. Semiconductor strain gages are used as a tension-sensing mechanism, and are placed in the hand to permit monitoring and controlling the power of finger forces.

The length of the fingers is 5.5 in. from first joint axis to fingertip. Urethane pads are attached to the fingertips to provide gripping friction. The structural parts are aluminum, and the weight of the hand is 2.5 lb.

The drive assembly weighs 12 lb. It uses dc motors with samarium-cobalt permanent magnets. The spur gear reduction is 25:1.

A simplified version of the three-finger hand is the Pennsylvania Articulated Mechanical Hand, PAMH or simply the Penn hand, which was developed at the University of Pennsylvania.[10] Professor Burton Paul was faculty adviser of the mechanical design of the Penn hand, which is shown in Fig. 3-15. The Penn hand has three fingers, each with two joints. One of the fingers is able to move above the base to oppose the other two, simulating the function of a human thumb. The approximate maximum fingertip speed is 2 in./sec, and the lifting capacity is about 1 lb.

To simplify installation and control, the fingers are moved by a system of flexible shafts instead of cables. The decision to use the flexible shaft

Fig. 3–15. Penn hand. (*From Abramowitz et al., Ref. 10*)

was made primarily to study friction, twisting, bending, and compliance of flexible shafts in robotic systems. There is one flexible shaft for each joint of each finger. The end of each shaft is a rigid extension with a nylon ball that pushes against a spring-loaded lever that moves the finger joint. This is illustrated in Fig. 3–16. The curve of the lever arm is such that at no point is there a bending moment acting on the extension of the flexible shaft.

Rovetta and Vicentini[11] of the Politecnico di Milano, Italy, described a multijoint mechanical hand, as shown in Fig. 3–17. Three fingers with three joints (or phalanges) and an elastic palm are used. In the prototype they are actuated by stepper motors, but dc servomotors could be used instead. The control system uses microcomputers and corresponding software for integrated action of the motors to optimize the grasping sequence. The prototype has been designed to grasp workpieces of arbitrary shape weighing up to 22.5 lb.

The step motor motions are reduced by a harmonic drive and a bevel gear that drives a shaft with a pulley. The pulley is connected with a traction

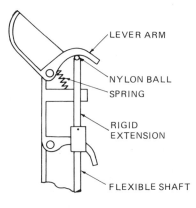

Fig. 3-16. Motion mechanism for Penn hand. (*From Abramowitz et al., Ref. 10*)

Fig. 3-17. Multifinger hand. (*From Rovetta and Vicentini, Ref. 11*)

wire that moves the fingers toward the palm. Springs with different mechanical characteristics are inserted in the finger joints to allow their motion in sequence.

The palm is an elastic system with variable stiffness. It constitutes a contact point that offers an increasing elastic reaction, thereby stabilizing the grasping operation.

Vacuum Cups

Vacuum cups have several advantages, among them the following:

- Only one side of the workpiece needs to be accessible for the gripper, as is the case when sheetmetal parts are stacked.
- They allow handling of fragile or vulnerable parts, such as glass plate.
- They can conform to contoured parts.

Whether one uses one, two, three, or even more of these cups depends entirely on the form of the object and on the lift force required.

Also, whether one selects vacuum cups or a gripper depends on the type of parts. I.S.I. Manufacturing Inc. makes the recommendation that parts with large, relatively smooth surface areas should generally be considered for vacuum handling. Examples are sheetmetal, wood, glass, plastics, cardboard, plate steel, and paper parts ranging in size from as large as an automotive roof to as small as a pack of cigarettes.

On the other hand, parts without large smooth, flat surface areas should be considered for mechanical grippers with fingers designed and contoured to fit the part configurations. Examples for this case are rounds, squares, and cylindrical and rectangular parts, including shafts, splines, hubs, gears, wood, and plastic objects.

Vacuum cups come in several shapes. Examples are shown in Fig. 3–18, as marketed by I.S.I Manufacturing Inc. The material is either neoprene or urethane. Polyurethane cups are usually considered more durable than neoprene, while the latter conforms better to uneven surfaces. In this illustration, (1) is an oval cup with a threaded insert that fits a 3/8-in. NPT pipe thread; (2) is a round cup with Vaclok Mount, a feature of vacuum cups made by I.S.I that prevents the cup from rotating on the thread; (3) is a Vaclok venturi assembly with Vaclok venturi mount and Vaclok bracket arm; and (4) is the outside appearance of an I.S.I venturi, model VB 38T.

The round vacuum cups come in sizes of 2, 3, 4, and 5 in. The 2-in. size is only 13/16 in. high. The low profile of this design allows for 60% sealing of the cup area when evacuated. Flexing is minimal. At a vacuum of 20 in. Hg, the cup has a lift capacity of 16 lb with a 2:1 safety factor. The larger

Fig. 3-18. Vacuum grippers. (*Courtesy of I.S.I. Manufacturing Inc.*)

sizes allow conforming to contoured parts and have lift capacities of 35, 62, and 98 lb, respectively.

The oval cups have applications in narrow areas. They flex in one direction only. One type is shallow with very limited flexibility. It measures $2\frac{1}{2}$ × $5\frac{1}{2}$ in. and has a lift capacity of 62 lb. The other is deep, has more flexibility, and lifts 90 lb.

Production of a vacuum requires either an extra vacuum pump with tank or a venturi. The latter is simpler and usually preferred. The VB38T Venturi offered by I.S.I. produces a vacuum of 20 in. Hg with plant air of 22 psi. Its air consumption is generally 5 to 9 ft³/min. It is essentially an air passage with a narrow section, called the throat, in the center. The approaches to the throat are carefully tapered on the entrance as well as the exit side. Since the flow velocity is very much increased in the throat, a pressure drop occurs in this section. Thus, tapping the throat lets air be aspirated through the tap and produces the vacuum.

End effectors with multiple cups are each fitted with their own venturi. This assures faster venturi evacuation time and eliminates dropping the part if the cup is damaged or the air line is severed. Venturis are very practical but do make some noise which can be objectionable. However, this can be overcome by using an I.S.I. silencer. Since it may be desirable to conserve energy, the venturi is usually cut off by a valve until the vacuum is needed. In this case, however, there is a short time delay involved to develop the necessary vacuum for lifting.

Electromagnets

While permanent magnets could be used to pick up objects, additional mechanisms or manual assistance would be needed to strip off the objects from the magnet once they arrived at their destination. This may be feasible in specialized situations, but in general it is the electromagnet rather than the permanent magnet that is useful as a robotic gripper.

With electromagnets the workpiece may be released by de-energizing the magnet or even by supplying a short pulse with reversed polarity. These magnets have strong pulling and holding force, but can be used only when the subject to be taken up is ferritic, well defined, and isolated from other magnetic materials that may be likewise attracted. When this is the case, the electromagnet merits consideration.

Holding value factors (i.e., the holding force) as given by manufacturers are usually determined on the basis of a smooth, ground surface of mild steel that is 2 in. thick or thicker. The stated values have a safety factor of two; but the holding force can decrease sharply when electromagnets are applied to very thin material or to rough, dirty, painted, rusty, or nonflat surfaces which cause air gaps.

Another consideration, when holding objects with electromagnets, is that in case of power failure the magnet will release the part, and injury or damage may result. Battery backup power packages are available to reduce this problem.

Figure 3–19 shows some typical magnets made by Industrial Magnets, Inc. (IMI), and Fig. 3–20 depicts an ASEA robot equipped with such a magnet.

IMI magnets operate on 12 or 24 V dc with the exception of rectangular magnets, which are also available for 110 V dc.

Round magnets offer concentrated holding power and high responsiveness. They are steel-encased and constructed with high-temperature potting material.

Round electromagnets from IMI come in sizes from $\frac{3}{4}$ in. to 12 in. in diameter. The holding force of a 5-in. magnet, which is 3 in. high, is 610 lb, requiring 48 watts.

Rectangular magnets deliver positive reach-out tailored to the material being handled. They come in six standard sizes with cross-sections of $1\frac{1}{2}$ ×

ROUND   RECTANGULAR

PARALLEL   MULTI-POLE

Fig. 3-19. Electromagnets. [*Courtesy of Industrial Magnets, Inc. (IMI)*]

$1\frac{1}{2}$ in. to 4 × 8 in. A rectangular magnet of $2\frac{1}{2}$ × $4\frac{1}{2}$ in., which is $1\frac{7}{8}$ in. high, has a holding force of 455 lb.

Parallel pole magnets deliver improved reach through air gaps. They can operate where parts have uneven surfaces or odd shapes by using custom-machined pole shoes.

Multi-pole electromagnets offer significantly greater performance and application flexibility than other types. They can have custom shoe configurations to fit heavy-duty applications. Standard sizes are available from 3 × 3 in. to 6 × 12 in. with a height of $3\frac{1}{2}$ in. A magnet of 4 × 8 in. has a holding force of 1100 lb.

REMOTE-CENTER COMPLIANCE

Often some compliance is needed so that the gripper gives when two parts have to be mated and are slightly offset. Thus, end effectors for loading round workpieces into a lathe chuck may be provided with a secondary pusher mechanism. After the lathe chuck has gripped the workpiece, the end effector spins around, the chuck jaws open again, and the end effector pushes the workpiece into its final position.

There is another method, which treats the loading operation the same way as a critical assembly task where a round peg has to be inserted into a

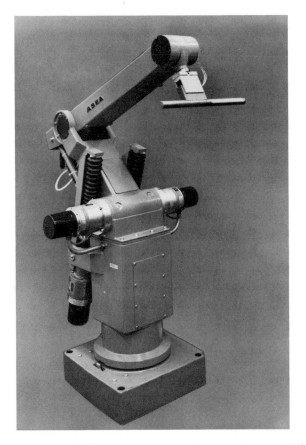

Fig. 3-20. ASEA robot with IMI electromagnet. [*Courtesy of Industrial Magnets, Inc. (IMI)*]

round hole. It makes a remote-center compliance (RCC) device part of the end effector.

Interfacing an RCC device between the wrist and the gripper gives the workpiece enough lateral freedom that it can be pushed all the way into the chuck of a machine tool the first time. Because of its inherent compliance, it permits the jaws to close concentrically.

The same is true in assembling parts. Small deviations in the relative position or angular orientation of the parts can prevent the robot from fitting the parts. Excessive forces can develop. Workpieces, fixtures, and the robot can be damaged. However, the use of end effectors with RCC can compensate for such misalignments.

The Charles Stark Draper Laboratory[12] began to research this concept in

1973. Their studies on the fitting of parts resulted in mathematical models, and mechanical devices were developed to confirm these models by experimental evidence. These devices used the RCC principles, which are based on the following assumptions:

- That the effective center of rotation is located at the point of contact between two objects that are in touch and to be assembled.
- That the mechanical structure is such that it is possible to insert the RCC device between the wrist and the gripper.
- That at least one of the parts to be mated has chamfered edges, whereby the chamfer has to be larger than the maximum allowable offset between the parts to be matched.
- That the conditions of maximum angularity, clearance, and coefficient of friction (see below) are met.

The typical case is the assembly of a round peg into a hole. Deviations in position and orientation may exist (1) because of lateral displacement between the axis of the peg and the axis of the hole, where both axes are parallel, as in Fig. 3–21a; and (2) because of angular deviations that destroy the parallel orientation, as in Fig. 3–21b. Both characteristics may occur simultaneously.

When an attempt is made to mate two parts with angular displacement, the first interference is a contact point between the peg and the hole. This is followed by a second contact point on the opposite side, as shown in Fig. 3–22.

To prevent jamming the peg, it is necessary that the second contact be established after the first lateral, sliding contact has penetrated to a depth that is at least equal to the product of the coefficient of friction multiplied by the diameter of the hole.

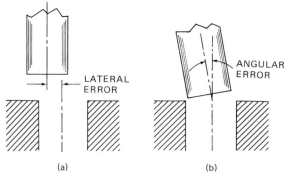

Fig. 3–21. Displacements between peg and hole.

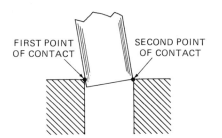

Fig. 3–22. Angular displacement and contact point.

Furthermore, the permissible maximum angle between the axis of the peg and the axis of the hole, A_{max}, has to be a function of the clearance between the diameters of the hole, D, and of the peg, d, as well as of the coefficient of friction, μ, so that:

$$A_{max} = \frac{D - d}{d\mu} \qquad (3.1)$$

The RCC device must be designed to rotate about the point at which the peg contacts the hole. This point is what is called the "remote center of compliance." An RCC device that is so constructed has the same effect as if the peg were pulled rather than pushed into the hole. It will prevent jamming and make the peg slide effortlessly into the hole.

An RCC device designed by the Charles Stark Draper Laboratory has been used for inserting ball bearings with a diameter of 1.6 in. into a seat with 0.0004 in. clearance in 0.2 sec. Initial position errors of 0.08 in. and several degrees of angular error could readily be tolerated.

Figure 3–23 shows the design principle as developed by the Charles Stark Draper Laboratory. It consists of rotational and translational parts. The upper three diagrams show only the RCC rotational part. It is free to swivel about the remote center, that is, the point of contact between the workpiece and the hole (not shown) into which it is inserted.

In the lower two diagrams, the RCC translational part is added on top of the rotational part. It provides lateral adjustments when parallel alignment at the remote center of the workpiece is required to enter it into the mating hole.

Figure 3–24 shows the embodiment of the principle in a commercial device. Here, the translational and rotational actions are combined (as described later), and a very compact unit results. It is a product of the Lord Corporation and considers angular misalignment not only in the direction

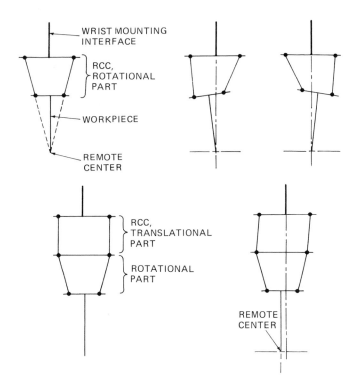

Fig. 3-23. Design principle of RCC device. (*From Whitney and Nevins, Ref. 12*)

Fig. 3-24. RCC device. (*Courtesy of Lord Corp.*)

of the pin (which Lord calls the "cocking effect") but also in the torsional sense about the axis of insertion. The latter is useful when, for example, the hole and pin are not perfectly round or have a square or other polygonal shape.

The compliance is obtained by stacking elastomer slices interspersed with metal shims. In compression, these sandwiched elements are much stiffer than in shear. By changing the type of elastomer, number of shims, and part geometry, the characteristics of this RCC device can be altered to fit the requirements of specific applications.

The action of the Lord RCC device is illustrated in Fig. 3-25. As pointed out above, a requirement is that the hole must have a chamfer that is larger than the maximum permissible offset. Lateral errors at the contact point between the pin and chamfer, as shown in (a), produce a contact force at the chamfer with a horizontal (lateral) component.

If the RCC device consists of rigid elements, as shown in (b), the horizontal force shifts the pin laterally until it slides into the hole.

In case of a displacement that is at an angle to the direction of insertion, as shown in (c), the pin will initially enter the hole. However, the edge of the pin will contact one side of the hole, and the edge of the chamfer will contact the other side of the pin. The result is a moment that turns the pin about the center of compliance, as seen in (d), and aligns the pin with the hole.

Thus, there are two degrees of freedom: the lateral and the angular motion of the pin. They are combined in the multidirectional compliance of the elastomeric components of the Lord RCC device, as illustrated in (e). Actually, there is a third degree of freedom, since the elastomeric component can also be twisted through a small angle.

Thirty different models are available. They differ in overall dimensions

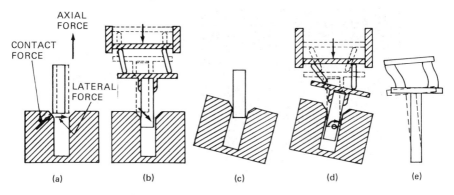

Fig. 3-25. Action of RCC device.

as well as in compression stiffness, lateral stiffness, cocking stiffness, torsional stiffness, and other characteristics.

As an example, model RCC55B01-BBHO-1 is designed for handling small parts with a PUMA robot (made by Unimation Inc.). It has a top plate of 3.5 in. to be mounted to the wrist of the robot, and weighs 8.3 oz. It is suited for longer parts, has a lateral stiffness of 100 lb/in., an axial or compression stiffness of 18,000 lb/in., a torsional stiffness of 75 in.-lb/rad, and a bending or cocking stiffness about an axis perpendicular to the axis of insertion of 4100 in.-lb/rad.

The permissible misalignment for which the compliance of this device can compensate in the lateral direction is ±0.20 in.; in angular cocking, ±1.4°; and in torsion, ±12°.

Another commercial extension of the RCC concept is the Astek Accommodator from Barry Wright Corporation, Inc., which is shown in Fig. 3–26. The design uses a set of six elastomeric shearing pads, which are quite stiff in compression but relatively soft in shear, like the above-described device. This allows sidewise motion during an assembly operation to correct for both lateral and angular misalignments. The basic model, AST-100, is also compliant in the torsional mode; that is, it permits a limited rotation about its axis. However, where torsional freedom is undesirable because torque is to be transmitted, an optional set of anti-rotation pins can be inserted.

The elastomer pads can be of different materials. Natural rubber provides the best mechanical properties, but is susceptible to environmental degradation. Silicone rubber provides comparable performance at some sacrifice in hysteresis. Neoprene is environmentally robust and is the elastomer usually chosen for all but the softest units.

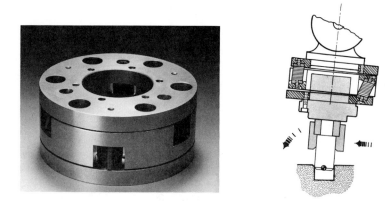

Fig. 3–26. Accommodator RCC device (*Courtesy of Barry Wright Corp.*)

Since the remote center of compliance is located at the tip of the peg, the design of the Accommodator, like any other RCC device, must consider the distance from the point of its attachment at the gripper to the tip of the workpiece it tries to insert. This distance is the projected center of compliance. The Accommodator is available for center-of-compliance projections of 3, 3.9, 4.9, and 5.9 in. Usually, deviations of 10 to 15% from these values are acceptable without deterioration of performance.

MOTION MECHANISMS

The joints of the robot must be moved, and there are a number of methods available to do this. The choice is determined by the robot's load carrying capacity, degrees of freedom, mechanical structure, speed, dynamic requirements, and so on.

The most direct way to produce motion is to put the motor, be it electric or fluid, directly on the joint. If the joint is connected to the base, this is no problem. If it is itself moved by a linkage, the inertia caused by the drive and its influence on the dynamic behavior of the robot must be considered.

Whatever actuator is chosen, the motion must be provided at a particular point of application. To accomplish this, transmissions, conversions from rotary to linear motion, and speed reduction frequently are necessary. The following discussion describes some of the most widely used elements for this purpose.

Transmissions

Cables. For smaller robots, steel cables can be used to advantage. Their application is often referred to as tendon technology, the proven method of using cable actuators in aircraft controls. When steel cables are used, possible stretching of the cables can result in positioning errors. To reduce such strain, engineers at the University of Utah have used woven synthetic fibers successfully. Hitachi Ltd. uses special metal wires that shrink when an electrical current passes through them.

A typical robot that uses steel cables for the transmission of motion to all main and wrist axes and even the gripper is the five-axis Alpha robot produced by Microbot, Inc. This is a low-cost programmable robot well suited for training purposes, research, and use in light-load material handling applications. It can carry loads of up to 1.5 lb at speeds of up to 20 in./sec. Its resolution is 0.010 in. and its repeatability ±0.020 in.

As Fig. 3–27 shows, the Alpha uses six cable drives. Drive gears and capstans for each motion are mounted independently of each other on a common shaft.

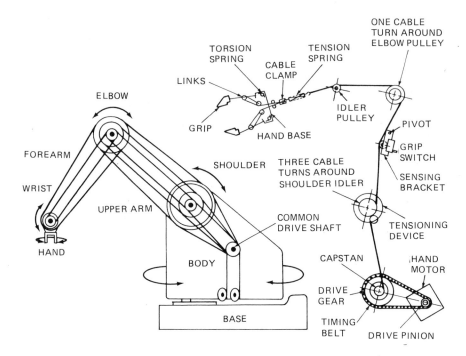

Fig. 3–27. Cable transmission. (*Courtesy of Microbot Inc.*)

The drive for the robot hand can be seen in detail on the right side of the illustration. It shows the cable drive for the hand that controls the gripping action. The drive pinion of the corresponding motor is connected by a timing belt to a drive gear and capstan mounted on the common drive shaft. The stainless steel cable is led from there to a three-turn idler on the shoulder shaft, over a grip switch—to sense gripper closure—to a one-turn idler on the elbow shaft, and then via an additional idler to the hand.

The cable drive system provides a base rotation of 330°, a shoulder bend and an elbow bend of 140° each, a wrist roll of 540°, and a wrist pitch of 180°.

Belts.   Timing belts are often called synchronous belts. They have no slippage like that of flat belts, but have teeth that fit into sprocket wheels, thus assuring synchronous speeds between the driving and driven ends of the transmission. Their teeth are trapezoidal.

Another type of belt has curvilinear teeth, as seen in the PowerGrip HTD belt from the Power Transmission Division, Uniroyal, Inc. It is shown in Fig. 3–28.

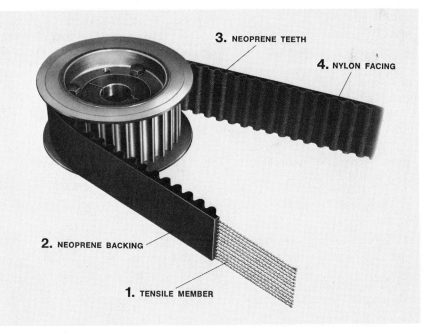

3. NEOPRENE TEETH

4. NYLON FACING

2. NEOPRENE BACKING

1. TENSILE MEMBER

Fig. 3-28. PowerGrip HTD belt. (*Courtesy of Power Transmission Division, Uniroyal, Inc.*)

The robot that was shown in Fig. 2-15, which is made by Citizens Robotics Corp., uses such belts for driving five of its axes. This robot can handle a 25-lb payload.

The PowerGrip HTD belts consist of a neoprene-rubber covering, non-stretch fiberglass tensile cords, and nylon-faced teeth that are profiled for uniform stress distribution. They have a high ratio of torque transfer to weight, which is an advantage for the relatively low-velocity, high-torque characteristics that are common when they are used in robotic drives. Other advantages are that they do not stretch with wear, are corrosion-resistant, and operate at reduced noise levels.

How well they stand up under the dynamic stresses of robotic application is not yet well documented. However, the fact that engines with timing belt drives for camshafts have been considered in the automotive industry has led to intensive and successful efforts at strengthening their design, and the technical world may have to get used to their ruggedness and reliability.

Roller Chains. Roller chains are widely used in robots and throughout the industry. They are well suited for short distances, where belt drives are not feasible. Precision construction provides efficient and quiet operation. Rated values are predicated upon adequate lubrication, but there are some self-lubricating types that have oil-impregnated sleeve bearings. They are capable of running at the same loads as true roller chains, but at the lower end of the speed range. Their service life lies between the service lives of nonlubricated and well-lubricated roller chains.

Figure 3–29 shows a chain drive as used by Unimation for the wrist axes of their Unimate Robot Series 2030. The piston rod of a hydraulic cylinder is connected to the roller chain, and an adjustable chain tensioning block keeps the chain at a suitable tension. Like the gear trains of robots, the chain would have to be lubricated at regular intervals.

Link-Rod Systems. For their IRB robot series, ASEA uses a patented link-rod system that drives the wrist motions from servomotors fixed to the rotating base. The base mounting of the motors keeps their mass at a distance from the moving parts. The link-rod system, the principle of which is shown in Fig. 3–30, gives an extremely accurate and reliable method of motion transmission. Consider disk *A* being rotated by the servomotor. The rods transfer the motion to disk *B*, which in turn transfers it to disk *C* to produce the wrist motion. By having the moving ends of the rods mounted on universal joints, the transmission provides a high degree of flexibility.

Rotary-to-Linear Motion Conversion

Rack-and-Pinion Drives. For converting rotary motion to linear motion, rack-and-pinion drives are simple, efficient, and economical. Their problem is load capacity. The entire force that the pinion transfers must be

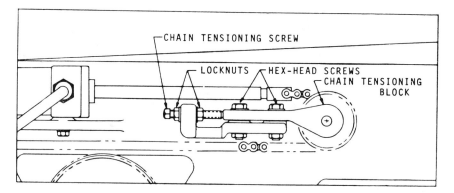

Fig. 3–29. Chain drive with chain tensioning block. (*Courtesy of Unimation, Inc.*)

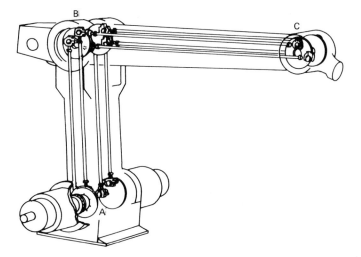

Fig. 3-30. ASEA's patented link-rod system. (*Courtesy of ASEA Robotics Inc.*)

supported by one pinion tooth at a given instant. Hence, the greater the load, the larger must be the pinion and the less is the reduction that is attainable. Rack-and-pinion drives are capable of resolutions of as small as 0.0005 in., but wear and backlash may develop in use.

Acme Screws.   The acme thread is a power transmission screw thread. The combination of acme screw and nut, as shown in Fig. 3-31, is one of the most common linear actuators. However, its efficiency is relatively small, since all motion between nut and screw thread implies sliding friction. Rolled-thread acme screws have an efficiency of about 30%, but their cost is also low, and they are used in some robots. The efficiency is improved when precision-ground threads are used, but even with precision-ground threads the major drawback is unavoidable friction.

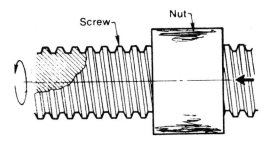

Fig. 3-31. Acme screw with nut.

One robot manufacturer chose a Teflon-coated lead screw with self-lubricating polymer nut, as used in the high-speed printer business. He found it to be robust and inexpensive, with zero backlash.

Another way to minimize friction and maximize durability is to use ball bearing screws.

Ball-Bearing Screws.   A ball-bearing screw, or ball screw for short, is what its name implies: it is a screw similar to an acme screw, but one that runs on ball bearings. The bearing balls provide the only physical contact between screw and nut. They replace the sliding friction of the conventional acme screw and nut with the smoother rolling friction of the balls.

Ball screws provide resolutions as fine as 0.00001 in., which is about 50 times finer than that of a rack-and-pinion drive. Figure 3–32 shows the design of a ball screw as produced by Beaver Precision Products, Inc. The thread is a precision-ground, hardened ball race. The nut consists of a series of bearing balls circulating in a similar race. The balls are carried from one end of the nut to the other by return tubes, so that they circulate in a con-

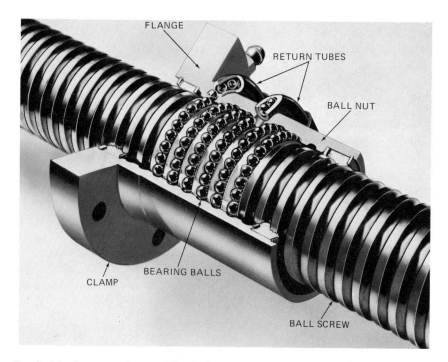

Fig. 3–32. Cutaway of a precision ball-bearing screw. (*Courtesy of Beaver Precision Products, Inc.*)

tinuous fashion. In a typical ball screw, like the one shown, two complete bearing ball circuits are used to increase the load carrying capacity of the screw, and the operating life. Unlike an acme screw, ball screw life is statistically predictable, based on B_{10} bearing ratings.

The efficiency of a ball screw is a nominal 90%. The accuracy is 0.0002 in. per foot of length of the ball screw.

There are two ways of using the ball screw. The most common permits the screw to rotate about its axis but prevents motion in the direction of the axis. This is the same as rotation of any typical drive shaft. The nut, however, is kept from rotating, but is permitted to move along the axis of the screw. The rotary motion of the screw is thus converted into linear motion of the nut.

The other way is sometimes called the inverted method. It rotates the nut but prevents it from traveling along the axis of the screw. In this case, the screw is prevented from rotating and travels lengthwise through the nut.

For maximum accuracy and minimum backlash, Beaver Precision Products offers preloaded ball screws. One configuration consists of two nuts face to face with springs between the faces tending to force them apart. This forces the balls into contact with the races, regardless of the direction of travel. Any backlash that could occur because of end play between the balls and races is effectively eliminated.

In addition to the double nut preload, other configurations include a double nut preloaded housing and a single nut unit with integral preload. Furthermore, the actual distance the screw travels during one revolution (i.e., the lead of the ball screw) can be specified for specific purposes. Thus, Fig. 3–33 shows a high-lead precision ball bearing screw with preload, designed by Beaver Precision Products specifically for the rapid transverse requirements of robot axes. It offers 2 in. of linear travel for each 360° of screw rotation. The lead accuracy is 0.0005 in. per foot.

Speed Reducers

Servomotors have low torque and high speed. The movement of robot joints requires high torque and low speed. Rare-earth magnets in dc servomotors are a significant improvement, since they offer higher torque and lower speed, but reducers still are needed in most applications.

Gear Reducers. The Merlin robot from American Robot Corp. uses gear trains to reduce the steps of its stepper motors to the small increments required for positioning of the robot joints. An outline of the robot is shown in Fig. 3–34. Payload of the robot is 50 lb. Speed is up to 8 ft/sec and repeatability ±0.001 in.

Fig. 3-33. High-lead precision ball-bearing screw designed specifically for the rapid traverse requirements of robot axes. (*Courtesy of Beaver Precision Products, Inc.*)

The gear reduction trains consist of two stages. The first one uses spur gears in a parallel shaft arrangement and provides a reduction of 4:1. The second stage rotates a spur gear against a face gear, so that intersecting shafts with a reduction of 12:1 can be used. The total reduction thus is 48:1. The face gear rather than bevel gearing was chosen because of the simplicity with which it allows preloading the gears for zero backlash.

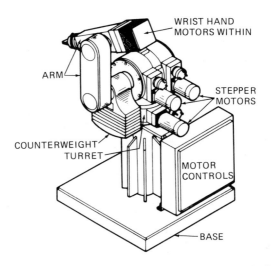

Fig. 3-34. Merlin robot. (*Courtesy of American Robot Corp.*)

Planetary Gearing. Planetary gearing is versatile and can be used for a number of different purposes. In robotic technology, however, it serves strictly for speed reduction.

A schematic of the basic planetary gear arrangement is shown in Fig. 3–35. A sun gear is located in the center and is surrounded by three to five planet (or pinion) gears. They are fastened to a spider-shaped planet (or pinion) carrier with pins on the end of its arms on which the planet gears sit. The planet gears are free to rotate on these pins. The ring (or internal) gear surrounds the planet gears. Since the ring gear has its teeth on the inside, and the planet gears fit between the sun gear and the ring gear, the planet gears can transmit motion from the sun gear to the ring gear, as long as the planet carrier is locked and forces the planet gears to rotate without changing the positions of their respective centers.

When a planetary arrangement is used as gear reducer, the ring is locked, and the planet gears with the planet carrier are free to orbit around the sun. The center of the planet carrier's spider is connected to the output. The sun gear sits on the input shaft.

The speed ratio, R, of input rpm to output rpm is given by:

$$R = 1 + \frac{N_r}{N_s} \qquad (3.2)$$

where N_r is the number of teeth inside the ring gear and N_s the number of teeth on the sun gear.

Thus, if there are four times as many teeth on the ring gear as on the sun gear, and the input speed is 3600 rpm, the output speed is 720 rpm.

By feeding the output into a second planetary arrangement of equal con-

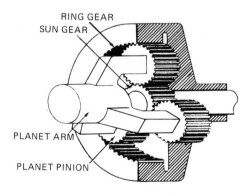

Fig. 3–35. Planetary gearing.

struction, one can reduce the output speed to 144 rpm. Thus, by connecting several planetaries in series, large reductions can be obtained. However, considering the number of gears involved in such a case, the efficiency of the planetary speed reducer suffers, and hence the torque transmitted is correspondingly diminished.

Planetaries are not free of backlash. Even the cycloidal drives, discussed below, though they are much better in this respect, may not be completely free of backlash, which hurts the dynamic response of robots. Manufacturers of robots have been known to reject cycloidal drives because of a backlash of 2 minutes of arc (about 0.03°).

Trochoidal and Cycloidal Drives. Gear reducers for robots should have zero backlash, high positioning accuracy, high efficiency, and high torsional stiffness, and should be small and compact. The Dojen orbital drive claims all this. Its producer is Dolan-Jenner Industries, Inc., a unit of the Barry Wright Corporation. The basic components and their assembly are shown in Fig. 3–36.

The dual-track cam is nested between the housing and the output shaft. It consists of two cams with trochoidal-shaped lobes on each track. The two cams are fastened together back to back. Each track cam mates with a corresponding set of rollers, one in the housing, the other on the inside of the output shaft. However, there is always one more roller on each set than the number of lobes on each of the matching tracks.

To obtain zero backlash, the rollers are preloaded against the cam track, assuring continuous contact between all the rolling elements without disengagement and engagement each time the direction of rotation is reversed. Rotation of the input eccentric sleeve causes the cam to orbit inside the sets of rollers.

Because of the difference in the number of lobes and rollers between each set, rotation of the orbital drive produces a continuous angular displacement of each set of rollers relative to every other set. The magnitude of the displacement is the ratio of input rpm to output rpm, which is expressed by the equation:

$$R = \frac{(N_1 - 1)N_2}{N_2 - N_1} \tag{3.3}$$

where R is the ratio of input rpm to output rpm, N_1 is the number of rollers on the housing, and N_2 is the number of rollers on the output shaft. If $N_1 = 12$, and $N_2 = 16$, then $R = 44$. Standard ratios for the Dojen drive are 224, 195, 168, 143, 104, 90, 77, 64, 55, and 44.

This means that a motor with 3600 rpm can be reduced to 25 rpm, when

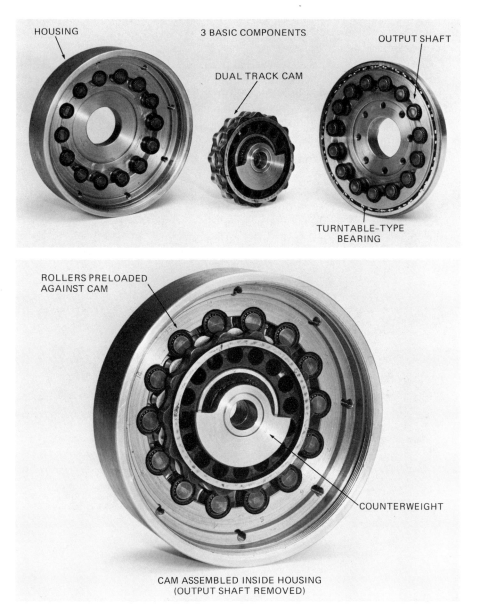

HOUSING

3 BASIC COMPONENTS

OUTPUT SHAFT

DUAL TRACK CAM

TURNTABLE-TYPE
BEARING

ROLLERS PRELOADED
AGAINST CAM

COUNTERWEIGHT

CAM ASSEMBLED INSIDE HOUSING
(OUTPUT SHAFT REMOVED)

Fig. 3-36.   Dojen orbital drive. (*Courtesy of Dolan-Jenner Industries, Inc., A unit of Barry Wright Corporation*)

$R = 143$. At the same time, the theoretical torque would be increased 143 times. For an efficiency of, say, 80%, the increase of torque would be less, yet it would still be 114 times.

Another representative of cycloidal drives is the Anti-Friction Drive (AFD), manufactured by Advanced Energy Technology Inc. As Fig. 3–37 shows, the orbiting member, which in the previous design was represented by the dual-track cam, is so designed that it does not mesh with the outer rollers. Instead, each outer ring contains two more lobes than the corresponding inner ring. Rollers occupy the spaces between inner and outer rings. The number of rollers is one less than the number of lobes on the outer ring.

The input shaft drives an eccentric that in turn causes the inner ring to orbit inside the corresponding outer ring. A second inner ring, which is rigidly connected to the primary inner ring, orbits inside a second outer ring that is rigidly connected to the output shaft.

The speed ratio of input rpm to output rpm for the AFD drive is given by:

$$R = \frac{1}{1 - (N_{o1} / N_{o2}) \, (N_{i2} / N_{i1})} \tag{3.4}$$

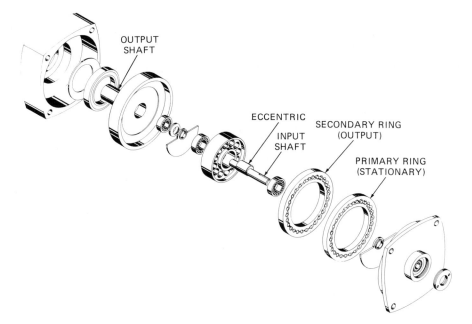

Fig. 3–37. Anti-friction drive. (*Courtesy of Advanced Energy Technology Inc.*)

where N_{o1} is the number of lobes on the primary outer ring, N_{o2} is the number of lobes on the secondary outer ring, N_{i1} is the number of lobes on the primary inner ring, and N_{i2} is the number of lobes on the secondary inner ring. Thus, if $N_{o1} = 36$, $N_{o2} = 28$, $N_{i1} = 34$, and $N_{i2} = 26$, the relation of input rpm to output rpm is 59.5.

Harmonic Drives. Figure 3–38 shows the construction of the harmonic drive produced by the Harmonic Drive Division of Emhart Machinery Group. Probably the most widely used of this type of gear reducers, it uses three concentric components. The outer ring is called the circular spline. It is stationary, with its inner surface shaped into gear teeth. The innermost part is an elliptical rotor, called the wave generator. Between the rotor and the outer ring is a flexible steel ring, the Flexspline. This element is under continuous deformation, shaping the ring into an ellipse rather than a circle. The axis of the ellipse rotates because of the revolutions of the elliptical rotor.

Fig. 3–38. The three principal parts of the harmonic drive. (*Courtesy of Harmonic Drive Division of Emhart Machinery Group*)

Since some of the teeth of the nonrigid Flexspline and the rigid circular spline are always in contact, and since the Flexspline has two teeth fewer than the circular spline, one revolution of the wave generator produces a relative motion between Flexspline and circular spline equal to two teeth. Thus, with the circular spline rotationally fixed, the Flexspline will rotate in the opposite direction to the input at a reduction ratio equal to the number of teeth on the Flexspline divided by two.

The relative rotation is shown in Fig. 3–39. The illustration follows the motion of a single Flexspline tooth for one-half of the input revolution of the wave generator. The tooth is fully engaged when the major axis of the wave generator is at zero degrees. When the major axis of the wave generator rotates to 90°, the tooth is disengaged. Full reengagement occurs in the adjacent circular spline tooth space when the major axis is rotated to 180°. This motion repeats as the major axis rotates another 180° back to zero, thereby producing a rotation of the Flexspline that is equal to a two-tooth advance per input revolution.

In these diagrams, the amount of Flexspline deflection has been exaggerated in order to demonstrate the principle. Actual deflection is much smaller than shown and is well within the material fatigue limits. Deflection thus is not a factor in the life expectancy of the drive.

In the typical robotic drive, a servomotor or stepper provides input through the wave generator. The circular spline is fixed, and the output, which is through the Flexspline, is a function of the ratio.

As an example, size 4M comes in standard ratios of 80, 100, 160, and 200 to 1. At a ratio of 160:1, a servomotor running at 3500 rpm will be reduced to an output speed of 21.9 rpm and provide a torque of 3300 lb-in. The motor would have to produce 1.55 hp.

Fig. 3–39. Principle of harmonic drive. (*Courtesy of Harmonic Drive Division of Emhart Machinery Group*)

REFERENCES

1. Konstantinov, Michael S. Jaw-type gripper mechanisms. *Proc. Fifth International Symposium on Industrial Robots,* Chicago, IL.: 1975.

2. Lundstrom, G., et al. *Industrial Robots—Gripper Review.* Bedford, England: International Fluidics Services, Ltd., 1977.

3. Engelberger, Joseph F. *Robotics in Practice.* New York: AMACOM, 1980.

4. Quingsen, H. A linkage mechanism for the concentric gripping of cylindrical components. *Proc. 12th International Symposium on Industrial Robots,* Paris, France, 1982.

5. Fitzpatrick, P. F., and Barto, J. J., Jr. Automated drilling and riveting system. *Proc. SME Conference on Applying Robotics in the Aerospace Industry,* St. Louis, MO, March, 1984.

6. Taylor, C. L., and Schwarz, R. J. The anatomy and mechanics of the human hand. *Artificial Limbs,* Vol. 2, May 1955, pp. 22–35.

7. Salisbury, J. K., Jr. Design and control of an articulated hand. *Proc. International Symposium on Design and Synthesis,* Tokyo, Japan, July, 1984.

8. ICAM—Robotics Application Guide. Technical Report AFWAL-TR-80-4042, Vol. II. General Dynamics Corporation, Ford Worth, TX, 1980.

9. Salisbury, J. K., and Craig, J. J. Articulated hands: force control and kinematic issues. *The International Journal of Robotics Research,* Vol. 1, No. 1, Spring 1982.

10. Abramowitz, Jeffrey D., Goodnow, John W., and Paul, Burton. Pennsylvania Articulated Mechanical Hand. *Proc. International Conference on Computers in Engineering,* American Society of Mechanical Engineers, Chicago, 1983.

11. Rovetta, A., et al. On development and realization of a multipurpose grasping system. *Proc. 11th International Symposium on Industrial Robots,* Tokyo, Japan, 1981.

12. Whitney, D. E., and Nevins, J. L. What is the remote center compliance (RCC) and what can it do? *Proc. Ninth International Symposium on Industrial Robots,* Washington, DC, 1979.

Chapter 4
HYDRAULIC CONCEPTS

ELECTRIC, PNEUMATIC, AND HYDRAULIC ROBOTS

Robots may be actuated by electric power, hydraulic power, or pneumatic power. The pneumatic type is limited to small payloads because of the low pressure common to pneumatic systems. Electric actuators are best suited for payloads of less than 200 lb. Compared with hydraulic systems, their repeatability is usually greater. They also require less housekeeping, and their noise level is considerably lower. However, where an explosion hazard exists, the electric actuator is generally avoided. For example, hardly any paint-spraying robots are electric because of that hazard.

Where high payloads are handled, the hydraulic system is in its element. While the use of hydraulic robots may be questionable with payloads from 100 to 200 lb, they definitely predominate in the heavy robot sector, where payloads of 200 lb or more are involved. However, there are exceptions.

Thus, the T^3586 Cincinnati Milacron, which has a load capacity of 225 lb, operates, as would be expected, with a hydraulic system. But the T^3566, with a payload of only 100 lb, is also hydraulically driven. The same is true of the Unimate Series 1000, 2000, and 4000. They are all hydraulically operated; but, while the Unimate 1000 has a load capacity of only 50 lb, the 2000 can carry 300 lb and the 4000 up to 450 lb. This illustrates that once a series is designed for hydraulic operation, even the lighter models of the series may have that type of drive.

Hydraulic actuators are smaller than electrical actuators for the same output power. This is one of their advantages, particularly at larger payloads. Furthermore, because hydraulic actuators have less mass, they are, from this perspective, dynamically superior to electric actuators. However, they also require control devices and connecting lines that can slow down their response. Thus, there is continuous discussion about which system performs better dynamically, a question that can only be decided when all the circumstances of a specific case are known.

One advantage of using a hydraulic system is in connection with transient overload peaks. Means for damping them out can more readily be provided with hydraulic and also with pneumatic systems, than with electric systems. Speed reducers are generally required with electric systems, and can be particularly vulnerable to such transient overload peaks.

Another advantage of hydraulic systems is their capability of storing energy for short-term peak demands in flow capacity. This becomes most apparent with larger drives. Thus, the Unimate 2000 momentarily requires as much as 60 gpm, which is supplied by a 17-gpm pump operating part-time loading an accumulator. Because of this extra power, hydraulic drives usually have an additional dynamic advantage over electric drives.

Hydraulic systems are complex, and a knowledge of fundamental concepts is necessary to understand them. The purpose of the following discussion is to give an overview of such fundamentals with particular emphasis on robotic technology.

CHARACTERISTICS OF HYDRAULIC FLUIDS AND FLOW

Pressure, Force, and Head

Force is a push or pull. Pressure is force per unit area. Pressure is commonly expressed in pounds per square inch (psi) or, when the International Standards are used, in bars or Pascals (Pa), which is another name for newtons per square meter. In fact, 1 bar $= 10^5$ Pa $= 10^5$ N/m^2.

However, industry in general seems to hold on to the customary U.S. units—that is, psi. One psi is the force that one pound exerts on one square inch of area.

Figure 4–1 shows a cylindrical vessel filled with water to a height of 60

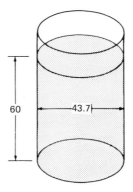

Fig. 4–1. Pressure, head, and density.

in. The inside diameter of the vessel is 43.7 in. This is equivalent to a cross-sectional area of 1500 in.^{2}. The volume of water is, therefore, 90,000 in.^{3}.

The density of water at room temperature is 62.3 lb/ft^3 or 0.036 lb/in.^{3}. Consequently, the water in the vessel in the figure weighs 3245 lb. This mass of water exerts a pressure against the bottom of the vessel of 3245/1500 = 2.16 psi.

All this can also be expressed by stating that pressure, p, is equal to the height or head of the liquid column, multiplied by the density of the liquid, ρ, or in mathematical terms:

$$p = h\rho \qquad (4.1)$$

It is important in using this equation that the head of the liquid and its density be expressed in the same units, such as inches for the head, and lb/in.^{3} for the density. The resulting pressure is then in lb/in.^{2}, or psi.

The pressure of 2.16 psi is actually the pressure of the liquid head plus the atmospheric pressure on the surface of the liquid minus the atmospheric pressure outside the containing tank.

If a hole were drilled in the bottom of the tank, water would rush out with an initial pressure of 2.16 psi. This could be indicated by a pressure gage such as those commonly used in fluid power. These gages measure the difference between atmospheric pressure on one side and fluid pressure plus atmospheric pressure on the fluid on the other side. They show what is known as *gage pressure.* They measure the hydrostatic pressure of the liquid column by canceling out the atmospheric pressure that is on it. This is done very simply by surrounding the measuring element, say a Bourdon tube, with the same atmospheric pressure.

Absolute pressure, on the other hand, includes the atmospheric pressure, which at sea level is 14.7 psi. The Bourdon tube would have to be enclosed in a vacuum in order to measure absolute pressure, or psia. In the above case, it would measure 16.86 psia: the hydrostatic pressure of 2.16 psi plus the atmospheric pressure of 14.7 psi.

In other words, 2.16 psi gage pressure is equal to 16.86 psi absolute pressure. (In Denver, Colorado, pressure would be about 14.4 psi absolute, but the gage pressure would be the same 2.16 psi.) If gage pressure is expressed by p_g and atmospheric pressure by p_a, then what is generally called pressure, p, is:

$$p = (p_g + p_a) - p_a \qquad (4.2)$$

where $(p_g + p_a)$ is the absolute pressure, and p is equal to the gage pressure, p_g.

Just as absolute pressure in customary U.S. units may be abbreviated psia, so gage pressure is equivalent to psig. Where it is understood which type of pressure is meant, the abbreviation psi suffices.

Transmission of Pressure

Fluid power operates by applying pressure to a load. It is based on Pascal's law, which simply states that pressure applied to an enclosed fluid is transmitted undiminished to every portion of the fluid and the walls of the containing vessel.

It follows that pressure exerts the same force on every square inch of a fluid power system. This is true as long as the fluid does not move. When it flows, other factors have to be taken into consideration, as discussed further below.

Pascal's law refers to quiescent conditions. In this case:

$$F = pA \qquad (4.3)$$

where F is the force in pounds, p the pressure in psi, and A the area in square inches.

The piston area of the cylinder in Fig. 4–2 is 12 in.^{2}. A pressure of 600 psi is applied against this piston. This pressure is capable of balancing a load of 7200 lb.

Figure 4–3 illustrates a further characteristic. Force F_1 applied to area A_1

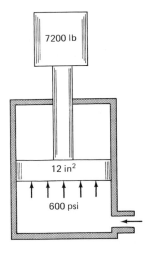

Fig. 4–2. Fluid pressure balancing load.

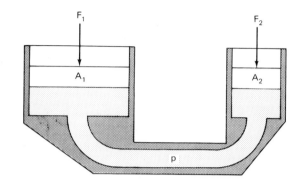

Fig. 4–3. Transmission of pressure and amplification of force.

produces a pressure in the fluid that requires a counteracting force, F_2, to keep the system in balance. According to Eq. 4.3, this pressure is:

$$p = \frac{F_1}{A_1} = \frac{F_2}{A_2} \qquad (4.4)$$

This illustrates what Pascal's law expresses, namely, that pressure can be transmitted undiminished to any point of the system and produce a force that is proportional to the area to which it is applied.

Let the effective area A_1 be 10 in.^{2} and area A_2 be 2.5 in.^{2}. If force F_1 is 40 lb, a pressure of 400 psi results. To keep both sides of the system in equilibrium, this must be balanced by a force F_2. Since A_2 is only 2.5 in.^{2}, it requires a force of $F_2 = 160$ lb to produce the same pressure, $p = 400$ psi. In other words, 40 lb force applied on one side can generate 160 lb on the other. This is the significance of Eq. 4.4.

Two practical conclusions can be drawn from these examples:

1. Fluid power not only permits easy transmission of pressure, but it is also readily convertible to almost any force, depending solely on the area on which it acts.
2. Tremendous forces can be produced with relatively small pressures, which should be a reminder of the safety aspects involved, namely, that the force that may act on some component in a hydraulic system can be far larger—and hence far more dangerous—than suspected.

The amplification of force that can be obtained by fluid power is by no means an amplification of the work the system is capable of producing. Work is the product of force times distance through which it is applied. This means that:

$$W = FL \qquad\qquad (4.5)$$

where work, W, is in in.-lb; force, F, is in pounds, and the distance over which the object is moved by the force, L, is in inches. Obviously, if the distance were expressed in feet, then the work would be in ft-lb.

Thus, let force F_2 in Fig. 4–3 push the piston with area $A_2 = 2.5$ in.^{2} downward by 3 in. As this is done, work of 3 in. $\times$ 1000 lb = 3000 in.-lb is expended. This work would displace a volume of liquid of 7.5 in.^{3}, which is the same volume that is added underneath A_1. Since $A_1 = 10$ in.^{2}, the resulting motion would be 0.75 in., and the work expended would be 0.75 in. $\times$ 4000 lb = 3000 in.-lb. This, of course, neglects friction losses, but illustrates that work, unlike force, cannot be increased without adding outside energy.

The transmission of pressure is only undiminished as long as the fluid is at rest. Flowing fluids create friction and hence losses.

Friction

The force needed to move a block of, say, 500 lb depends on a number of factors, some of which were mentioned in Chapter 1 with respect to the static and dynamic behavior of robots. They are included here with specific reference to hydraulics. We will consider coulomb friction, static friction, and viscous friction.

Coulomb Friction.   This is also called dry or sliding friction (see page 13). It is the force, F_1, required to move an object over a surface. This force is proportional to the vertical component of the weight, w, of the sliding object multiplied by the coefficient of sliding friction, K, which is an empirical factor determined by the material, its smoothness, and the lubrication between object and surface. For horizontal sliding, this can be expressed by:

$$F_1 = Kw \qquad\qquad (4.6)$$

Static Friction.   This is also called stiction, or break-loose friction. It is the force required to start moving an object over a surface against which it presses with a certain force. Static friction is due to minute asperities between the surfaces. Pressure applied on the object may even cause a welding together of these asperities and thus increase break-loose friction.

The magnitude of static friction is expressed by the coefficient of static friction, which is determined experimentally and is subject to variations.

Values given in tables for sliding as well as for static coefficients of friction can only be used as a rule of thumb.

Viscous Friction. This is caused by the shearing of fluid in narrow clearance spaces. Pistons, pumps, and motors that have internal leakage produce viscous friction. In many cases, viscous friction is desirable.

Sometimes, dashpots are used to provide time delays in the 1-sec to 2-min range. Here, a sealed cylinder is filled with a viscous fluid, and a piston is pushed through it. The clearance between the piston and the cylinder wall is usually so small as to be negligible, but a small metering hole in the piston permits the fluid to pass from one side to the other. The velocity, v, with which the piston can move through the liquid is proportional to the force, F_2, that is applied from the outside to move the piston through the fluid, and is inversely proportional to the damping constant, c, caused by the viscous friction of the fluid moving through the metering hole. This means that $v = F_2/c$. Obviously, the fluid in a dashpot must maintain a rather constant viscosity when temperature changes occur; otherwise the velocity, and hence the time to complete its motion, will change.

An essential factor in the case of the dashpot is the force, F_2, because the preceding equation can be written:

$$F_2 = cv \qquad (4.7)$$

In a hydraulic system, where sliding parts sometimes must move through more or less entrapped oil, the same force is active, and it can reach considerable magnitude. But since it is a force that is proportional to velocity, it is a characteristic that also lends itself to damping unstable systems, and it may well be used for this purpose, particularly in servovalves.

Consequences of Friction. Friction is essential. Without friction we could not walk, cars would not roll, and so on. Viscous friction is beneficial because it provides damping. However, there are many negative effects that require reducing friction to an absolute minimum.

Let a steel block of 500 lb slide on a horizontal cast iron table. Assuming a coefficient of static friction of 0.5, a force of 250 lb is needed to start moving the block.

If the coefficient of sliding friction in this case is assumed to be 0.3, then the force required to keep the block moving would be 150 lb.

If the steel block is moved by a hydraulic cylinder, the friction of the piston against the cylinder walls and seals must be added. Furthermore, if the piston rod bends under the load, be it ever so slightly, the friction force increases considerably. When the load is moved upward, the corresponding gravitational forces must be added.

One should understand all these factors to appreciate the excessive amount of energy that can be lost unless all precautions are taken to minimize friction. In addition, these various sorts of friction cause heat, which can further increase the friction if materials of different thermal expansion are used. It can also increase the temperature of the hydraulic fluid beyond its permissible limits. Furthermore, friction produces abrasion, which shortens the life of components.

ACCELERATION

Equation 1.1 stated Newton's second law in the form $F = ma$; that is, force equals mass times acceleration. If mass is expressed as weight in pounds, acceleration in in./sec², and force in pounds, then:

$$F_a = \frac{wa}{386} = 0.0026 \ wa \qquad (4.8)$$

which is in accordance with Eqs. 1.1 and 1.2.

Suppose that the above steel block has to be brought from rest to a velocity of 4 in./sec within 0.2 sec; then the acceleration, a, is 20 in./sec². Since the block weighs 500 lb, the force required to produce this acceleration, as calculated by Eq. 4.8, is 26 lb.

Let the steel block be accelerated by the force of a hydraulic cylinder with a piston with an effective area of $A = 3$ in.². The question would be: What hydraulic pressure is needed to provide acceleration of 20 in./sec² for 0.2 sec? According to Eq. 4.3, pressure times area is force. The total pressure for a given area that is required to accelerate the block once it starts moving, can be written:

$$P = \frac{0.0026 \ wa + cv + Kw}{A} \qquad (4.9)$$

Here, the term $0.0026 \ wa$ is the force component according to Eq. 4.8; the force component due to viscous damping is cv, as given in Eq. 4.7; and Kw is the force component due to sliding friction, as given in Eq. 4.6.

Generally , static friction need not be considered, since it is active only at break-loose and not during motion. Here, it is assumed that the block has begun to move. Static friction may not be of great significance in the overall picture, since at the point of break-loose there would be neither sliding nor viscous friction. In other words, if there is enough force for significant acceleration, there is also enough force to start the motion to begin with. Only in exceptional conditions (and these should usually be remedied by suitable lubrication, etc.) may it be found that static friction is unduly high and that allowance must be made in the hydraulic pressure to take care of this condition.

Consider the case of the sliding steel block and its characteristics, as used in the previous examples. It may also serve to illustrate the application of Eq. 4.9. To begin with, the magnitude of the term 0.0026 wa in Eq. 4.9 is 29 lb, as was already calculated with respect to Eq. 4.8.

Furthermore, the magnitude of cv is the damping constant times velocity. It is a force component primarily caused by the hydraulic fluid being displaced in the cylinder and in its connections. The velocity is that of the load, which is the same as that of the piston. Since it was previously assumed that the steel block changes velocity during the acceleration from 0 to 4 in./sec, one can assume that this change is linear, and hence the average velocity is 2 in./sec. Thus with a damping constant of, say, 7.5, one obtains $cv = 15$ lb.

The sliding friction, Kw, is primarily caused by the steel block unless the piston rod is subject to a bending moment from the load. Such a condition would lead to rapid deterioration of piston seals and other parts, and must be eliminated. In a properly designed system, sliding friction of the piston can be neglected. Hence, Kw in Eq. 4.9 is, as previously calculated, 500 × 0.3 = 150 lb.

Inserting all these values in Eq. 4.9, and assuming that the effective area of the piston is $A = 3$ in.^{2}, gives $p = (26 + 15 + 150)/3 = 63.7$ psi.

In the preceding, it has been assumed that the load moves horizontally and that no significant pressure losses occur in the connecting lines. The vertical lifting of a load by a robot, however, would have to consider the additional weight of the load itself and the friction of the joints. Thus, a robot by Cincinnati Milacron may operate with 2250 psi, not only to deliver the force required for the payload but also to cover the above mentioned losses plus others, such as those due to friction in robot joints, connecting hydraulic lines, valves, and so on.

THE CONTINUITY EQUATION

Flow rate expressed in pounds of fluid flowing at some velocity v through a conduit with cross-sectional area A is equal to $\rho A v$. In this expression, ρ is the density of the fluid in lb/in.^{3}. Let a hydraulic fluid have a density of 0.03 lb/in.^{3}. It flows through a conduit with a cross-sectional area of 0.8

in.^{2} at a velocity of 100 in./sec. This means that the flow rate, expressed in pounds of fluid per second, is 2.4 lb/sec.

What happens if the conduit gets narrower or wider? Consider an aluminum rod that is extruded through a die to a smaller diameter. Obviously, if one pound of aluminum is to be extruded every second, this "rate of flow" remains the same, independent of the diameter: 1 lb/sec is fed into the die, and 1 lb/sec must come out. What happens, however, is that the velocity must increase during the extrusion—and the increase must be in inverse proportion to the cross-sectional area of the rod. Otherwise, the passage would be choked.

The rod will probably increase in temperature as it passes through the die. This will reduce its density and expand the volume. Again, the velocity has to increase to take care of that extra length of outgoing rod. This influence of thermal expansion is negligible in hydraulic fluids, although in pneumatic systems it becomes a significant factor.

In other respects, the hydraulic fluid behaves like the aluminum rod. The weight of fluid flow remains the same, independent of changes in density, velocity, or area. In algebraic terms:

$$\rho_1 A_1 v_1 = \rho_2 A_2 v_2 = \rho_3 A_3 v_3 = \text{constant} \qquad (4.10)$$

This is known as the continuity equation.

However, the density of hydraulic fluids practically remains the same (contrary to pneumatic system where the fluid is air); consequently $\rho_1 = \rho_2 = \rho_3$, etc., and, hence:

$$A_1 v_1 = A_2 v_2 = A_3 v_3 = \text{constant} \qquad (4.11)$$

Figure 4–4 shows a pipe section of three different diameters, A, B, and C. Let the cross-sectional area of A be 0.2 in.^{2}; of B, 0.04 in.^{2}; and of C, 0.8 in.^{2}. If the flow enters A with a velocity of 100 in./sec, it must, according to Eq. 4.11, increase its velocity in B to 500 in./sec, and then decrease it in C to 25 in./sec.

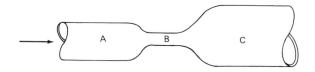

Fig. 4–4. Pipe section with changing diameters.

VOLUMETRIC FLOW

The die through which the aluminum rod in the above example is extruded will have a diameter of 0.5 in. This is equivalent to an area of about 0.2 in.^{2}. Let it pass through this area with a velocity of 4 in./sec. The volume of aluminum that passes through the die is the product of velocity and cross-sectional area, which is 0.8 in.^{3}/sec. The same is true of a fluid. Its volumetric flow is the product of the velocity and the cross section of the conduit through which it flows.

Hydraulic flow, in customary U.S. units, is expressed in gallons per minute (gpm). One gallon equals 231 in.^{3}. And since it is expressed in terms of minutes, while velocity is expressed in seconds, a conversion factor of 60 must be used. Consequently:

$$Q = 0.26 \, vA \qquad (4.12)$$

where Q is flow rate in gpm, v is flow velocity in in./sec., and A is cross-sectional area in in.^{2}.

FLOW THROUGH RESTRICTIONS

More than 300 years ago, Torricelli observed liquids discharging through orifices in the bottom of vessels, as shown in Fig. 4–5. He found an analogy between the motion of falling bodies and the flow of liquids through an orifice. Galilei had already established that the velocity of a falling body is:

$$v = \sqrt{2 \, gh} \qquad (4.13)$$

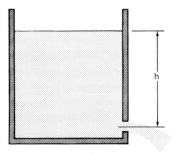

Fig. 4–5. Torricelli's theorem.

If g is the acceleration due to gravity, which is generally assumed to be 386 in./sec², and h is the height from which it falls, in inches, then v is velocity in in./sec.

Torricelli applied this concept to liquids. It became known as Torricelli's theorem, which simply states that "the liquid velocity at an outlet discharging into the free atmosphere is proportional to the square root of the head."

According to Eq. 4.1, $p = h\rho$. This permits substituting p/ρ for h in Eq. 4.13, which then reads $v = \sqrt{2gp/\rho}$.

Furthermore, by substituting in Eq. 4.12 and considering that $g = 386$, Eq. 4.13 can be written as:

$$Q = 7.2\ A\ \sqrt{\frac{p}{\rho}} \qquad (4.14)$$

The equation applies to idealized conditions. In actual systems, the flow rate through an orifice diminishes because of pressure losses in the passage through the restriction. To make allowance for these losses, the discharge coefficient C_d is introduced, and Eq. 4.14 is written as:

$$Q = 7.2\ C_d A\ \sqrt{\frac{p}{\rho}} \qquad (4.15)$$

Various types of restrictions are shown in Fig. 4–6. The one on the left is a sharp-edged orifice. In this case, the discharge coefficient, C_d, can usually be assumed to be 0.65. If the orifice is rounded, as shown in the restriction in the middle, C_d is usually between 0.70 and 0.75. For a bell-shaped nozzle, as shown on the right, a discharge coefficient of $C_d = 0.93$ can be expected.

Consider a diameter d in Fig. 4–6 of 0.2 in. This makes $A = 0.0314$ in.².

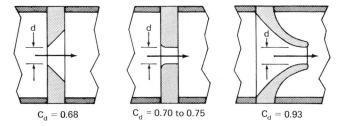

$C_d = 0.68$　　　$C_d = 0.70$ to 0.75　　　$C_d = 0.93$

Fig. 4–6. Shapes of flow restrictions.

Let $\rho = 0.03$ lb/in.^{3} and $p = 10$ psi. The result, according to Eq. 4.15, would be a flow rate of about 2.68 gpm for the sharp-edged orifice, 2.89 to 3.10 gpm for the rounded orifice, and 3.84 gpm for the bell-shaped orifice.

Valve ports that modify flow are restrictions that may be considered orifices. A discharge coefficient of $C_d = 0.625$ is generally assumed for these ports. Thus, a valve that is to pass 10 gpm at a pressure drop from supply to load of 50 psi must, at a fluid density of 0.03 lb/in.^{3}, have a port area of 0.054 in.^{2}.

The above equations ignore the viscosity of the fluid. Actually, this is a simplification that, while used in practice, can introduce some error. Thus, it does not show that sharp-edged orifices, although the most inefficient, are the ones least sensitive to changes in viscosity. Viscosity and its variations due to temperature changes will be discussed later. However, because the pressure drop across a sharp-edged orifice is the least affected by changes in viscosity, it is often used in flow control valves to measure pressure drop and thereby keep the flow constant.

ENERGY AND POWER

Energy is capacity for doing work. Inversely, work is the manifestation of energy. A fluid has energy by virtue of its pressure. Through work, the energy reveals its presence. Both work and energy can be expressed in in.-lb. If a 5-lb weight is lifted 10 ft (120 in.), the work done on it is 600 in.-lb. This is the energy stored in the weight.

There are two basic forms of energy: potential and kinetic. Potential energy is due to position or condition. Let the 5-lb weight be lifted by a hydraulic cylinder, and then let the supply line to the hydraulic cylinder be closed by a valve. Now, the energy stored in the system has potential energy of 600 in.-lb. If the valve is opened, the weight pushes the fluid out, and the potential energy becomes kinetic energy.

Energy can be converted. It cannot be created or destroyed, but it can be dissipated, usually as heat, which may then be referred to as energy loss—at least for all practical purposes.

Power is the rate of doing work. It is expressed by:

$$P = \frac{W}{6600\, t} \qquad (4.16)$$

where P is horsepower (hp), W is work in in.-lb, and t is time in seconds. This equation shows that producing the same work in half the time requires twice as much power.

Lifting a load requires not only pressure but also a certain flow rate. Energy and work are functions of pressure, but power is a function of pressure and flow rate. To obtain the necessary equation, only three steps are necessary:

1. Substitute Eqs. 4.3 and 4.5 in Eq. 4.16 and write:

$$P = \frac{pAL}{6600\ t} \qquad (4.17)$$

2. Consider that L/t (distance per time) is velocity and hence:

$$P = \frac{pAV}{6600} \qquad (4.18)$$

3. Consider that Eq. 4.12 can be written $v = 3.85\ Q/A,$ and substitute this in Eq. 4.18. The result is very nearly:

$$P = 5.83 \times 10^{-4}\ (pQ) \qquad (4.19)$$

Suppose, it is necessary to lift a load of 750 lb, plus 50 lb for friction and so forth, to a height of 60 in. in 2 sec. Neglect acceleration and deceleration, and let the piston area be 2 in.^{2}, which means the pressure necessary to develop the force of 800 lb amounts to 400 psi.

The work to be done is 800 lb multiplied by 60 in. Hence, $W = 48,000$ in.-lb. It has to be done within 2 sec., which according to Eq. 4.16 requires 3.64 hp.

The necessary flow rate can now be obtained from Eq. 4.19. Inserting the values for P and p gives 15.6 gpm.

DENSITY AND SPECIFIC GRAVITY

Weight density, ρ, is weight per unit volume. Mass density, ρ_m, is weight density divided by the gravitational constant. Hence, $\rho_m = \rho/g$. In engineering, it is customary to refer to weight density simply as density.

Specific gravity of a liquid is the ratio of its density at 60°F to the density of water at the same temperature. (Physicists use 4°C as the reference temperature.) Since pressure generally has an insignificant effect upon the density of liquids, temperature is the only condition that must be considered in defining the basis for specific gravity.

The density of water at 60°F is 0.0361 lb/in.^{3}, while the density of a typical mineral oil at the same temperature is 0.0316 lb/in.^{3}. Hence, the specific gravity of this oil is 0.875. This may be written sp gr 60/60°F = 0.875. Some typical densities and specific gravities of hydraulic fluids are given in Table 4-1.

Density of hydraulic fluids may be expressed in other units. One system that is frequently encountered is the API (American Petroleum Institute) scale. The units are commonly expressed in "degrees API" or "API gravity at 60°F." To convert from sp gr 60/60°F (Table 4-1) to degrees API, one uses the formula:

$$\text{deg API} = \frac{141.5}{\text{sp gr } 60/60°F} - 131.5$$

For example, the API reading of SAE 30 (Table 4-1) would be 26 deg API.

The API system is not used for specific gravities above 1.000. Above this point, the Baumé scale is used, as follows:

$$\text{deg Baumé} = 145 - \frac{145}{\text{sp gr } 60/60°F}$$

Thus, the phosphate ester in Table 4-1 would have a density of 31.3 deg Baumé.

Densities, and thus, specific gravities, vary only slightly with temperature. The temperatures a robotic hydraulic system is concerned with do not exceed 150°F, and usually 120°F is the upper limit. In this temperature range, changes in density are rather small and can usually be ignored. Errors will hardly exceed 4% when the temperature reaches 120°F and 7% when the temperature reaches 150°F.

Table 4-1. Densities and specific gravities of some hydraulic fluids.

FLUID	DENSITY AT 60°F		SP GR 60/60°F
	LB/FT^3	LB/IN^3	
Mineral oil—SAE 10	54.64	0.0316	0.875
SAE 30	56.02	0.0324	0.898
Water-in-oil emulsion	57.14	0.0331	0.916
Synthetic silicone	60.07	0.0348	0.963
Water–glycol base	66.12	0.0383	1.060
Undiluted phosphate ester	71.49	0.0414	1.146
Phosphate ester	79.54	0.0460	1.275

EXPANSION

The density of hydraulic oil, like that of other fluids, decreases with increasing temperature, with a decrease in density corresponding to an increase in volume. Such an expansion of oil can have serious consequences. When it is confined to a space without air pockets, an increase of temperature and consequent expansion leads to either leakage or bulging, and eventually to the bursting of restraining walls. But even when air space is provided and there is no vent, large forces can develop, as will be shown.

Expansion can be expressed by an equation that is sufficiently accurate for most engineering applications, as follows:

$$V_2 = V_1 \, [1 + B(T_2 - T_1)] \qquad (4.20)$$

where V_2 is the larger volume in in.^{3}, V_1 is the smaller volume, $(T_2 - T_1)$ is the temperature increase in degrees Fahrenheit, and B is the coefficient of thermal expansion. For most hydraulic fluids, an average value of $B = 5 \times 10^{-4}$ is a relatively safe assumption.

If the hydraulic fluid can only expand in length and the cross-sectional area is the same throughout, as it is in a reservoir, a pipe, and so on, then Eq. 4.20 can be written:

$$L_2 = L_1[1 + B(T_2 - T_1)] \qquad (4.21)$$

where L is the length of the liquid column.

Let a closed reservoir of rectangular shape be $20 \times 8 \times 12$ in. high and be filled with hydraulic fluid at 60°F up to a level of 11 in., so that a 1-in. air cushion remains on top. The tank is closed, and the oil heats up to 120°F. According to Eq. 4.21, this expansion would increase the height of the oil level by 0.33 in. Consequently, the air cushion that initially was 1 in. is now compressed to 0.67 in.

If the air was at atmospheric pressure (i.e., 14.7 psia) to begin with, then its absolute pressure would now increase to $p_a = 14.7 \, (1/0.67) = 20.9$ psia. This is an increase of pressure of 6.2 psi, which results in a force pushing against the top of the tank equal to $6.2 \times 20 \times 8 = 992$ lb.

COMPRESSIBILITY

Hydraulic fluids are often considered incompressible. This is a valid assumption in many applications. However, minute as the compressibility is, it can become a factor, since the natural frequency of servo systems is inversely proportional to the square root of compressibility. Practical use of the compressibility of liquids is made in liquid springs but not in robots.

Compressibility leads to a reduction of volume under pressure. It can be approximately expressed by:

$$V_2 = V_1 (1 + Cp) \qquad (4.22)$$

where V_2 is the larger and V_1 the smaller volume in in.^{3}; p is the pressure in psi; and C is the compressibility coefficient in in.^{2}/lb.

Instead of compressibility, the term bulk modulus is often used. The latter is simply the reciprocal of compressibility.

Compressibility coefficients change with temperature, pressure, and type of fluid. Exact magnitudes are not available. They vary between 2×10^{-6} and 7×10^{-6} in.^{2}/lb. For practical purposes, it is customary to use a compressibility coefficient of 4×10^{-6} or 5×10^{-6} in.^{2}/lb. A compressibility factor of 4×10^{-6} in.^{2}/lb is equivalent to a compression of 0.4% for every 1000 psi pressure.

Where the cross-sectional area is fixed, it is only the length of the liquid column, L, that is affected, and thus takes the place of V in Eq. 4.22. Consider a cylinder that has a 12-in. stroke. Let the piston rod be fully extended. A load is applied that increases the fluid pressure from zero to 2250 psi. Since the cross-sectional area remains the same, only the length of the fluid column changes. Substituting L for V in Eq. 4.22 and inserting a compressibility coefficient of 4×10^{-6} in.^{2}/lb, the length of the oil column in the cylinder is found to change from 12 in. to 11.893 in. In other words, this piston will move slightly more than 0.1 in. when the load is applied.

VISCOSITY

Motion of objects creates friction. This also applies to fluids in motion. Viscosity is defined as a measure of the internal friction or, similarly, of the resistance the pressure has to overcome in order to cause the fluid to flow.

There are two kinds of viscosity: absolute (also called dynamic viscosity) and kinematic. Viscosity without further reference means, at least in this text, kinematic viscosity.

The absolute viscosity of a liquid is determined experimentally by means of a viscosimeter, the principle of which is illustrated in Fig. 4-7. Here, a cylinder rotates inside a stationary cylindrical container (the container could also be made to rotate around a fixed cylinder—the effect on the test would be the same). Cylinder and container walls are concentric and have a very narrow gap between them of width L. The space is filled with the liquid to

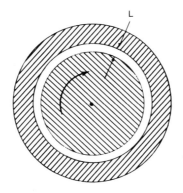

Fig. 4-7. Schematic cross-section of viscosimeter for measuring absolute viscosity.

be tested. The force, F, necessary to rotate the inner cylinder at a given velocity is measured.

The liquid in contact with the moving cylinder surface is found to have the same velocity as that surface, while the liquid adjacent to the inner wall of the stationary cylindrical container is at rest. The velocity of intermediate layers of the liquid increases uniformly from one wall to the other. Thus, there is a velocity gradient or shear rate which is equal to v/L.

The force that is applied to rotate the inner cylinder tends to drag the liquid around. We can visualize the liquid as successive concentric layers with the rotating cylinder exerting a shearing force on them. If A is the area of the liquid over which the force is applied, the ratio F/A is the shearing stress exerted on the liquid.

In sum, the force depends on:

- The linear velocity, v, of the rotating cylinder.
- The surface area, A, of the rotating cylinder in contact with the liquid.
- The distance, L, or width of the interspace between cylindrical container and rotating cylinder.
- The absolute viscosity of the fluid, μ.

The absolute viscosity, the liquid characteristic that is to be determined, is proportional to FL/Av. This relation contains the shearing stress, F/A, as well as the shear rate, v/L. Thus, absolute viscosity is the ratio of shearing stress to the shear rate of the fluid.

If, for example, the force were 1.25 lb, the interspace 0.002 in. wide, the area of contact 7 in.^{2}, and the velocity 30 in./sec, then the absolute viscosity would be 0.000048 lb sec/in.^{2}. Units expressed in these dimensions are called Reyns.

As seen from this example, the Reyn yields uncomfortably low numbers. Therefore, it became customary to use the centipoise (cP) as the basic unit of absolute viscosity. One centipoise is approximately equal to the absolute viscosity of water at 68°F; that is, one centipoise is equal to 1.45×10^{-7} Reyns. The symbol for cP is μ. The following equation can thus be written:

$$F = 1.45 \times 10^{-7} \frac{\mu A}{L} \qquad (4.23)$$

Inserting the same values as used above would give a viscosity $\mu = 82$ cP.

Equation 4.7 is $F = cv$, where c is the damping constant. Comparing this with Eq. 4.23, the damping constant is the equivalent of $c = 1.45 \times 10^{-7} (A/vL) \mu$, which illustrates the relationship of damping and absolute viscosity. For damping to become of significant magnitude, the width of the passage, L, must be very narrow.

This sort of viscous damping occurs in narrow passages of valves, cylinders, pumps, and so on. It has the advantage of increasing the stability of a system, but the disadvantage of requiring force, hence dissipating energy, and of being dependent on viscosity, which changes with temperature. It should also be noted that friction factors are significant in such narrow passages, and temperature could be considerably higher locally than in the rest of the system, thus causing degradation of hydraulic fluid at an apparently normal overall temperature.

In many of the equations concerning fluid power, viscosity occurs as the ratio of absolute viscosity to fluid density, and this ratio (for unknown reasons) is called kinematic viscosity.

In practice, kinematic viscosity is usually measured by an instrument quite different from the one used for absolute viscosity. Kinematic viscosity is determined by means of the Saybolt Universal Viscosimeter, which is sensitive to viscosity as well as density. It has the virtue of simplicity, since it measures the time required for 60 cm^3 of hydraulic fluid at 100°F to flow through an orifice of specific size. The number of seconds is the Saybolt viscosity. The exact method is described in ASTM Standard D 88–35, "Viscosity by Means of the Saybolt Viscosimeter." The resulting kinematic viscosity is expressed in SUS (Saybolt Universal Seconds), which is also written SSU.

Kinematic viscosity (or viscosity for short) is usually expressed in either SUS or centistokes (cSt). SUS is experimentally established, but the centistoke is a mathematical quantity that is equal to the absolute viscosity divided by the specific gravity of the fluid. Thus:

$$\nu = \frac{\mu}{s} \qquad\qquad (4.24)$$

where ν is the viscosity in cSt, μ is the absolute viscosity in cP, and s is the specific gravity of the hydraulic fluid.

Furthermore, the equation:

$$\text{SUS} = 4.6 \, \nu \qquad\qquad (4.25)$$

can be used to convert centistokes to SUS. Although it is not an exact conversion, the results are close enough for practical results with hydraulic fluids. Thus, by means of Eqs. 4.24 and 4.25 useful relations between the three most commonly used expressions of absolute and kinematic viscosities are obtained.

FLUID FLOW

To pass flow through a conduit a pressure differential between the inlet and the outlet is necessary. This can be caused by elevation (head) or by a pump. The pressure differential is needed because of pressure losses caused by inertia, wall friction, and viscosity. The factors that determine the required pressure differential for a given flow depend to a large extent on whether the flow is laminar or turbulent.

Laminar and Turbulent Flow

As long as flow through a circular conduit may be visualized as many concentric molecular layers of fluid, the flow is laminar.

As laminar flow moves through a conduit, the outermost layer rubs against the conduit wall. This boundary layer friction is much higher than that between the outermost layer and the next layer. The result is that the outermost layer clings to the wall and lets the next layers do the flowing. There is some friction on each successive layer, but since each layer is in motion, the absolute flow velocity of each layer increases gradually toward the center.

A profile may be drawn, as shown in Fig. 4–8, illustrating how velocity increases from zero along the wall of the conduit to maximum in the center. The average flow velocity that would result from these different velocities of the concentric layers would be about 50% of the maximum which takes place in the center of the pipe.

As the flow rate increases, the flow changes gradually from laminar to turbulent. With increasing turbulence the curve shown in Fig. 4–8 flattens

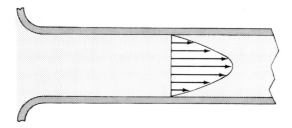

Fig. 4–8. Velocity profile of laminar flow.

out more and more, making the flow velocity more equal throughout the cross section. At the beginning of turbulence, the average velocity may be about 70% of maximum. It increases with the Reynolds number until the average velocity settles down to about 82% of maximum velocity.

Reynolds Number

When the flow under certain given conditions increases beyond a critical flow velocity, it becomes turbulent. The otherwise straight boundaries between concentric layers begin to dissolve. The velocity at which this occurs depends on the diameter of the conduit and the density and viscosity of the fluid. All these factors are combined in the Reynolds number, R_e, by the equation:

$$R_e = 645 \frac{dvs}{\mu} \qquad (4.26)$$

where s is the specific gravity of the fluid, v is the velocity of the fluid in in./sec, d is the diameter of the conduit in inches, and μ is the absolute viscosity in cP.

Substituting Eq. 4.24 in Eq. 4.26 gives:

$$R_e = 645 \frac{dv}{\nu} \qquad (4.27)$$

The important function of the Reynolds number is that it gives an indication of whether or not a flow is turbulent. Above a Reynolds number of 4000, flow will be turbulent. Below 2000, it is laminar. Between 2000 and 4000, the flow pattern is unpredictable.

The lower limit of 2000 applies to a sufficiently long and straight conduit. If there are abrupt bends in the line, turbulence may result with Reynolds numbers as low as 1200. After the flow has passed the bend, the turbulence will gradually smooth out again.

Turbulent flow requires more pressure to supply the same flow rate through the same conduit than does laminar flow. Turbulence produces added friction, internally within the fluid as well as with respect to the conduit wall. The total friction effect is a function of the roughness of the wall and of the Reynolds number.

Because of these pressure losses, turbulent flow is avoided as much as possible, and the Reynolds number in conduits is kept well under 2000. Hydraulic fluids may vary in their viscosity between 20 and 80 cSt at 120°F. At lower temperatures, the viscosity increases, and, consequently, the Reynolds number decreases. The lowest value of 20 cSt could be chosen to substitute in Eq. 4.27, which then reads $R_e = 32\,dv$. If the Reynolds number is to be kept at a maximum of 1600, then dv must not be greater than 50. For an inside diameter of the conduit of 1 in., the maximum velocity of the fluid would then be 50 in./sec. Since a diameter of 1 in. corresponds to a cross-sectional flow area of 0.79 in.^{2}, the maximum flow through this pipe would be about 10 gpm.

If the diameter were 2 in., the velocity would have to be only 25 in./sec. But since the area would be 3.14 in.^{2}, laminar flow would be assured for up to 20 gpm.

PRESSURE DROPS

Darcy's Formula

The general equation for pressure drop is known as Darcy's formula. It may be written as:

$$p = 1.3 \times 10^{-3}\,\frac{fL\rho v^2}{d} \qquad (4.28)$$

where p is the pressure drop in psi; f, the friction factor; L, the length of the conduit in inches; ρ, the density in lb/in.^{3}; v, the velocity in in./sec; and d, the diameter in inches.

Darcy's formula in this form applies equally for laminar and for turbulent flow. If the flow is laminar, the friction factor is expressed by:

$$f = \frac{64}{R_e} \qquad (4.29)$$

By combining Eqs. 4.27 and 4.29, the friction factor can be eliminated, and Darcy's formula can be written with a slight approximation:

$$p = 1.3 \times 10^{-4} \frac{L\rho v \nu}{d^2} \qquad (4.30)$$

It should be noted, however, that in this form Darcy's formula applies *only* to laminar flow.

Consider the above example, where $R_e = 1600$, the viscosity $\nu = 20$ cSt, the velocity $v = 50$ in./sec, and the diameter $d = 1$ in. A conduit that is 12 ft long, that is, where $L = 144$ in., and density of the hydraulic fluid, ρ, equal to 0.0316 lb/in.^{3} are also assumed. The pressure drop in this case would be $p = 0.59$ psi.

Resistance Coefficients

Laminar flow can be maintained in straight pipes by using suitable dimensions. However, when flow encounters an obstacle, turbulence is usually unavoidable. Valves, bends, and fittings are such obstacles. Turbulence must not only be expected for the flow through the valve, but also for some distance in the downstream conduit until the flow smoothes out. Even upstream some of the turbulence may be reflected and produce an increased pressure drop.

To account for these losses, a resistance coefficient, K, is introduced. It is an empirical factor and like many empirical factors should not be considered much more than a rule of thumb. The resistance coefficient takes the place of L/d in Eq. 4.30, which then reads:

$$p = 1.3 \times 10^{-4} K \frac{\rho v \nu}{d} \qquad (4.31)$$

Consider, for example, a conventional swing check valve. It has a resistance coefficient of $K = 135$. If the density is, as in the previous example, 0.031 lb/in.^{3}, the velocity 50 in./sec, the viscosity 20 cSt, and the diameter 1 in., then the pressure drop caused by the check valve is 0.54 psi, or practically the same as in the above 12-ft-long pipe.

The above equation also applies to exits of pipes from the reservoirs of hydraulic systems. Figure 4-9 shows various configurations of such exits and their corresponding resistance coefficients. It illustrates that the pressure loss in the case of an inward projection is almost 20 times greater than that of a well-rounded connection.

Flow through Orifices

An orifice is a flow restriction of negligible length. Figure 4-10 shows one form of such a restriction. The flow, in passing through the orifice, obvi-

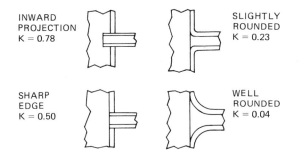

Fig. 4-9. Values of K for different exit configurations.

ously must contract to the diameter of the orifice. An important phenomenon, however, is that it continues contracting after passing through the orifice, before spreading out again.

While the flow contracts and then expands again, it becomes turbulent. The point of its narrowest cross-sectional area, A_2, is about 62% of the orifice area, A. Hence:

$$A_2 = 0.62A \qquad (4.32)$$

The location of A_2 is generally referred to as the vena contracta. There is also the cross-sectional area of the flow upstream from the orifice, designated A_1.

Equation 4.11 stated that $v_1 A_1 = v_2 A_2$, which may also be written $v_1 = v_2 A_2 / A_1$. This means that the upstream flow velocity, v_1, is equal to the velocity through the vena contracta multiplied by the ratio of the area of the vena contracta to the area of the upstream conduit.

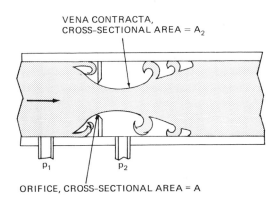

Fig. 4-10. Flow through orifice.

The law of conservation of energy establishes that energy cannot be lost and cannot be created, although countless transformations from one form of energy into another can occur. This also applies to the fluid energy upstream and downstream from the orifice. As stated before, even what is commonly called "energy loss" is converted into some other form—usually heat. It is merely lost for doing useful work but is maintained in a global sense (disregarding entropy effects).

Bernoulli's theorem for hydraulics is based upon this concept of conservation of energy. The theorem includes a number of terms, some of which can be neglected in considering a robotic hydraulic system. In its simplified form, it may be written:

$$\frac{v_1^2}{2q} + \frac{p_1}{\rho} = \frac{v_2^2}{2g} + \frac{p_2}{\rho} \tag{4.33}$$

where p_1 and v_1 refer to the conditions upstream from the orifice, and p_2 and v_2 to those at the vena contracta. The term g is the gravitational constant, which is 386 in./sec².

Substituting in Eq. 4.33 the previously established relationship $v_1 = v_2 A_2 / A_1$ and solving for v_2, we obtain the flow velocity through the orifice:

$$v_2^2 = \frac{772 \, (p_1 - p_2)}{\rho[1 - (A_2/A_1)^2]} \tag{4.34}$$

If the orifice diameter is 50% of the conduit diameter, then, in accordance with Eq. 4.32, the relation $(A_2/A_1)^2$ equals 0.024, which is so small that it can be neglected. Hence, for orifice diameters that are 50% or less of the conduit diameter, Eq. 4.34 can be simplified and written:

$$v_2^2 = \frac{772}{\rho} \, (p_1 - p_2) \tag{4.35}$$

Flow through valve ports, jet pipes, and flapper nozzles, as described in the next chapter, is considered orifice flow. If there were no energy losses (conversions into heat), the upstream pressure would be fully recovered downstream. However, the energy losses are considerable, as shown in Table 4-2. In other words, for small orifice openings practically the entire pressure differential is a permanent loss. This is the case in valve ports and flapper nozzles; hence, for all practical purposes, $(p_1 - p_2)$ can be considered the permanent pressure loss, p.

If we insert this value, as well as Eqs. 4.12 and 4.32, into Eq. 4.35, then it can be written:

Table 4-2. Pressure losses in orifice flow.

ORIFICE DIAM/PIPE DIAM	PRESSURE LOSS IN PERCENT OF $(p_1 - p_2)$
0.5	0.75
0.4	0.84
0.3	0.91
0.2	0.96
0.1	0.99

$$Q = 23 \, A \sqrt{p} \qquad (4.36)$$

where Q is flow in gpm, A is the cross-sectional area of a small orifice, and p is the pressure drop across the device that contains the orifice. If pressure gages were applied at p_1 and p_2 in Fig. 4-10, the difference in their readings would be equal to p in the above equation.

There are a number of approximations involved in arriving at this equation. The fact is, however, that it gives excellent practical results.

Figure 4-10 illustrates that a pressure gage in the pipe wall at the point of the vena contracta, P_2, would read a minimum. The amount of pressure decrease would be a function of the flow velocity. Therefore, measuring the pressure differential between the upstream pressure and the pressure at the vena contracta is used as one method to measure flow rate. This phenomenon occurs because the pressure in a flowing fluid consists of a static and a dynamic component. The latter is determined by the flow velocity. At high velocity, most or all of the static pressure is converted into dynamic pressure. In straight hydraulic lines, velocities are usually kept below 800 in./sec, and, under these conditions, dynamic (or kinetic) pressure can be ignored. In orifices, however, and in particular in the vena contracta, velocities increase to a point where the effect is very noticeable. At small valve openings, which are equivalent to orifices, the difference between static and dynamic pressure is a factor that must be considered, as will be shown in the next chapter.

SUMMARY

The most important equations of this chapter can be summarized as follows:

$F = pA$ (Pascal's law)

$W = FL$

$F_1 = Kw$ (coulomb friction)

$F_2 = cv$ (viscous friction)
$F_a = 0.0026\ wa$ (acceleration force)

$$p = \frac{0.0026\ wa + cv + Kw}{A}$$

$A_1 v_1 = A_2 v_2 = A_3 v_3 = $ constant (only valid when ρ is constant)

$Q = 0.26\ vA$

$$Q = 7.2\ C_d A \sqrt{\frac{p}{\rho}}$$

$$P = \frac{W}{6600\ t}$$

$P = 5.83 \times 10^{-4}\ (pQ)$
$V_2 = V_1\ [1 + B\ (T_2 - T_1)]$
$V_2 = V_1\ (1 + Cp)$

$$\nu = \frac{\mu}{s}$$

$\text{SUS} = 4.6\ \nu$

$$R_e = 645\ \frac{dv}{\nu}$$

$$p = 1.3 \times 10^{-4}\ \frac{L\rho v \nu}{d^2} \qquad \text{(Darcy's formula for laminar flow)}$$

$$p = 1.3 \times 10^{-4}\ K\ \frac{\rho v \nu}{d}$$

$$Q = 23\ \sqrt{p} \qquad \text{(for orifice flow through valves, etc.)}$$

where:

$A = $ Area, in.^{2}
$a = $ Acceleration, in./sec^2

B = Coefficient of thermal expansion, $°F^{-1}$
C = Compressibility coefficient, $in.^2/lb$
C_d = Discharge coefficient
c = Damping constant, lb sec/in.
d = Diameter, in.
F = Force, lb
K = Coefficient of sliding friction or resistance coefficient for pressure losses in fittings
L = Length, in.
P = Power, hp
p = Pressure, psi
Q = Flow rate, gpm
R_e = Reynolds number
SUS = Saybolt Universal Seconds, sec
s = Specific gravity, sp gr
T = Temperature, °F
t = Time, sec
V = Volume, $in.^3$
v = Velocity, in./sec
W = Work, in.-lb
w = weight, lb
μ = Centipoise, cP
ν = Centistoke, cSt
ρ = Density, $lb/in.^3$

Chapter 5
HYDRAULIC SYSTEMS AND COMPONENTS

GRAPHIC SYMBOLS

Hydraulic and pneumatic circuits and components are usually represented with the aid of graphic symbols. These symbols were established by the International Organization for Standards (ISO), a worldwide federation of a variety of national standards institutes. The symbols are contained in International Standard ISO 1219. The appendix of this book shows its most essential graphic symbols.

Another important standard in the graphic representation of hydraulic circuits is ANSI B93.9, issued by the American National Standards Institute. It establishes "Symbols for Marking Electrical Leads and Ports on Fluid Power Valves."

From this standard, only the symbols most frequently used in connection with the ports of hydraulic devices are listed below:

- *Symbols A and B* identify the working ports of valves when there are no more than two such ports. Where there are more than two, the symbols K_1, K_2, K_3, etc. are used to identify the working ports.
- *Symbol D* identifies a hydraulic drain port, which usually is connected to an unrestricted line.
- *Symbol F* identifies the port of a flow control valve, from which the controlled flow exits.
- *Symbol P* identifies the pressure inlet port. It may appear as PA, PB, P1, etc., where A, B, and 1 indicate the working port which can connect to the pressure supply.
- *Symbol T* identifies the outlet port, usually connected to a tank (reservoir). Similar to symbols for the combinations with P, the symbols TA, TB, T1, etc. may be used.
- *Symbol X or Y* may be used for either a hydraulic pilot drain port, a hydraulic pressure supply port, or a hydraulic pilot control port. The affixes A, B, 1, etc. may also be used with X or Y.

140

SYSTEMS

Figure 5-1 shows the application of these graphic symbols to the hydraulic drive of the shoulder axis of a robot. The robot here is the model T^3566 of Cincinnati Milacron. Its appearance corresponds with that of Fig. 2-1.

In addition to the shoulder-axis drive system discussed below, there are five more drives on the same robot. Together they cover the six axes of the robot. While all six axes operate with the same control scheme in principle, their drive systems differ as follows:

- Base axis: servovalve rated 15 gpm; rotary actuator with 240° rotation and displacement of 25 in.^{3}/rad (160 in.^{3}/rev).
- Shoulder axis: servovalve rated 15 gpm; rotary actuator with 90° rotation and displacement of 40 in.^{3}/rad.
- Elbow axis: servovalve rated 15 gpm; linear actuator with 12-in. stroke, 1.38-in. rod diameter, and 2.50-in. bore.
- Pitch axis: servovalve rated 5 gpm; rotary actuator with 180° rotation and displacement of 3.8 in.^{3}/rad.
- Yaw axis: servovalve rated 5 gpm; rotary actuator with 180° rotation and displacement of 3.8 in.^{3}/rad.
- Roll axis: servovalve rated 5 gpm; rotary actuator with 270° rotation and displacement of 12 in.^{3}/rad.

A prime mover (electric motor), hydraulic pump, reservoir, heat exchanger, and other accessories are common to all six axes. They are discussed further below with respect to Fig. 5-2.

The pressure line in Fig. 5-1 delivers a hydraulic pressure of up to 2250 psi from the pump. As long as the shoulder axis is not moving, the servovalve is in the center position as shown, and flow from the pump is returned through the servovalve ports via the return line to the reservoir.

The servovalve is mounted directly at the actuator. This minimizes the fluid volume in the connection between the valve and the actuator, thereby improving the dynamic response.

The main stage of the servovalve is actuated by the hydraulic pressure of a pilot valve, which receives an electric signal and amplifies it by hydraulic means, as will be discussed later. The electrical input signal comes from the robot's computer control system. It is the signal that initiates shoulder motion.

Capacities of servovalves for Cincinnati Milacron's robots vary according to size and application of the robot. In this particular model, as already mentioned, the servovalve for the shoulder motion is rated at 15 gpm. Other models have shoulder drives with servovalves rated at 25 gpm.

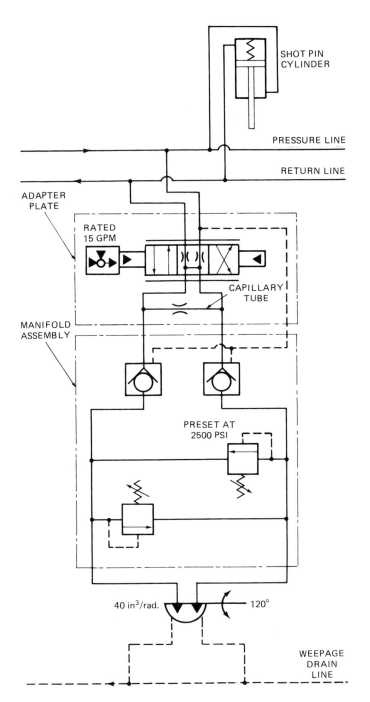

SHOT PIN
CYLINDER

PRESSURE LINE

RETURN LINE

ADAPTER
PLATE

RATED
15 GPM

CAPILLARY
TUBE

MANIFOLD
ASSEMBLY

PRESET AT
2500 PSI

40 in³/rad.

120°

WEEPAGE
DRAIN
LINE

Fig. 5-1. Hydraulic circuit of shoulder axis of T³586 robot. (*Courtesy of Cincinnati Milacron*)

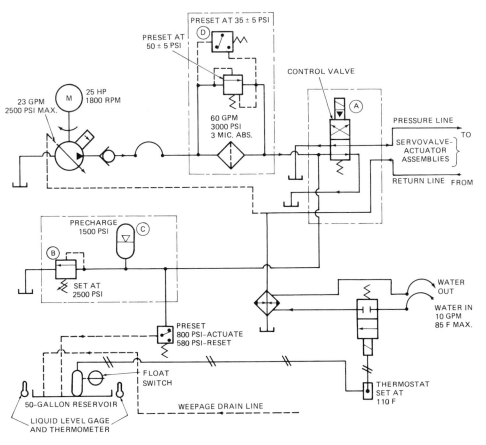

Fig. 5–2. Hydraulic power supply for T³586 robot. (*Courtesy of Cincinnati Milacron*)

There are two reasons for the different sizes of servovalves . One is that flow requirements are dictated by the desired velocity of the actuator and by the rated load carrying capacity of the robot. The latter dictates the size of the actuator for a given supply pressure; and the larger the effective area of the actuator is, the more flow is required for a given speed.

The other reason is that the smaller the servovalve is, the better is its dynamic response and the less its possible null shift. Null corresponds with the servovalve's center position as shown in the illustration. It is defined as the position where, at zero electrical input, the servovalve supplies zero control flow at zero load pressure drop. Null shift may occur with changes in supply pressure, temperature, and other operating conditions. It means that the servovalve loses its exact center position.

Returning to Fig. 5–1, a cross-over capillary tube is shown after the servovalve. Its purpose is to provide effective damping of the actuator motion and to stabilize the servovalve in very high-gain systems.

The manifold assembly contains two pilot-controlled nonreturn or check valves. As long as pressure is available, they are kept open and play no role in the functioning of the system. However, when the system is shut down, and the pressure is released to the reservoir, the check valves keep the pressure on the actuator and prevent a sudden drop of the robot arm. Actually, internal leakage will cause a slow but safe motion in the direction of gravity.

Each robot has a "home" position, a point of departure for all motions. To prevent the robot arm from overshooting this position, as it gradually moves downward after system shutdown, a shot pin mechanically blocks motion beyond this point. The shot pin is positioned by the shot pin cylinder, shown in the illustration. Under operating conditions, the shot pin is retracted by the hydraulic pressure under the cylinder piston. With no pressure present, the mechanical spring on top of the piston pushes the piston forward, and with it the shot pin.

To protect the components and structure of the robot from high shock loads or excessive pressures due to external loads when the valve is in neutral, two relief valves are provided. Shock loads could possibly increase the line pressure, perhaps beyond safe limits. Hence, if it exceeds 2500 psi on either side, one of the relief valves opens and releases the pressure.

At the end of the circuit is the rotary actuator, which is capable of moving the shoulder axis through 90°. The actuator requires 40 in.^{3} of hydraulic fluid per radian of motion. This is equivalent to 0.272 gallon for the total motion of 90°. Since the servovalve delivers 15 gpm, the shoulder can go through its full swing in 1.1 sec. This is with a fully open valve and disregards acceleration and deceleration. The fact is, however, that the shoulder motion can reach a velocity of 13.8 rpm. Considering the weight that is being swung around, the dynamic problem of bringing this mass at this velocity to a smooth stop becomes obvious.

To prevent leakage through the shaft seals of the actuator, an interspace is provided between the system pressure on one side of the seal and atmospheric pressure on the other. These interspaces are connected to the reservoir by weepage drain lines. The lines consist of $\frac{1}{4}$-in. plastic tubing.

The motion of the rotary actuator is provided with a resolver for position feedback and a tachometer for velocity feedback (see Fig. 1–15). However, these feedbacks are electric and so are not shown in the hydraulic diagram.

The pump that delivers the flow to all six servovalves is rated at 23 gpm and 2500 psi max., as shown in Fig. 5–2. It is driven by a 25-hp motor. While the maximum rating of the pump is 2500 psi, the operating pressure is reduced to 2250 psi.

Equation 4.19 stated that $P = 5.83 \times 10^{-4} pQ$. This would mean that this pump delivers about 30 hp. An electric drive motor of 25 hp is used because the motor is rated for continuous load, while the duty cycle for maximum power is only part-time.

It is conceivable that, at a given moment, the system would have to supply maximum hydraulic flow to all six servovalves. This could momentarily call for a supply that would exceed the pump capacity. It is one of the purposes of the accumulator to supply such peak demands.

The accumulator is precharged with gas at a pressure of 1500 psi. The gas is separated from the hydraulic fluid by a bladder. As the hydraulic fluid enters the accumulator at a pressure of 2250 psi, it compresses the bladder and with it the gas until it, too, reaches 2250 psi. The precharge ensures that the reserve flow from the accumulator never drops below 1500 psi.

Other important purposes of the accumulator are (a) to dampen spikes in fluid pressure, thus stabilizing the system, and (b) to maintain fluid pressure should a momentary pressure drop occur.

The pump has a pressure compensator. The principles of the device are described further below. It suffices here to say that it enables the pump to maintain a pressure of 2250 psi during idle, and thus to maintain it on the accumulator also. An additional safety measure is the pressure relief valve, B. It opens when the system exceeds 2500 psi.

The filter is rated at 60 gpm, 3000 psi, and 3 microns absolute. The designation "absolute" means that no spherical solid particles with a diameter of 3 μm (1.2×10^{-4} in.) or more are supposed to pass through. Such fine filtration is required to safeguard the proper operation of the servovalves. For further protection, additional filters are provided for the pilot stages of each of them.

As they accumulate more and more debris, filters increase their resistance to flow. This leads to an increased pressure differential across the filter, thus providing a means to signal the need for filter change. The sensor used for this purpose is the pressure-actuated electric switch, D. When the pressure differential exceeds 35 $\pm$ 5 psi, a contact on the pressure switch closes. The switch is wired into the computer control system of the robot, and part of the control system is a video screen. In case of an excessive pressure differential, an alert message appears on the screen, calling attention to the filter condition.

If the filter is not changed in spite of the warning, and the pressure differential continues to increase, then a relief valve, which is set at 50 $\pm$ 5 psi pressure differential, goes into action. It causes the flow to bypass the filter, and a message to this effect is flashed on the screen.

The control valve, A, is shown in the shutdown condition. The pressure line from the pump and the accumulator is connected directly to the reservoir, as is the pressure line to the servovalves. The entire system is no longer under pressure.

The control valve is, as the symbol indicates, a pilot-operated solenoid valve. It is actuated by an electric signal that makes the system start up. At this point, pressure builds up in the pressure lines, and the servovalves take over.

A pressure-actuated electric switch is provided between the line that leads to the accumulator and the reservoir. It is adjusted to actuate at 800 psi and to reset at 580 psi. Another of the functions tied into the robot's computer control, it serves two purposes:

1. When activated, the system is set at its lowest gain. When it reaches the 800-psi mark, full gain is applied. If the pressure does not begin to build up within 2 sec, something is wrong, and the pressure switch signals the computer to shut down the system.
2. If, during operation, the pressure falls below 800 psi, indications are that a line is broken. In this case, the pressure switch action also causes shutdown of the system. The system cannot be restarted until the pressure has increased to 580 psi.

Owing to internal friction of various sorts, the temperature of the hydraulic fluid increases. Good repeatability of the accuracy with which the robot is to move to its targets demands that the maximum temperature of the hydraulic fluid be kept between 110°F and 120°F. This is the purpose of the water-cooled heat exchanger in the return line. When the fluid temperature reaches 110°F, the thermostat actuates a valve that admits water flow through the heat exchanger. If the fluid temperature rises above 120°F, the hydraulic system will shut down when the robot arm returns to the home position. An error message will appear on the video screen, indicating overheating of the hydraulic fluid.

The reservoir has a capacity of 50 gallons. It not only contains the hydraulic fluid but also serves as a base for the pump and motor. The reservoir has a float switch that, like other functions already mentioned, is monitored by the computer. If the level goes too low, a warning signal will indicate the condition.

It becomes apparent in this description of a typical hydraulic robot system that there are quite a number of components involved. The principles of their operation in this and other systems will be discussed in the following pages.

VALVES

Basic Valve Concepts

Valves are used to regulate and stop the passage of flow. Basic concepts that are most frequently encountered in valves for the hydraulic control of robots are summarized below:

Seating Action.   These valves use a plug, such as a ball or truncated cone, to open or close a seat in the body of the valve. They are often used to provide quick action. While lacking the precision for regulating flow that is characteristic of valves with shear action, they are lower in cost and can open and close flow passages with a minimum of stroke. They are often called poppet valves, and the plug is referred to as a poppet.

The needle valve, which is also a valve with seating action, is distinguished by a stroke that is intentionally long, and a port opening that can be regulated very accurately for small flows. It is frequently used as a manually adjustable restriction for controlling flow.

Shear Action.   These valves modulate flow by means of an element that slides across the flow path. The shear action is generally produced by one of three elements. One device is a ball that rotates on an axis normal to the flow path. It is provided with porting that matches porting in the body of the valve in such a way that it can regulate the flow rate and/or change the direction of flow through the valve. In regulating the flow, it may also reverse it or stop it completely.

Instead of a ball, a plug may be used that rotates on an axis normal to the flow path. The porting of the valve body and the plug has the same effect as the porting of the rotating-ball design. Both the rotating ball and the rotating plug are rarely used in the systems under consideration and will not be discussed in detail.

A third type of element that uses the concept of a valve with sliding action is a spool, sometimes called a plunger. The spool slides in the valve body, and by covering and uncovering ports it directs and regulates flow. It is this design that is most frequently used for flow control valves of all sorts in robotic hydraulic systems.

Flapper Action.   These valves combine a flapper and a nozzle. They do not regulate flow so much as they control the back pressure of the nozzle. Since the flapper requires very little positioning force, these valves are generally used as a pilot for the main stage of a larger valve. They play the role of an electrohydraulic amplifier. The flapper is moved by a small torque motor or equivalent toward a nozzle through which flow passes. In moving toward the nozzle, the flapper throttles the flow and increases the back pressure between the nozzle and a fixed orifice that is located further upstream. Usually, the flapper is positioned between two nozzles with their respective fixed upstream orifices. While the flapper opens one nozzle, it closes the other, thus creating a differential pressure that is then applied across the spool of a valve with shearing action.

Jet Action. Jet action is generally used the same way flapper-nozzles are used, that is, for pressure-producing electrohydraulic amplifiers. In this case, however, it is the fluid-carrying jet pipe and not the flapper that is being controlled. The jet pipe issues a steady fine stream of hydraulic fluid under pressure. In its center position it directs this jet between two fixed receiver orifices. However, by deflecting the jet pipe toward one or the other orifice, it creates a pressure differential in the dead-ended passages that lead from the receiver orifices to both sides of the main spool. The magnitude of the pressure differential is almost proportional to the amount of deflection.

The following sections discuss the construction details of typical valves according to the basic designs mentioned above.

Valves with Seating Action

The typical valve with seating action uses a poppet in the form of either a ball or a truncated cone. Both forms of poppets are shown in Fig. 5-3. The poppet is kept against a seat by a spring. As long as the force produced by the hydraulic pressure applied under the poppet does not exceed the spring force, the poppet closes the passage. When the pressure produces a force that exceeds the spring force, the passage opens.

The angle of the truncated cone should differ from the angle of the seat, so that there is only a line-to-line fit between the poppet and the seat. If

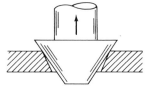
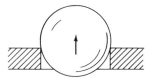

TRUNCATED CONE   BALL

Fig. 5-3. Poppet shapes.

this were not so and the slopes of poppet and seat were identical, several things would be likely to happen:

1. The two walls would tend to cling to each other, and break-loose friction would be high.
2. Upon opening of the valve, the walls would remain parallel, which in turn would lead to an increased pressure drop. By comparison, line-to-line closure, as shown in Fig. 5–3, forms an annular sharp-edged orifice on opening.
3. The flow path between the parallel seat and poppet would produce capillary flow at small openings with a rate that, as stated in Chapter 4 for narrow passages in general, is sensitive to viscosity and, hence to temperature changes. This, too, would be undesirable.

Hence, the preferred design is a line contact between the poppet and the seat, at least if both elements are metal. However, if the seat is made of an elastomer and the poppet of metal (or vice versa), then the resiliency of the elastomer increases the contact area.

The choice between a metal-to-metal seat and a metal-to-elastomer seat also applies to ball poppets. The elastomer gives tight closure but may deteriorate and requires periodic renewal. Sometimes, valves combine a metal-to-metal seat and an elastomer seat, one backing up the other.

Valves with seating action can close, throttle, or fully open the flow passage. In opening it they create, as noted above, an annular orifice between the poppet and the seat. The size of the annular orifice determines the flow rate for a given pressure differential across it. A truncated cone can be made quite long with a very small cone angle. It may actually approach or become a needle valve. The transition between the two is not defined. The smaller the cone angle is, the larger the motion can be made for a given increase of flow rate. Hence, a small cone angle can be practical for the fine metering of flow and is used, as mentioned before, as a manually adjustable restriction in some flow controls.

For a given flow rate, the valve requires a pressure differential that increases with throttling. This is expressed by Eq. 4.15, which states that $Q = 7.2\, C_d\, A\sqrt{p/\rho}$. At a seat diameter of 0.5 in. and an annulus width of 0.0185 in., the flow area is 0.028 in.^{2}. Let the flow rate, Q, be 6 gpm, the discharge coefficient, C_d, be 0.627, and the density of the fluid, ρ, be 0.032 lb/in.^{3}. The pressure differential required for this case would be 72 psi. The energy equivalent, according to Eq. 4.19, is $P = 5.83 \times 10^{-4}\,(pQ)$. Hence, 0.25 hp or 188 watts are required to make the flow pass the metering port of the valve. It is not only that this is energy lost for moving the actuator, but also that this loss becomes heat that must be removed in one way or another.

The pressure drop between higher and lower pressure takes place at approximately the narrowest point of the passage. Pinpointing exactly when this transition takes place can be difficult, particularly when the plug is only slightly removed from the seat. And yet this transition point and its shifting with the opening of the valve are a critical phenomenon.

Figure 5-4 shows a cone-shaped poppet in the closed as well as in the barely open position. The narrowest part of the passage in the "cracked" position is indicated by a dashed line. At this point, the diameter of the poppet is D_2, which, as shown, is smaller than the seat diameter, D_1. The passage indicated by the dashed lines is approximately the point of transition between upstream and downstream pressure.

In the closed position, the hydraulic pressure acts against an area that is proportional to the square of the diameter D_1. In the cracked position, the area is indicated by D_2. Thus, the area against which the pressure acts, and hence the force on the poppet, is larger in the closed than in the open position.

Consider a pressure relief valve that is to open to atmospheric pressure at 2500 psi. The valve will normally be closed by a spring above the plug. For a port area of 0.8 in.^{2} and a pressure of 2500 psi, the force pushing the poppet up is 2000 lb. The spring pushes downward on the poppet with an equal force.

At the moment when the hydraulic force exceeds that of the spring, the valve opens, and the poppet area on which the pressure is active decreases slightly. Not only does the change of area permit the spring force to push the poppet back, but the slight compression of the spring further increases the counterforce on the poppet. So the poppet is pushed back again and closes the valve. This lets the force caused by the hydraulic pressure increase again, and the cycle is repeated. The result is a chattering pressure relief valve. The effect is the same whether the plug is a truncated cone or a ball.

Hence, pressure relief valves are, and must be, designed with suitable damping to prevent chattering.

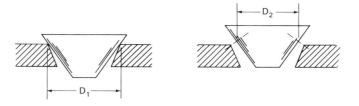

Fig. 5-4. Metering orifice of cone-shaped poppet.

Valves with Shear Action

The majority of valves with shear action use spools as the flow-controlling element. The basic principle of such a spool valve is shown in Fig. 5-5. There are four essential ports in the body of the valve. The supply pressure is connected to P. Any supply flow would enter through this port and pass on through either load port A or B. In the position shown, the pressure port is closed, and the pressure is dead-ended. The fourth port is T, which is the return connection to the reservoir.

It should be noted here, that "port" may refer either to the outside connection or to the variable opening inside the valve.

The load may consist of a hydraulic cylinder and some external force that the piston of the cylinder must move. If the spool is displaced toward the right, flow will pass from P to A to the cylinder. The return flow will pass through B to T.

Inversely, if the spool is displaced toward the left, flow will pass from P to B to the cylinder. The return flow will pass through A to T.

The ports in the valve body have about the same width as the lands of the spool that close them. There may be underlaps and overlaps, as discussed later, but these are minor deviations. The ports of the valve body are cut all around the spool land. Without it, pressure acting at right angles to the spool would push the spool against the sleeve with such force that the resulting friction probably would prevent any spool motion.

The advantage of the spool valve lies primarily in its balancing of forces that act upon it. In any spool position, it appears that the sum of the forces that act in one direction is equal to the sum acting in the other. While this is true to a certain extent, it is not *entirely* true. The fact is that flow forces, which are generated when flow passes through a valve, have to be accounted for. Flow forces oppose the opening of a valve.

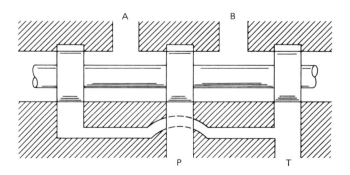

Fig. 5-5. Principle of valve with shearing action.

Figure 5-6 shows only ports P and B of the valve schematic in Fig. 5-5. The spool here has been shifted slightly to the left, and flow passes from P to B. The opening is exaggerated in the illustration. The actual opening is supposed to be minimal.

The opening at P creates an annular passage, the so-called metering port, through which flow rushes from the pressure side to the load side. Since the area of the metering port is small, the velocity is extremely high. This means high kinetic pressure and consequently a lowering of the static pressure, particularly in the vicinity of the metering port outlet. This makes the static pressure against the spool less on the left side than on the right. The net effect is a force acting to move the spool to the right and close the passage.

There are in reality two metering ports, one from P to B, the other from A to T. Both generate flow forces; both are acting to move the spool back to close the valve. As a rule of thumb, the total flow force is

$$F = 2.2 \times 10^{-2} Q \sqrt{p} \tag{5.1}$$

where F is the force acting to close the valve, in pounds, p is the total pressure drop across the valve, in psi, and Q is the flow rate through either port, in gpm. If there is a difference between flow *to* the load (P–B) and flow *from* the load (A–T), the average must be taken. A valve that passes 80 gpm with a pressure drop of 500 psi would, in accordance with Eq. 5.1, oppose opening with flow forces that amount to almost 40 lb. This is the main reason why valves with shearing action, except for the very smallest flow rates, are designed as two-stage valves, with a pilot valve and a main valve.

Other forces that must be overcome are friction forces caused by contact between the spool and the sleeve. Even the most minute inaccuracies in machining the spool and the sleeve can prevent the spool from floating in

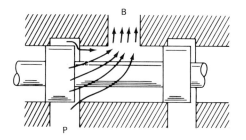

Fig. 5-6. The cause of flow forces.

the sleeve while still preventing excessive leakage across the lands. Dimensional inaccuracies—out of perfect roundness, for example—force the spool against the wall, particularly when hydraulic fluids carry silt or other particles that lodge between the spool and the sleeve.

Such conditions may produce hydraulic lock, that is, practically prevent the spool from moving at all. Balancing grooves cut circumferentially in the spool lands greatly alleviate this situation. They also reduce viscous forces opposing spool motion. Experiments have shown that the force required to position a valve spool may be reduced by about 94% by providing three circumferential grooves in each land and by more than 97% by seven grooves.[1]

Hydraulic valve spools move in sleeves or bores and have essentially metal-to-metal contact with a hydraulic film as interface. Very accurate fitting is necessary to minimize leakage, and yet tolerances and free motion require some clearance. As a rule of thumb, minimum and maximum clearances for different nominal bore diameters, d, are as shown in Table 5-1. To calculate the worst case of leakage, it may be assumed that the spool rubs against one side of the wall of the sleeve. In this case, the passage through which leakage can take place has the sickle-shaped form of an eccentric annulus. The flow through such an area is 2.5 times more than it would be with the spool floating in the center of the sleeve, and can be expressed by the approximate equation:

$$Q = 4.3 \times 10^4 \; \frac{pdb^3}{v\rho L} \tag{5.2}$$

where Q is the leakage flow in gpm, p is the pressure drop across the leakage path in psi, d is the diameter of the sleeve in inches, b is the clearance between centered spool and sleeve in inches, v is the viscosity in cSt, ρ is the density of the fluid in lb/in.^{3}, and L is the length of the leakage path in inches.

Table 5-1. Clearances between bore and centered spool

| D, IN | CLEARANCE, IN. $\times 10^{-4}$ | |
	MINIMUM	MAXIMUM
0.50	2	7
0.75	3	9
1.00	4	12
2.00	8	18

Thus, if the leakage path is 0.16 in. long with a pressure differential across it of 1500 psi, and if the sleeve diameter is 0.8003 in. and the spool diameter 0.7995 in., making $b = 0.0004$ in., and if the viscosity is 23 cSt and the density 0.0316 lb/in.[3], then the maximum leakage is about 0.03 gpm.

The most important application of valves with shear action is in servo-valves, which are dealt with later under a separate heading. Our next two topics are two forms of pilot valves, flapper-nozzles and jet pipes, which are generally used as part of servovalves.

Flapper-Nozzles

A flapper always works with a nozzle. The action is illustrated in Fig. 5–7, which shows a single-nozzle as well as a twin-nozzle design. As stated above, the typical use of such flapper-nozzles is as the pilot stage for the main spool of servovalves.

Pressure is admitted at P, and flow passes through a fixed orifice and then through the nozzle to the tank (reservoir). As the flapper approaches the nozzle, the back pressure between the orifice and nozzle increases. With the single-nozzle design, back pressure is pilot pressure X, and with the twin-nozzle design, it is pilot pressures X and Y, which are applied against the end faces of the main spool.

With the single-nozzle design, the pilot pressure is applied against one end face of the main spool, which is biased by a spring so that spring force and pilot pressure balance. This has two disadvantages. One is that the position of the main spool shifts with changes in pilot supply pressure. The other is that in case of pilot pressure failure, the spring will push the main spool hard over to one end, since it no longer has a counterforce. This position is generally undesirable in case of failure.

With two nozzles, two different back pressures are produced. One in-

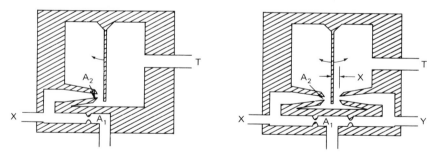

SINGLE-NOZZLE DESIGN   TWIN-NOZZLE DESIGN

Fig. 5–7. Flapper-nozzles.

creases while the other decreases. The resulting pressure differential is applied across the main spool for positioning.

The flapper is generally positioned by a torque motor (see pages 167–169), which is not shown in the illustration. The torque motor rotates the flapper about a pivot through a very small arc. The motion is so small that it can be considered a straight-line motion. Frequently, a feedback force is applied from the motion of the main spool back to the flapper, but this is another feature that will be described later. The feedback has several advantages. One is to eliminate springs across the main spool.

The back pressure between the fixed orifice and the nozzle can be determined when the areas of the two restrictions are known. The area of the first restriction, A_1, is simply that of the fixed orifice. If the orifice has a diameter d_1, then $A_1 = \pi d_1^2/4$.

The area of the second restriction, A_2, is that of an annulus formed by the circumference of the nozzle and the gap between the flapper and the nozzle. The gap in the center position is indicated as x in Fig. 5-7. The magnitude of x will change with the position of the flapper. If the diameter of the nozzle is d_2, then the circumference of the nozzle is πd_2. Hence, the area of the exit between the nozzle and the flapper is $A_2 = \pi d_2 x$.

Let the supply pressure be designated p_s, and the back pressure p_c. The efflux from the flapper-nozzle discharges to the tank through connection T. The pressure here may be assumed to be atmospheric, or zero psi. For static conditions, the flow rate through A_1 is the same as that going through A_2. This leads, on the basis of Eq. 4.36, to the result that:

$$A_1 \sqrt{p_s - p_c} = A_2 \sqrt{p_c}$$

Expressing area A_2 in terms of diameter d_2 and gap x, one can also write:

$$(d_1/4) \sqrt{p_s - p_c} = d_2 x \sqrt{p_c}$$

which in turn yields:

$$\frac{p_c}{p_s} = \frac{1}{1 + 16x^2 (d_2^2/d_1^4)} \qquad (5.3)$$

If there were no flapper, the only restriction would be the nozzle itself, with an area of $\pi d_2^2/4$. This means that $\pi d_2^2/4$ is the maximum area of the variable restriction, and when x becomes large enough to make $\pi d_2 x = \pi d_2^2/4$, which is equivalent to $x = 0.25\, d_2$, this maximum opening is reached. Beyond this, there is no longer control. This leads to a general rule that limits the gap x to a practical maximum of $0.2\, d_2$.

Suppose the diameter of the fixed orifice as well as that of the nozzle is 0.03 in. Using the maximum gap of 0.006 in. for this case and inserting values in Eq. 5.3 would result in a ratio of back pressure, p_c, to supply pressure, p_s, of 0.61. This means that, for example, at a supply pressure of 1000 psi and a maximum gap of 0.2 d_2, the back pressure, p_c, would be 610 psi. This 61% of the supply pressure is the smallest back pressure attainable when the diameters of the fixed orifice and of the nozzle are the same.

By means of Eq. 5.3, back pressures can be calculated for any gap that is smaller than 0.2 d_2, and the results can be plotted to produce a graph as shown in Fig. 5–8.

It can also be calculated from Eq. 5.3 that by making the orifice diameter smaller than that of the nozzle, the minimum back pressure can be made lower. This permits producing a greater pressure differential for a given flapper motion, which is desirable. However, because of the smaller diameter of the fixed orifice, the likelihood of plugging by particles carried in the oil is increased. Nevertheless, the diameter of the fixed orifice is sometimes chosen slightly smaller than that of the nozzle.

The force required to move the flapper is small but not negligible. It is primarily due to the static pressure, p_c, and the momentum transferred from the flow velocity to the flapper. Both forces are proportional to the cross-sectional area of the nozzle opening. This is one of the reasons why the nozzle opening is chosen as small as is feasible. Another reason is that the flow passing through the flapper-nozzle represents a loss that one wants to minimize to the extent possible.

The forces exerted on the flapper change not only because of changes in

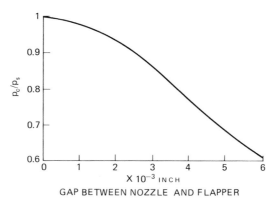

Fig. 5–8. Characteristics of flapper-nozzle.

back pressure in accordance with Fig. 5–8, but also because of a transient phenomenon that is described below.

A change in flapper position produces main spool motion. This is its purpose as pilot valve. The displacement of the spool requires a certain pilot flow to the back chamber of the spool that must pass through the fixed orifice. This means more flow passing through the fixed orifice than through the flapper-nozzle restriction. A transient drop of back pressure is the consequence, though the flapper position does not change.

Similarly, flow returns from the other end of the main spool and passes through the nozzle. Here, the consequence is an increase in back pressure. The combined transient change in back pressures of the two nozzles results in a potential for instability that is prevented partly by inherent viscous damping and partly by very high spring rates and mechanical feedback (to be discussed).

The nozzle, as described here, has a sharp edge at the exit of the fluid. The width of this edge should be less than the minimum gap between the flapper and the nozzle. With a broader edge of the nozzle exit, characteristics of the flapper-nozzle change drastically. The annulus becomes a flow path with laminar flow characteristics. Pressure drop becomes unpredictable, and changes with viscosity. Sometimes erosion caused by fluid rushing over the initially sharp edge dulls it sufficiently to produce these erratic conditions.

Jet Pipes

Of the four basic valve designs discussed here—seating action, shear action, flapper action, and jet action—the last is the only one that does not act by throttling. Instead, it directs flow by means of a jet pipe. The principle is illustrated in Fig. 5–9.

Although the valve design is described here only in generic terms, it should be mentioned that the only significant producer of such servovalves today seems to be Atchley Controls, Inc.

The jet pipe is rigidly fastened (a) to the armature of a torque motor or equivalent positioning mechanism and (b) to a flexure tube that bends when the armature rotates the jet pipe through a very small angle.

A narrow jet stream of, say, 0.008 in. diameter leaves the jet pipe. It is directed toward two receiver ports. These ports have a slightly larger diameter than the jet, and are separated from each other by a very narrow ridge.

When the jet pipe is in its center position, it straddles the two receiver ports, and the two pressures at X and Y are the same. Suppose, the supply

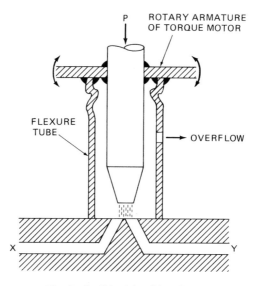

P   ROTARY ARMATURE
    OF TORQUE MOTOR

FLEXURE
TUBE

OVERFLOW

X

Y

Fig. 5-9. Principle of jet pipe.

pressure is 1000 psi and the recovery 90%. In this case, the pressure in each receiver port is 450 psi.

The space between the jet pipe and receiver ports must always be sur-rounded by a pool of the hydraulic fluid that passes through the jet. This prevents air from being drawn into the receiver ports, a condition that could easily lead to erratic behavior and dynamic imbalance. As indicated in the illustration, an overflow can maintain this fluid pool and drain it contin-uously to the tank.

As with the flapper-nozzle, the primary application of the jet pipe is as a pilot for servovalves. (This was not always so. The jet pipe has a long history as a single-stage control valve.) Comparison between flapper-nozzles and jet pipes reveals their advantages and disadvantages, some of which are the following:

- The opening of the jet pipe typically is 0.009 in., while that of a typical nozzle is 0.030 in. But the smallest flow passage of the jet pipe is still 0.009 in., while that of the flapper-nozzle is the width of the annulus, that is, the gap x, which in the center position of the flapper of a twin-nozzle design is 0.003 in., and in the hardover position is only 0.006 in. Thus, somewhat larger particles can pass through the jet pipe than through the flapper-nozzle. Besides, it is argued that the flapper-nozzle arrangement has four restrictions that can clog up—two of these being the fixed orifices, though they have a relatively large passage, and two the gaps between flapper and nozzle—while with the jet pipe there is only one restriction, the jet pipe tip—unless one wants to consider the two receiving ports as additional, albeit slightly larger, restrictions.

- The jet pipe is capable of producing a maximum pressure differential under static conditions of 90% of supply pressure, although some jet pipes have much lower recovery pressures than that. The maximum pressure differential that a flapper-nozzle with equal diameters for orifice and nozzle can achieve is somewhat less than 40%.
- The jet pipe is more difficult to design, particularly since the moving element carries the fluid. It is comparatively easy to make the flapper the moving element.
- Raymond D. Atchley[2] has pointed out that erosion on the flapper and nozzle has more severe consequences than on the jet-pipe tip and receiver; hence life expectancy is greater for the latter.
- The jet pipe has to contend with backsplashing of fluid against the jet-pipe tip, which can cause erratic behavior. On the other hand, it is not subject to the pressure of impinging fluid as the flapper is.
- The jet pipe requires one filter, while the flapper-nozzle requires two—one upstream of each fixed orifice. As one of these filters begins to get clogged, an imbalance between the two back pressures results that produces erratic behavior of the main stage. Clogging of the jet pipe merely diminishes the available power, gradually slowing down the response of the main stage.

These contentions have to be taken with caution. In the first place, while it is true that many of the flapper-nozzle servovalves use two filters, it is also true that this is merely a matter of convenience. The Moog servovalves, for example, which are described further below, use only one filter, which is built as a hollow tube. The flow enters from the outside and is distributed between the two fixed orifices.

Furthermore, it is also true that a minute obstruction at the exit of the jet pipe can play havoc with the jet and, more likely than not, destroy performance completely. Incidentally, this is true also of the erosion mentioned above.

Be that as it may, flapper-nozzles as well as jet pipes have excellent performance records, and academic claims are all too often negated by practical results.

Rating of Valves

Hydraulic valves are usually specified by the flow rate they are capable of passing. A single specification of 25 gpm, however, is rather meaningless, unless the pressure drop across the valve at this particular flow rate is given, as well as the viscosity and mass density of the fluid.

An example of a comprehensive and accurate rating system is that used by Vickers Inc. Consider their DG4V–3–0–B valve. Its rated flow capacity is given as 5 gpm, and its maximum flow rate is 10 gpm. At the same time, it is stated that these flow rates are valid for a hydraulic fluid of 0.030 lb/in.^{3} density and 21 cSt kinematic viscosity. Furthermore, reference is made to a flow chart. This chart lists the various pressure drops within the valve at the rated flow capacity. They are given for the flow from the pressure port to the A-port of the load (P to A) and from the B-port to tank (B to T) as well as for P to B and A to T, and also for the center condition where in this particular valve all flow goes directly back to the tank (P to T). The pressure drops (p) at rated flow (Q) are: P to A: 26 psi; A to T: 26 psi; B to T: 26 psi; P to T: 29 psi; P to B: 26 psi.

Total pressure drop, when the valve is fully open, as seen from this listing is $2 \times 26 = 52$ psi. Sometimes pressure drops in the various passages differ from each other much more than in the above case.

On the basis of the tabulated values, which are $Q = 5$ gpm, $p = 52$ psi, $\rho = 0.030$ lb/in.^{3}, and $\nu = 0.21$ cSt, additional information is given as follows:

- Pressure drop, p_x, for any other flow rate, Q_x, is approximately $p_x = p\,(Q_x/Q)^2$.
- Pressure drop, p_x, for any other density, ρ_x, is approximately $p_x = p\,(\rho_x/\rho)$.
- Pressure drop for any other kinematic viscosity will change approximately as follows:

Viscosity in cSt:	14	32	43	54	65	76	86
Percent of p:	93	111	119	126	132	137	141

Thus, if the flow rate is 6.5 gpm instead of 5, the density 0.034 lb/in.^{3} instead of 0.030, and the viscosity 32 cSt instead of 21, then the total pressure drop with the valve fully open will increase from 32 psi to $32 \times (6.5/5)^2 \times (0.034/0.030) \times (111/100) = 68$ psi.

Directional Control Valves

Valves with shear action are generally directional control valves. They have as their primary function the control of direction and/or the prevention of flow. The modulation of the flow is often included. Servovalves are directional control valves with highly accurate flow modulation and superior dynamic behavior. They control flow as a function of some variable input signal.

In robotic hydraulics, directional control valves usually operate with a sliding spool. Such valves can be classified in various ways. One is to divide them into straightway, three-way, and four-way valves. This refers to the main valves. Pilot valves are ignored in this connection.

Straightway valves either pass or interrupt flow from P to A. The thermostat-controlled valve in Fig. 5-2 represents this category. P here is the "Water in" connection, and A is the connection to the heat exchanger.

Three-way valves differ from straightway valves. They connect port A either to pressure P or to the reservoir or some equivalent connection T. Three-way valves inherently have three ports, disregarding possibly separate pilot valve ports.

Four-way valves connect two working ports, A and B, in either one of two different modes. One is from P to A and from B to T, the other from P to B and from A to T. They are represented by the servovalve in Fig. 5-1 and the control valve in Fig. 5-2.

Another way of classifying directional control valves is by the number of positions the spool can assume. A four-way, three-position valve, for example, has one position for P to A and B to T, and another position for P to B and A to T. A third position, which is the center or neutral position, may either block both working ports or connect them both to the tank, or produce some other nonactive condition.

Servovalves are essentially three-position valves. However, they change between these positions gradually. Thus there is an infinite number of intermediate points at which the spool can be positioned. The servovalve in Fig. 5-1 is a typical example.

A two-position valve is represented by the control valve in Fig. 5-2, where one position connects P to the pressure line and the return line to T, and the other position reverses these connections.

ISO 1219, the standard on graphic symbols, refers to a "directional control valve 3/2" or a "directional control valve 5/2" (see the Appendix). This indicates a three-way valve that has two positions or a five-way valve (which is rarely used) with two positions. This method of designation has found acceptance in Europe but has hardly been used in the United States.

Figure 5-10 shows the four most commonly used center conditions for three-position valves with shear action. The symbols on the right show only the center position of the valves, which is the relevant one for this discussion. When the spool is moved to the left, the supply pressure, P, is connected to working port A, in all cases, and working port B is connected to T. Shifting any of the spools to the right connects P to B and A to T.

Four-way valves are shown, but they can readily be converted to three-

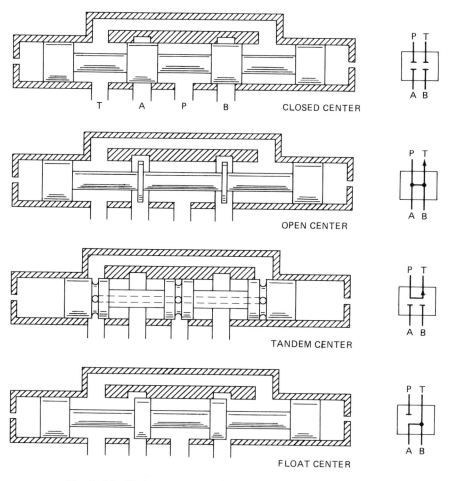

Fig. 5-10. Various center conditions for three-position valves.

way valves by plugging one of the working ports. The diagrams are not intended to be dimensionally realistic. Their purpose is solely the demonstration of principles.

The closed-center configuration on top blocks all flow passages in its center position. The load, which is connected to A and B, can be firmly locked in place—except for leakage.

If a working port is blocked, excessive pressure can be produced by an external load that acts on the cylinder piston or fluid motor shaft exerting force or torque. In the circuit of Fig. 5-1, the two relief valves that are preset at 2500 psi protect against such an event.

The closed center is obtained by overlap. This means that the spool lands are wider than the ports they control. When the spool is neutral, the pump either discharges over a relief valve, or its delivery goes to zero because of pressure compensation (described later).

In either case, full supply pressure is available when the spool valve is only slightly opened. With a constant-displacement pump, the pressure drop across the slightly opened valve is significant. If the load is to be metered and demands a pressure that approaches the relief valve setting, most of the flow will go over the relief valve. Yet some flow will pass through the control valve and move the load, be it ever so slowly.

Closed centers, because of the overlap, cause a dead band, or at least a very diminished increase of flow in response to valve motion. This affects the dynamic performance, so servovalves in particular are designed to approach zero overlap. With an ideal zero lap, there is no leakage when the spool is centered, but the slightest displacement produces load flow. This is hardly attainable. One has to consider that the total movement of a spool in a servovalve may be in the order of ± 0.01 in. For a dead band of not more than 2%, the maximum overlap would have to be 0.0002 in. This is a tolerance that makes for costly machining. Even if it were done this way, tolerances between the spool diameter and the sleeve diameter would still permit leakage flow. Hence, zero lap is more an ideal to be approached than a condition that can be realized.

The open-center configuration in Fig. 5-8 interconnects all ports. The lands are narrowed sufficiently to make sure the working ports are never closed. In the center position, pressure is connected back to the tank. The pressure drop across the ports is kept to a minimum. Typically, this may be 15 psi for a hydraulic valve with a capacity of 35 gpm. Pump pressure will drop to a correspondingly low value. The working ports are connected to the tank so that the load is free to float.

With the open-center valve, full pressure is not available until the passage from the pressure to the reservoir is completely cut off.

The tandem-center configuration connects P to T in the same manner the open-center valve does, but blocks the working ports A and B. This is accomplished by connecting P to T via a passage through the axis of the spool. When the spool is shifted to the left or to the right, the passage through the spool is blocked, and normal connection to the working ports is obtained.

Tandem-center valves act like open-center valves, insofar as full pressure availability is obtained only after the spool has moved sufficiently to cut off the passage from P to T.

The float-center configuration connects both working ports to T, permitting the load to float. Like the closed center, it is used where pressure must be available for other loads, when the valve is in its center position. It is also used with pressure-compensated pumps that reduce flow to zero or with prime movers that automatically shut down or are de-clutched from the pump whenever the spool goes to its center position.

Servovalves

The servovalve is the essential element in the control loop that moves the robot. All the other elements in the hydraulic system merely support the action of the servovalve. Its task is to convert a relatively small electrical signal with a power of, say, 1.5 watts into a hydraulic output power of some 5000 watts. This makes the servovalve an electrohydraulic power amplifier of very high gain. Since it is part of a control loop, its dynamic behavior is of the greatest importance.

Hydraulic systems can be designed either to supply constant flow and vary the pressure or to have constant pressure and vary the flow. The latter is the conventional system for servovalves. It is illustrated in Fig. 5–2, where a variable-displacement, pressure-compensated pump is used. It will deliver flow as required by the load which is controlled by the servovalve. Pressure will remain constant.

Servovalves are usually rated at 3000 psi maximum supply pressure, but servovalve *capacity* is generally rated at 1000 psi pressure drop across the valve while the full, rated signal current is applied.

Control flow is the flow passing through the valve at any given pressure drop and signal current. The relation of control flow, rated flow at 1000 psi pressure drop, and pressure drop across the valve is the characteristic square-root relationship:

$$Q_c = Q_r \sqrt{\frac{p_v}{1000}} \qquad (5.4)$$

where Q_c is the control flow, Q_r is the rated flow, and p_v is the pressure drop across the valve.

It can be assumed that the servovalve returns flow to the tank at a pressure that approaches atmospheric relative to the high operating pressures. Under these circumstances, the supply pressure, p_s, minus the load pressure, p_l, equals the pressure drop across the valve. This permits writing Eq. 5.4 in the form:

$$Q_c = Q_r \sqrt{\frac{p_s - p_l}{1000}} \tag{5.5}$$

This equation is graphically represented in Fig. 5–11, where the ratio of control flow to rated flow, Q_c/Q_r, is plotted against the supply pressure, p_s, minus the pressure drop across the load, p_l. The graph also shows the changes in control flow in proportion to reduced signal currents.

The rotary actuator in Fig. 5–1 is rated at 80 in.^{3}/rad. This is equivalent to 126 in.^{3} or 0.54 gallon for the full rotation of 90°. The rated flow of the servovalve is 25 gpm, and the supply pressure in this case is 2250 psi.

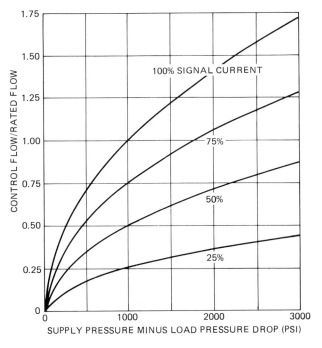

Fig. 5–11. Control flow vs. load pressure drop.

It is customary to limit the pressure across the load to two-thirds of the supply pressure. Hence, the maximum load would be 1500 psi. At this pressure drop across the load, the flow rate, according to Eq. 5.5, would be 21.6 gpm.

Since 0.54 gallon is required to move the actuator through its 90°, it would be theoretically possible to complete the motion in 0.54/21.6 = 0.025 min or 1.5 sec. In practice, the movement would be slowed somewhat by acceleration and deceleration.

Instead of Eq. 5.5, one can also use Fig. 5–11, which gives practically useful but less accurate results. At a supply pressure of 2250 psi and a load pressure drop of 1500 psi, the difference between the two pressures is 750 psi. At 100% signal current the graph would show that the control flow is 0.86 times the rated flow, which would give a result that is for all practical purposes sufficiently close to the 21.6 gpm obtained above. Reducing the signal current to 50% would halve the control flow.

The Moog Series 78 Servovalve made by Moog Inc. is an example of a two-stage design with a flapper-nozzle as pilot. It is available at rated flows of 20, 30, and 40 gpm at 1000 psi pressure drop across the valve. This rating is for petroleum-base hydraulic fluids with a density of about 0.032 lb/in.^{3}. The valve can accommodate viscosities from 60 to 450 SUS at 100°F.

The no-load flow characteristic of these servovalves is plotted in Fig. 5–12. It shows flow gain, linearity, and symmetry.

Flow gain is the slope of this curve. The steeper the slope is, the more will the flow increase for a given increment in input current; that is, the greater the flow gain will be. As seen in the graph, flow gain changes somewhat over the operating range of the valve. Flow gain without any specific reference to the region of such a graph is considered nominal gain. It is given by a straight line drawn from the point of origin to the point that corresponds to 100% of rated current.

However, it is also important to distinguish between gain in the null region and gain in the region beyond. The graph shows that in the null region, that is, at less than 5% of rated current, flow gain may be anywhere between 50 and 200% of nominal gain. This variation is unavoidable because of tolerances in a closed-center valve, as mentioned before. However, it may make a difference in the dynamic behavior. A valve that homes in with a 50% gain is likely to slow down the robot arm response for small (i.e., low-flow) motions, but it will probably prevent it from overshooting. On the other hand, a gain in the null region that is 200% of nominal gain will make the robot arm home in faster but could produce overshoot for small motions at low flow. In other words, 200% gives less damping.

A similar effect may result from the difference in the slope of the gain

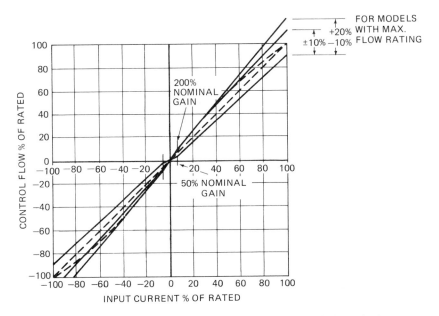

Fig. 5-12. No-load flow characteristics (*Courtesy of Moog, Inc.*)

curve due to manufacturing tolerances. It can change from valve to valve between −10 and +10%, and in the large valve even to +20%.

Hysteresis is primarily caused by electromagnetic characteristics of the torque motor. For the Moog Series 78 it is rated at less than 3% of rated current.

Hysteresis should not be confused with threshold, which is the minimum change of input current required to produce a change in servovalve output. Threshold is normally specified as the current increment required to revert from a condition of increasing output to a condition of decreasing output. It is more serious and, hence, should be much less than hysteresis. In the case of the Moog valves under consideration, it is less than 0.5% of rated current. This means that at a rated current of 20 mA, for example, after increasing the flow to a certain value, a reduction of current of 0.1 mA may be required to start the valve toward decreasing the flow. This may be the unavoidable, though small, dead band caused by the servovalve in the robot control system.

Threshold cannot be shown in the flow curve, but symmetry can. The flow curve in Fig. 5-12 shows a symmetrical valve. Symmetry is measured as the difference in nominal flow gain of each side, expressed in percent of the greater flow gain. For the Moog Series 78 it is less than 10%. The only effect this may have is a slight change in dynamic response similar to nonlinearities in the flow characteristics.

The torque motor and its operation are illustrated in Fig. 5–13. It is based on the interaction of two magnetic fields. One is produced by the permanent magnet, the other by two coils with direct current flowing through them. The coils are wound around a pivoted armature. The armature is movable through a small angle, while the coils are stationary. Firmly connected to and moving with the armature is the flapper.

The coils use a series or parallel connection. Each of the two coils is wrapped around one of the arms of the armature. Their polarity is such as to produce a north pole on one end of the armature and a south pole on the other. At no current flowing, the armature has no magnetic poles. If the current is reversed, the poles of the armature are reversed.

Both ends of the armature are located in corresponding gaps of the permanent magnet, which has its north poles above the armature and the south poles below. The north pole of the armature is attracted by the south pole of the permanent magnet, and vice versa. Figure 5–13 shows the left end of the armature to be north. This results in a counterclockwise rotation of the armature, albeit through a very small angle. A counteracting force is provided by the flexure-mounted armature. It establishes a force balance that produces an armature position proportional to the signal current.

If the current in both coils were reversed, the left end of the armature would become south. In either direction, there is considerable torque exerted on the flapper. The larger the current that passes through the two coils, the greater is the torque and with it the angle of armature and flapper rotation.

A feedback spring is provided by the flexible extension of the flapper. It ends, as shown in the illustration, in a ball that fits inside a hole machined vertically into the shaft of the spool.

When the armature pivots counterclockwise, pilot pressure is applied against the right end face of the spool. It displaces the spool to the left,

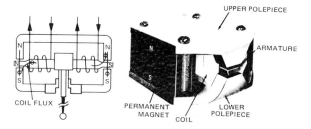

Fig. 5–13. Torque motor. (*Courtesy of Moog, Inc.*)

and the feedback spring exerts a restoring force on the flapper. Continued motion of the spool will increase the force until the flapper is brought back to its neutral position, and motion of the spool ceases. This force balance produces a position of the spool that is very closely proportional to the input current to the torque motor coils.

In order to reduce threshold, a small-amplitude dither signal may be used. It is recommended that its frequency be 200 to 400 Hz and less than 20% of the rated current amplitude. This gives no discernible motion of the spool and yet allows the spool to respond to the slightest change of input signal.

The jet pipe, though different in its operation (described above), would, when applied as the first stage of a servovalve, fit a pattern that is very similar to the flapper-nozzle valve. The same type of torque motor would be used, and feedback of the spool position would be applied to the jet pipe rather than to the flapper.

ACTUATORS

Linear Actuators

One of the advantages of hydraulics is that linear motion can be easily obtained by means of cylinders. Figure 5–14 shows the construction of a cylinder manufactured by Parker Hannifin Corporation. The cylinder itself consists essentially of a tube covered at the right by the cap and at the left by a head. Built into the cap is the fluid connection to one side of the piston (the cap end port) and a cushion, which is described further below. The head carries the bearing through which the piston rod can pass to the outside. It also is provided with a fluid port (the rod end port) that carries fluid from and to the rod side of the piston.

Tubing, cap, and head are held together by four tie rods that traverse the length of the cylinder. They are prestressed at assembly to hold the ends of the cylinder against the tubing.

If heavy loads are involved or piston speeds are in excess of 20 ft/min and the piston goes through full stroke, cushions, such as shown in the illustration, are generally indicated.

When the piston moves toward the head end, full flow enters at the cap end through the opening in the center of the cylinder bore, and the fluid displaced on the other end of the piston leaves through an annulus between the piston rod and the head end. The piston rod has a section with an enlarged diameter that enters the annulus and gradually closes it, thereby reducing the flow through it. The effect is smooth deceleration and damping. A bypass is provided with an adjustable restriction to provide the desired degree of deceleration.

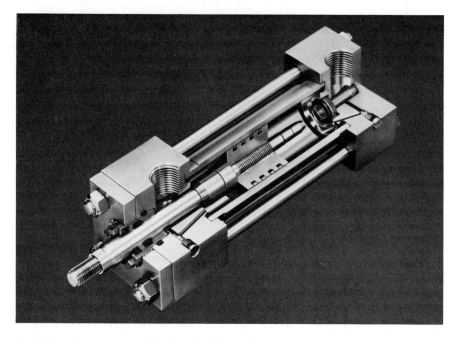

Fig. 5-14. Construction of Parker Hannifin cylinder. (*Courtesy of Parker Hannifin Corporation*)

Similarly on the cap end, when the piston moves in the opposite direction: a plunger extension of the piston rod enters the center outlet and gradually blocks the flow through it. Another bypass with an adjustable restriction provides the regulation needed.

Figure 5-15 shows the piston assembly of a Vickers hydraulic cylinder. The piston is slipped over the piston rod, and the follower nut tightens the

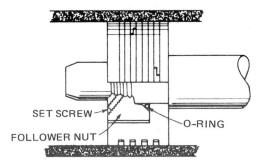

Fig. 5-15. Piston assembly of Vickers hydraulic cylinder. (*Courtesy of Vickers Inc.*)

piston against the shoulder, with the O-ring preventing leakage around the piston rod. A set screw locks the assembly. The illustration shows a metallic step seal ring. These seals are available as an option and have the advantage of a very long life. A disadvantage, however, is that they permit some leakage, usually between 1 and 3 in.^{3}/min. Elastomer seals virtually eliminate leakage, but their life is more limited.

The rod bearing is another critical part of the cylinder. It must prevent leakage, cause no significant friction, keep the piston rod moist and lubricated, and be durable. A seal and a wiper are provided for these purposes, as shown in the generic representation of Fig. 5–16. The gland may be of bronze or of ductile iron with a high graphitic iron content to give a good bearing surface for the piston rod. It contains an oil seal and, generally, a wiper for the piston rod. The purpose of the wiper is to prevent dirt from getting between the gland and the rod. Such dirt may be picked up while the rod is outside the cylinder. Elastomers are often used for wiper material, as they have good abrasion resistance; however, leather has lower friction. Leather can absorb lubricant, in this case the hydraulic fluid, and release it as needed. It is often impregnated to control its porosity so it will seal fluids, yet retain its self-lubricating property. A metal scraper is also sometimes used as a wiper.

With robots, boots are often used instead of wipers. They consist of corrugated covers or bellows that are fastened to the cylinder head and rod, moving accordion-like with the rod. They may also provide a wrapping of the hydraulic actuator, as is the case with the DeVilbiss/Trallfa TR-4500 robot shown in Fig. 5–17. Since this robot is used for spray finishing, particularly good protection is required.

There are different opinions about the relative position between the seal and the bearing. Having the seal on the outside provides good lubrication

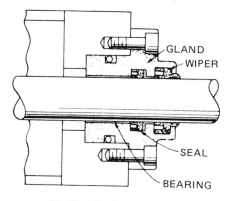

Fig. 5–16. Cylinder gland.

Fig. 5-17. TR-4500 spray finishing robot. (*Courtesy of The DeVilbiss Company*)

for the bearing, which has full contact with oil in the cylinder. Having the bearing on the inside reduces lubrication but brings the bearing closer to the connected load and thus improves support. Sometimes the seal is in the middle, with the bearing surface before and after.

It is advantageous to have a long bearing surface to support some side loading without making the piston bear the lever force of the side load. In a fully extended piston, this force cannot be entirely eliminated, but the long bearing surface helps to approach the ideal of a free-floating element that serves only to push and pull and to seal.

Piston seals, gland seals, and wipers produce friction, which increases with the pressure differential across them. Thus, in Fig. 5-16 pressure applied in the cup of the seal spreads the ring more firmly against the rod, increasing sealing force but also friction. Where low friction is important, a tank connection may be provided at the end of the bearing immediately before the seal. This is similar to the weepage drain line for the rotary actuator in Fig. 5-1. Having a close fit between the bearing and the rod reduces internal leakage to a minimum that can now pass from the cylinder to the tank. With this arrangement, the seal is exposed to a minimum pressure differential at all times, thus reducing friction.

Robots that are used for spray finishing applications, such as the DeVilbiss/Trallfa TR-4500 in Fig. 5-17, are normally programmed by manually leading the spray gun at the end of the robot arm through its desired cycle. During this teaching operation, the motions are programmed and afterward repeated automatically as often as desired. In the teaching mode, the hydraulic system is not pressurized, but the piston in the cylinder must move with the manual motions of the spray gun. Because of pressure losses in conduits, valves, and so on, this could mean a considerable manual effort. To avoid this, a unique split-piston design is used. Figure 5-18 shows the split-piston concept under operating, pressurized conditions. The piston consists of three segments that are sandwiched together. The piston rod is hollow, and moves in and out of a center rod along the axis of the cylinder. Under pressure, all three piston segments are held together as one unit.

Figure 5-19 shows the cylinder in the teaching mode. The cylinder has been depressurized, and the piston rod has been manually moved to its extreme position. This has forced the hydraulic fluid out of the cylinder, and the space has filled with air. Since there is no pressure, the sandwich is no longer kept together, and front and rear split-piston segments have separated as shown. The manual force required is now reduced to a minimum, and programming can proceed with ease.

Once teaching has been completed, hydraulic pressure is restored. The

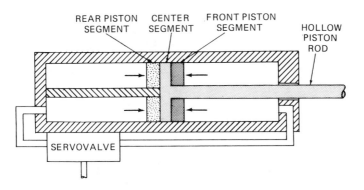

Fig. 5-18. Split-piston in operating condition. (*Courtesy of The DeVilbiss Company*)

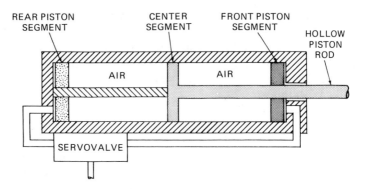

Fig. 5–19. Split-piston in depressurized condition. (*Courtesy of The DeVilbiss Company*)

two split-piston segments rejoin the center segment, and the unit again operates like a conventional linear actuator.

The force that can be exerted with any linear actuator is considerable. Consider a piston diameter of 3 in. With a supply pressure of 2000 psi, the piston rod can be pushed out with a force of over 14,000 lb. On the other hand, the effective areas on both sides of the piston are unequal. For example, a 3-in.-diameter piston with a $1\frac{3}{8}$-in. rod has an area of about 7.1 in.^{2} on the cap end, but only 5.6 in.^{2} on the opposite end. This means that with 2000 psi pressure available, the maximum force is 10,200 lb in the retracting direction, while in the pushing direction it is 14,000 lb. This is not very relevant for robotic applications because with a retracting piston, the weight of the arm would probably assist in the motion.

Rotary Actuators

Electrohydraulic actuators by Bird-Johnson Co. and Moog Inc. were described in Chapter 2 as wrist drive systems. At this point, they deserve another look, as hydraulic elements driven by a servovalve.

Figure 5–20 shows the internal hydraulic circuit of the electrohydraulic wrist made by Moog. It uses a torque motor with flapper-nozzle and mechanical position feedback. The rotary actuator is equipped with a vane that divides the chamber into two compartments, one connected with the A-port of the valve, the other with the B-port. Since this is a closed-center valve, the rotor is maintained in place when the valve is in neutral. When the spool moves, connecting one working port to pressure and the other to the tank, the rotor will move in the corresponding direction.

The illustration shows the rotor at right angles to the output shaft. This was done for better visualization. Actually, the rotor is part of the output

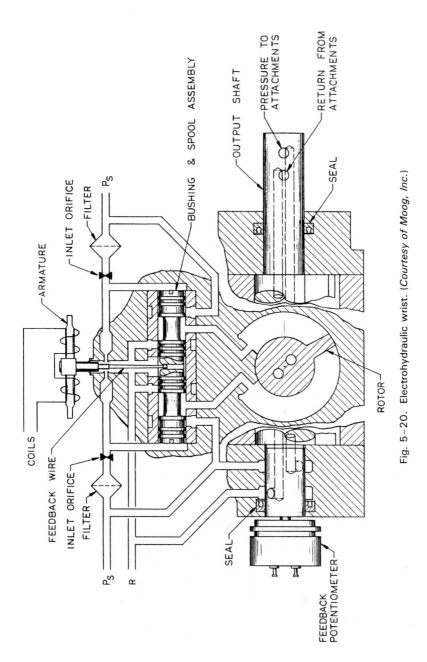

Fig. 5-20. Electrohydraulic wrist. (*Courtesy of Moog, Inc.*)

shaft and turns with it. The two openings in the shaft part of the rotor connect the fluid supply and its return through the length of the output shaft to the wrist attachments. Also operated by the shaft is a feedback potentiometer, which transmits an electric signal that opposes the input signal to the torque motor coils. As the potentiometer signal approaches the point where it zeroes out the input signal, the valve returns to its neutral position. The result is that for each input signal, the output shaft rotates a corresponding amount.

Moog Inc. is primarily a producer of servovalves rather than rotary actuators. For Bird-Johnson, the opposite is true; it is primarily a producer of rotary actuators. The Bird-Johnson units are for higher power than those of Moog Inc., and consequently the rugged Bird-Johnson actuators can be used to advantage not only for wrist action but also for the three main axes of the robot.

The Bird-Johnson actuator shown in Fig. 5–21 is called the Hyd-ro-ac, and is produced in both single-vane and double-vane designs. The double-vane actuator can produce twice the torque of a single-vane actuator. However, the available travel is only half, and, for the same flow rate, its velocity is half. In the Cincinnati Milacron robot, five Hyd-ro-ac units are

Fig. 5–21. Hyd-ro-ac rotary actuator. (*Courtesy of Bird-Johnson Company*)

used, three of them aluminum and two cast-iron units. Their angular travel ranges from 100° ±5° (double vane) for the shoulder to 280° ± 5° (single vane) for the roll axis. Double-vane units that can deliver torques of up to 741,000 in.-lb at 3000 psi supply pressure are available. Single-vane units would deliver half that much torque.

Another type of actuator converts linear action into rotary output. An example is the rack-and-pinion rotary actuator from Flo-Tork, Inc., illustrated in Fig. 5-22. Among other applications, it is used to rotate the base and elbows of robots. The hydraulic piston here drives a rack that in turn rotates a pinion gear as part of the rotating output shaft. The design is available with single as well as with double racks. Rotations of 90°, 180°, and 360° are standards, but any rotation from 0° to 1800° can be produced.

The model 7500, for example, has a stall torque of 3750 lb-in. when the pressure differential across the actuator is 1500 psi. Its weight is 25 lb, and the overall dimensions are approximately 11 × 4 × 4.5 in. At a maximum rotation of 180°, it displaces 8.92 in.^{3}.

Primarily because of the rack-and-pinion drive, there is a backlash, which is less than 1°. This can be eliminated in a double-rack design by using only one rack to develop torque and using the other rack to produce a counterforce at all times. However, this method cuts the available torque in half.

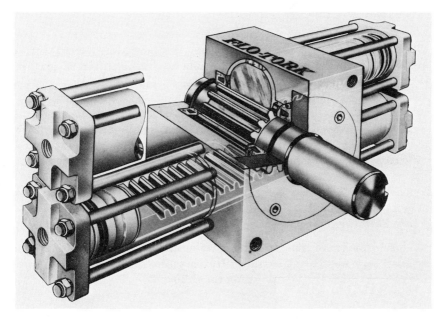

Fig. 5-22. Rack-and-pinion rotary actuator. (*Courtesy of Flo-Tork, Inc.*)

Another method used is a spring-loaded block that applies a constant engaging force on the rack into the pinion and thus prevents backlash without loss of torque.

Still another way to eliminate backlash is to load the rack with a hydrostatic load cell behind the rack. Experience has shown, however, that, even without any of these methods, the standard models have less backlash than expected and so may be used without such adjustment.

PUMPS

Most pumps can be used as hydraulic motors with minor modifications. As hydraulic motors they could be used as actuators for robotic joints, but they are rarely used for this purpose because of their inherently high speed. Hence, only pump designs are discussed here.

The pumps used in practically all hydraulic control systems are positive-displacement pumps. This means pumps that carry with each revolution one or several individual "compartments" filled with oil from the low-pressure side, and force the oil so carried into the high-pressure side. The form of the compartments determines whether they are gear, vane, piston, or any other category of pumps.

Pumps used with servovalves require either a variable-displacement pump or a fixed-displacement pump with accumulator and unloading valve. The time in motion of a robot is generally relatively small. Usage of a fixed-displacement pump with accumulator and unloading valve is in this case a less efficient method, since the pump operates at all times at maximum displacement even when not producing useful work. On the other hand, the fixed-displacement pump is a far simpler design and requires less maintenance. Unimation, for example, opted for a fixed-displacement vane pump.

The variable-displacement pressure-compensated pump reduces the flow automatically to the minimum required at any given time, providing greater efficiency. The Cincinnati Milacron system shown in Fig. 5–2 is based on use of a variable-displacement pump.

Pump Performance

The performance of a pump under varying conditions depends on its characteristics. A number of characteristics must be considered, the ones that are of particular significance being the following:

- Rated pressure and rated speed.
- Volumetric displacement and overall efficiency under different pres-

sures and, in case of variable-displacement pumps, under different volumetric conditions.

• For variable-displacement pumps, the response and recovery characteristics.

In determining these characteristics, two publications are frequently referred to. One, from the Society of Automotive Engineers (SAE), is "Recommended Practice SAE J745c—Hydraulic Power Pump Test Procedure." The other is the standard approved by the American National Standards Institute (ANSI), numbered B93.27 and entitled "Method of Testing and Presenting Basic Performance Data for Positive Displacement Hydraulic Fluid Power Pumps and Motors."

Rated Pressure, Rated Speed, and Life.   Both rated pressure and rated speed of a pump as given by its manufacturer should be qualified, but this usually is not done. The qualification is necessary to answer this question: What is the life expectancy when the pump runs continuously at rated pressure *and* rated speed?

Bearing life is usually the limiting factor of a pump, and since bearings have a life expectancy inversely proportional to the third power of the load, it follows that the life of a pump can be increased substantially by running it below its rated pressure. For example, a pump rated at 3000 psi with a life expectancy of 500 hours could be expected to last 45,000 hours at 1000 psi. This approach, of course, must be used with great caution. The concept of extending the life of the pump by running it below the rated conditions makes it more likely that life expectancy will be influenced by chance occurrences rather than by bearing life.

On the other hand, this same concept is often used in determining the life expectancy of a specific pump design. Here, the pump is run *above* its rated pressure in an accelerated test until it fails. Extrapolations from the test results determine long-time performance under rated conditions. Experience has shown that such extrapolations are justified.

Any statement by the manufacturer regarding life expectancy at rated pressure and/or speed is based on very clean fluid. Any contamination leads to deterioration, due either to overheating of the bearings or to rapid wear of pump parts.

Even the term life expectancy requires definition. Frequently, the so-called B-10 life is used, which means that during the given life expectancy not more than 10% of all pumps of the particular type will break down.

Volumetric Displacement.   Volumetric displacement is the volume of liquid that passes through a pump during one revolution of the unit's shaft.

Volumetric rating according to SAE J745c requires that the pump be operated at 1000 rpm with 100 psi differential pressure at 120°F using a mineral-based hydraulic fluid with a viscosity of 95 to 115 SUS at this temperature and 50 to 54 SUS at 180°F. The ANSI standard is similar.

Overall Efficiency.   The ratio of hydraulic output power to mechanical input power is overall efficiency. The hydraulic output power expressed in horsepower is given by Eq. 4.19, $P = 5.83 \times 10^{-4}$ (Qp). The mechanical input power, P_{in}, is the horsepower delivered by an electric motor or other prime mover. Hence, overall efficiency, η, expressed in percent, is given by:

$$\eta = 5.83 \times 10^{-2} \frac{Qp}{P_{in}} \qquad (5.6)$$

A pump requiring 40 hp to deliver 25 gpm at a pressure differential of 2250 psi across the pump thus has an efficiency of 82%.

Overall efficiency has two components of quite different nature. One refers to friction in the pump of various origins and the torque it produces on the input shaft. This part of the overall efficiency is often referred to as the torque efficiency. The other component is volumetric efficiency. It refers to the internal leakage in the pump, which is whirled around in the pump, diminishing delivery and producing heat.

Response and Recovery.   Response and recovery time as well as rate of pressure rise and drop are of particular interest with variable-displacement pumps because they show the dynamic response of the pump to step changes. Figure 5–23 shows two oscilloscope recordings as given in the above-mentioned SAE J745c. One gives the response, the other the recovery characteristic. To obtain these readings, a circuit is used as shown in Fig. 5–24.

The deadhead pressure, which is the pressure developed by a variable-displacement pump when delivery is zero, is set at the recommended maximum pressure. The relief valve is set to limit maximum pressure to 125% of deadhead pressure. The manual shutoff is opened to maintain 75% of deadhead pressure.

The solenoid-operated shutoff valve is then closed, with the solenoid assuring rapid action. The pressure transducer is connected to an oscilloscope that displays the graphs, as shown in Fig. 5–23. The slope of the curve at the moment when the instantaneous pressure equals the deadhead pressure is the rate of pressure rise in psi/sec. Also the response time is determined, namely, the time interval between the occurrence of the pressure at which

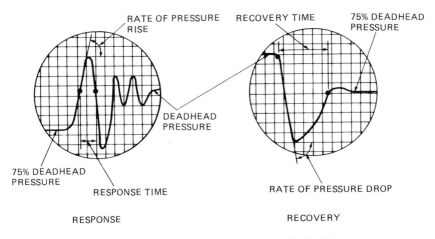

Fig. 5-23. Oscilloscope recordings from SAE J745c.

the rate of pressure rise was measured and the return of the oscillating pressure to the same pressure.

After these readings are taken, the solenoid valve is opened rapidly. The corresponding display now shows a return to the initial condition of 75% deadhead pressure. Here, the rate of pressure drop and the recovery time are determined in the same way as they were when the valve was closed.

Typical results are shown in Table 5-2.

Performance Data. Pump manufacturers usually supply performance graphs of pump types they furnish. The way they should be presented, according to SAE J745c, is shown in Fig. 5-25. For the fixed-displacement pump, the various parameters are plotted as a function of pump rpm at various pressures. Test conditions should be the same as those mentioned for volumetric displacement.

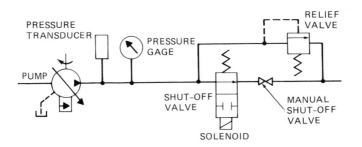

Fig. 5-24. Typical test set-up for response and recovery from SAE J745c.

Table 5–2. Variable-displacement pump response data.

SPEED, RPM	DEAD HEAD PRESSURE, PSI	RESPONSE, MSEC	RECOVERY, MSEC	RATE OF PRESSURE RISE, PSI/SEC	RATE OF PRESSURE DROP, PSI/SEC
2000	2000	55	75	20,000	20,000
1000	1000	75	120	20,000	20,000

The left plot of Fig. 5–25, which refers to a fixed-displacement pump, shows, for example, at 1600 rpm and 1000 psi, a horsepower loss of approximately 4.9 hp with a horsepower input of 38.4 hp and a delivery of 57 gpm at 87% overall efficiency. The horsepower loss is the difference between power input and power output.

The right plot shows the performance of a variable-displacement pump at various volumetric deliveries and pressures. As seen from these graphs, the pump can deliver slightly more than 20 gpm at 1000 rpm, 35 gpm at 1500 rpm, and 45 gpm at 2000 rpm. For example, at 20 gpm and 1500 psi, horsepower loss amounts to 8.3 hp, horsepower input to 26.4 hp, and overall efficiency to 91%. This is as close as one can read from a graph.

These graphs are part of the afore-mentioned "Recommended Practice" and considered only as examples, for which purpose they serve well. However, as an exercise it is interesting to detect some inconsistencies beyond the normal numerical uncertainties inherent in graphical representation. The horsepower output, according to Eq. 4.19, is $20 \times 1500 \times 5.83 \times 10^{-4} = 17.5$ hp. It should, however, be equal to the horsepower input minus horsepower loss—which appears from the graphs to be $26.4 - 8.3 = 16.1$ hp. On the basis of these data and of Eq. 5.6, the efficiency, as shown in the graph seems to be too high.

Types of Pumps

The three most common pump types are gear, vane, and piston pumps.

Gear Pumps. Basic elements of a gear pump are shown in Fig. 5–26. Two gears are meshed and rotate against each other. The shaft of one gear passes through a bearing in the housing to the outside and connects to a prime mover, usually an electric motor. It becomes the driver, while the other gear is the driven gear or idler.

The meshing of the gear teeth prevents hydraulic fluid from being carried between the gears. However, between the housing wall and the teeth of each gear, fluid can be carried from inlet to outlet.

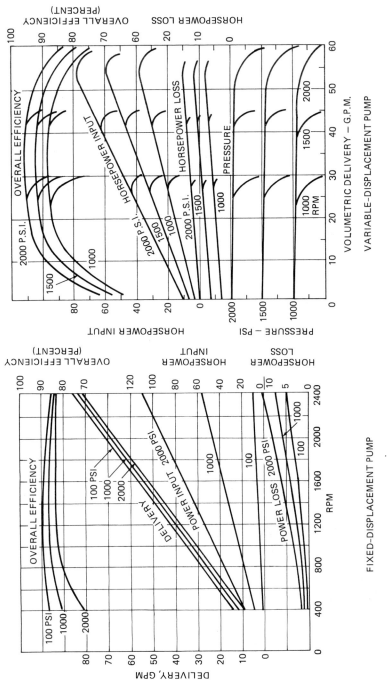

Fig. 5–25. Performance data of hydraulic pumps according to SAE J745c.

VARIABLE-DISPLACEMENT PUMP

FIXED-DISPLACEMENT PUMP

183

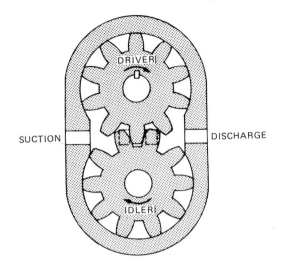

Fig. 5-26. Concept of gear pump.

As the gears rotate against each other, they separate on the inlet (or suction) side, draw fluid into the interspaces, carry it to the outlet side, and force it through the outlet (or discharge) port.

As gears begin to mesh on the discharge side, a remnant of the fluid tends to get trapped between the tooth tip of one gear and the tooth root on the other. Without providing some release of the resulting pressure, very high forces would result, and the fluid would somehow be squeezed through clearances, accompanied by considerable noise. One of several methods for relieving entrapped fluid is to provide shallow recesses in the housing, as shown by the small dotted rectangles in Fig. 5-26.

Vane Pumps.   The basic design of a vane pump is shown in Fig. 5-27. A cylindrical rotor with sliding vanes in radial slots rotates inside an elliptical ring. The rotor axis is eccentric to the ring. Since the vanes are free to slide in their slots, they move outward by centrifugal force and/or hydraulic pressure and/or springs, and a number of chambers are thus formed between the vanes and the ring. These compartments change their volume continuously as the vanes follow the inner contour of the ring.

The compartments between vanes decrease in volume when they enter the discharge side. As a result, they squeeze out the fluid against the load, which is connected to the pump. The discharge and suction sides of the pump are sealed from each other at any time by at least one vane. Delivery of the

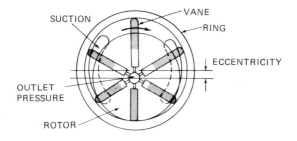

Fig. 5-27. Concept of vane pump.

vane pump is proportional to the eccentricity between the center of the ring and the center of the rotor, as indicated in the illustration.

An adjustable eccentricity results when the center of the rotor is made movable relative to the center of the ring. This is accomplished by a ring that can float in one direction. It is a concept that leads to a variable-displacement pressure-compensated pump, as illustrated in Fig. 5-28. The unbalanced reaction forces that result from the system pressure on one side of the rotor and the suction pressure on the opposite side are taken advan-

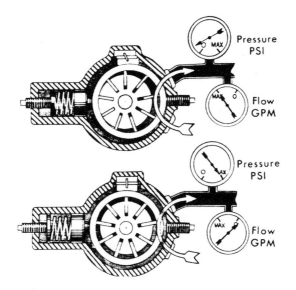

Fig. 5-28. Concept of variable-displacement vane pump. (*Courtesy of Racine Hydraulics Division, Dana Corporation*)

tage of and used to control delivery of the pump as a function of output pressure.

The resulting thrust of the ring is resisted by a spring. In the upper diagram, pressure is low, and the force of the spring pushes the ring against the stop, so that maximum delivery is obtained. As pressure increases, reaction forces push the ring to the left, decreasing eccentricity and, hence, delivery. At a certain maximum pressure, flow is reduced to zero. The pressure at which flow begins to decrease can be altered by adjusting the screw on the left. The ratio of flow change to pressure change depends on the spring rate.

A vane pump must be produced very accurately. Only if all moving parts are finished to exacting tolerances can the volumetric efficiency be kept high. This refers specifically to the rectangular vanes and the slots in which they move. As they slide out of the slot, the vanes are cantilevered and temporarily subject to a large pressure differential across the vane. Considerable frictional forces are thus generated between the vane and the slot. The hydraulic fluid used must be not only clean but of low viscosity for the necessary lubrication. On the other hand, viscosity should not be too low, since it has a significant influence on volumetric efficiency.

Designs of fixed-displacement vane pumps can be balanced as shown in Fig. 5–29. Here, two discharge ports and two suction ports, instead of one of each, are provided. They are, of course, internally connected to obtain only one outlet and one inlet to the pump. All forces that act on the pump rotor and shaft are now balanced, extending bearing life considerably. This design is not applicable to the typical variable-displacement vane pump, which is generally limited to a somewhat lower pressure rating than the balanced fixed-displacement pump, although the SV-10, for example, a variable-volume vane pump for 7.5 gpm from Racine Hydraulics Division, is rated at a relatively high 2000 psi.

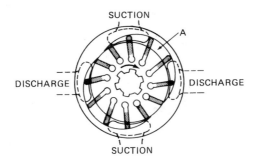

Fig. 5–29. Concept of balanced vane pump.

A recent development of Vickers Inc. provides hydraulic balancing even of variable-displacement vane pumps. These designs are rated for continuous duty at pressures to 3000 psi and speeds to 1800 rpm. Their rated flow capacity is 54 gpm (at 1200 rpm and 2500 psi).

Piston Pumps.   Compared with vane pumps, the piston pump is easier to manufacture, as tolerances usually can be kept closer in bored holes and pistons than in flat surfaces, particularly when they are recessed. Forces can be relatively well balanced, and operating pressure as well as efficiency can be high. Among the many types of piston pumps, the bent-axis and the in-line designs are probably the most widely used.

The operating principle of the bent-axis pump is shown in conditions (a) through (c) of Fig. 5–30. They correspond to several tilt angles, α, between the drive shaft and the cylinder block. In (a), the drive shaft, pistons, and cylinder block all rotate about a common axis. The tilt angle is zero (i.e., nonexistent). Obviously, in this condition there is no reciprocating motion of the pistons and, hence, no pumping. Displacement, like the tilt angle, is zero.

In (b), the cylinder block axis forms an angle with the drive shaft axis. The radius from the cylinder block axis to the centerline of the piston is indicated by r. As the entire unit rotates, the piston is forced into a reciprocating motion relative to the cylinder block. The amount of this motion is indicated by x.

When the tilt angle α increases farther, as shown in (c), stroke x must also increase. The length of the stroke, and thus the displacement of the pump, is proportional to the tangent of the angle α (tan α), and can be expressed by:

$$x = 2r \tan \alpha \qquad\qquad (5.7)$$

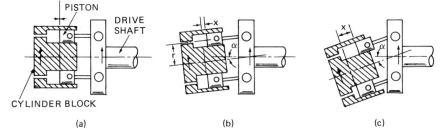

(a)   (b)   (c)

Fig. 5–30. Concept of bent-axis piston pump.

For example, if tilt angle α is 12° and radius r in Fig. 5–30 is 1.2 in., the stroke x is 0.51 in.

An actual design of a variable-displacement bent-axis pump would connect the cylinder block to the drive shaft by a rod with universal joints at both ends. Pistons and piston rods would still be connected by the same basic method as shown, but they would be relieved from carrying the thrust loads produced by hydraulic forces in the pump.

The bent-axis pump can be built with either a fixed or an adjustable tilt angle. In the latter case, it becomes a variable-displacement pump. The same is true of the in-line pump, the principle of which is shown in Fig. 5–31.

There are several closely related design concepts of the in-line pump. The one shown is typical. The cam plate here is fixed. If the angle α of the inclined shoe plate were adjustable, it would become a variable-displacement pump. Equation 5.7 applies equally for this case.

The shoe plate rotates against the cam plate, separated from it by a thin film of hydraulic fluid as lubricant. The shoe plate is part of the rotating group that includes the drive shaft, cylinder block, and pistons. The shoes consist of flexible sliding links connected to the shoe plate.

To provide pressure compensation for a piston pump, special valving is needed. The purpose is to prevent the pump, when all servovalves are closed, from discharging its flow over a pressure relief valve and thus generating heat and wasting energy. Pump stroke and, hence, delivery are reduced to zero when the system pressure rises to the setpoint of the pressure compensator. Since the setpoint pressure is below the relief valve setting, the latter serves only as a safety backup.

Figure 5–32 illustrates how the pressure compensator operates. It is mounted to the pump case as an integral part of the pump. The stem of the pressure compensator is connected to the mechanism that adjusts the angle α, that is, the displacement of the pump. The stem in the illustration

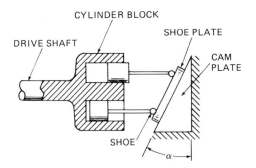

Fig. 5–31. Concept of in-line piston pump.

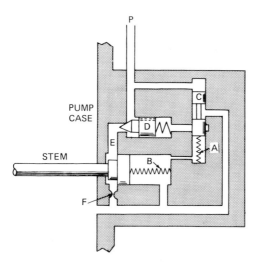

Fig. 5-32. Principle of pressure compensator.

should move to the right when the pressure exceeds a maximum, which is determined by spring A. The spring has an external adjustment (not shown) to set the pressure at which the stem is to move.

Under normal conditions, spring B pushes the piston and, hence, the stem to the left. The output pressure of the pump, which is the system pressure, is connected at P. If this pressure produces a force on spool C that exceeds the force of spring A, the spool moves downward. This connects the back chamber of plug D to the tank, since the pump is at tank pressure.

Plug D has a hole drilled through it leading from the supply pressure side to its back chamber to equalize the pressure across the plug when the back chamber is dead-ended. With spool C in the downward position, the back chamber is connected to the tank. Very little flow passes through the hole in plug D, and the pressure differential that now exists across it moves it to the right, opening a passage from the supply pressure to the tank through restriction F. The restriction produces a pressure differential across the stem piston, making the stem move to the right and thereby reduce the pump displacement.

When pressure drops again, spool C moves upward, dead-ending the back chamber of plug D. The pressure across D equalizes, because of the hole drilled through it, and the spring closes the passage. The pressure in E drops to tank pressure, and the stem piston moves to the left, acting to restore the original conditions—but not quite doing so.

The moment when the displacement of the pump suffices to provide the very small flow passing through restriction F, the system arrives at a balanced condition, where pump delivery is minimal and full supply pressure is maintained. The energy required to drive the pump is also minimal, since it is proportional to the product of the flow rate and the pressure.

LEAKAGE

Leakage usually occurs at the connection between the hydraulic components and the connecting members between the components. But, as James Mockler[3] says, hydraulic leakage can be virtually eliminated with existing technology.

The Dryseal American Standard Taper Pipe Thread, or NPTF, as covered in ANSI Standard B 1.20.3, has been used in hydraulic equipment for many years. The sealing action of NPTF threads is accomplished by the metal-to-metal interference of the tapered threads. While connections that use this thread seal well when the installation is made with great care, they tend to leak more than other types. When the threads are too loose or have been tightened too many times, leakage will occur.

For optimal prevention of leakage, the preferred method is to use either straight thread O-ring ports according to SAE Standard J514g or, in case of larger ports, the four-bolt split flange for tube, pipe, and hose connections, according to SAE Standard J518c.

The threads for either method are straight, not tapered, and are provided with an O-ring seal as shown in Fig. 5–33. Tightening the locknut and washer against the face of the boss compresses the O-ring and assures a leak-tight seal.

The aerospace industry has changed completely to fittings that seal with an O-ring, and has proved that leakage can be prevented even under high vibrational, atmospheric, and mechanical stresses.

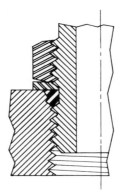

Fig. 5–33. O-ring port.

RESERVOIRS, FILTERS, AND MAINTENANCE

Hydraulic systems are sometimes criticized because they (a) leak and (b) fail, thus causing shutdowns. Both situations can be avoided by good housekeeping and maintenance. General Motors, for example, credits its maintenance program with 100% uptime in its robotic hydraulic systems.

Not only can pumps and other components be severely affected by contaminants in hydraulic systems, but servovalves are particularly sensitive to any particles that are not properly filtered out. Under such conditions servovalves get erratic or stuck, and shut down the system.

Maintenance starts with storing the hydraulic fluid. Generally, it is shipped in 55-gallon drums, which should be stored under a roof at least, preferably in a closed room. They should lie on their side across wooden runners to protect them from rusting and from dirt collecting on top. Before a drum is opened, its top is cleaned to prevent dirt from falling into the fluid. After the required amount is withdrawn, bungs are tightened with a wrench that is tapped firmly with a mallet.

Hoses, piping, and so on, used in filling a hydraulic system must also be clean, and the hydraulic fluid must be filtered before it enters the system.

Although the reservoir of a system serves primarily to store and supply hydraulic fluid, it also helps to separate contaminants and dissipate heat generated within the system.

Figure 5–34 shows a typical reservoir as used by Vickers Inc. It is designed in accordance with ANSI/B93.18, an American Standard for "Non-Integral Industrial Fluid Power Hydraulic Reservoirs." This means that its structural details include the following:

- Supporting legs to raise the reservoir a minimum of 6 in. above the floor to facilitate handling and draining and to improve heat dissipation.
- A rigid baffle plate that separates returning hydraulic fluid from that entering the pump. It permits slower and more complete circulation, facilitates precipitation of some contaminants, releases air, and dissipates heat.
- Cleanout openings for cleaning the reservoir without disturbing any components or piping.
- A reservoir bottom shaped to facilitate complete drainage.
- Provision of two filler openings and two fluid level indicators, each located in close proximity to a filler and visible during filling.
- Provision of a breather or filler–breather combination with a filter having a rating of not more than 40 microns, to maintain approximately atmospheric pressure during changes in fluid level.
- A heavy plate on top for mounting the pump, electric drive motor, and

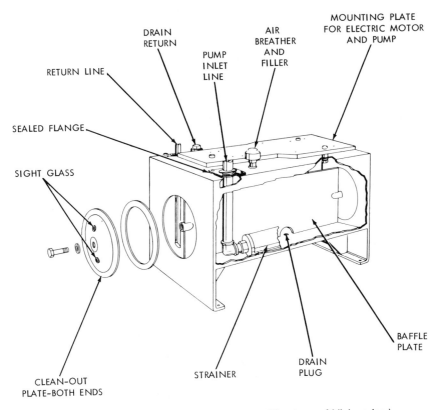

DRAIN RETURN

AIR BREATHER AND FILLER

MOUNTING PLATE FOR ELECTRIC MOTOR AND PUMP

PUMP INLET LINE

RETURN LINE

SEALED FLANGE

SIGHT GLASS

BAFFLE PLATE

CLEAN-OUT PLATE–BOTH ENDS

STRAINER

DRAIN PLUG

Fig. 5–34. Reservoir for hydraulic system. (*Courtesy of Vickers Inc.*)

associated components, that is rigid enough to prevent distortion and subsequent shaft misalignments between the pump and drive.
- A capacity that is equal to three times the fluid volume the pump can deliver in one minute, plus 10% to provide a free surface to the atmosphere for thermal expansion of air and air separation from the fluid.

The pump inlet line is usually provided with strainers, either one or several in parallel, which are immersed in the reservoir. They consist basically of a fine mesh wire screen or some equivalent. They do not provide as fine a cleaning action as filters but offer less flow resistance. This is important for the pump inlet, since pressure drop in this line must be avoided to prevent cavitation and thereby excessive noise, loss of efficiency, and rapid wear of active parts.

In mounting the pump to the prime mover, a flexible coupling is required.

Angularity and offset alignment between shafts of the pump and prime mover must be aligned within 0.003 in. total indicator reading, in order to prevent undue bearing and shaft seal wear. Filters are used on the outlet side of the pump as shown in Fig. 5–2, and possibly additionally in the return line. Their filtration rate is much finer than that of strainers. Their purpose is to remove contaminants that consist primarily of minute metal or metal oxide particles that result from wear of moving parts in the system, primarily of the pump. There is also sludge formation resulting from oil decomposition, which is often caused by hot spots in the system. And, finally, there are all the foreign contaminants or dirt that may be introduced in the system.

Filters are rated by the size of the particles they are capable of retaining, the size being expressed in microns. One micron is one micrometer (μm), and 1 μm = 0.000039 in. A filtration rating of, say, 5 μm can be deceiving, unless clearly specified whether this is an absolute rating or not. Even then, its practical significance can be questioned.

Absolute filtration rating, which is the most restrictive rating, is defined by the National Fluid Power Association as "the diameter of the largest hard spherical particle that will pass through a filter under specified test conditions."

It is the nature of contaminants and filter design that a better definition cannot be offered, but it does leave unanswered what a "hard" particle is (a soft one can do just as much damage), and to what extent the specified test conditions are applicable to actual working conditions. The diameter of a spherical particle, of which the definition speaks, does not apply to particles that have the same diameter in one dimension but are considerably longer in another—although such elongated particles will usually be trapped with depth filters, where the fluid has to pass through a tortuous path of thick filter media, particularly when they are "hard" particles. In any case, the micron rating of a filter should be taken primarily as a relative measure, in the sense that it means that a 5-micron filter is better than a 10-micron filter, and so forth.

The micron rating may also refer to the so-called efficiency, which is defined as "a measurement of the average size of the pores of the filter medium"—and the stress here should be on "average."

One particular effect of contaminants that may especially affect servovalve action is so-called silting. This is caused by tight passages between moving surfaces, such as the valve spool in its sleeve, that themselves act as filters. Extremely small particles may accumulate there and increase friction.

All filters and strainers require a regular maintenance schedule to ensure their efficiency and minimize the pressure drop they cause in the system.

With proper maintenance malfunction may be blamed on the hydraulic system rather than on insufficient care.

Before a hydraulic system is put into operation, all air should be purged from the system. Cylinders must be worked back and forth several times to remove entrapped air. It may also be necessary to bleed the air from fittings of bleed petcocks at the highest points or at cylinders. Air trapped in a hydraulic system will cause spongy operation.

SUMMARY

Equations of this chapter are repeated here for ready reference.

Flow forces:

$$F = 2.2 \times 10^{-2} Q \sqrt{p}$$

Leakage flow through eccentric clearance between spool and sleeve:

$$Q = 4.3 \times 10^4 \frac{pdb^3}{v\rho L}$$

Flapper-nozzle characteristics:

$$\frac{p_c}{p_s} = \frac{1}{1 + 16 x^2 (d_2^2/d_1^4)}$$

Rated vs. control flow of servovalve:

$$Q_r = Q_t \sqrt{\frac{1000}{p_s - p_l}}$$

Overall pump efficiency:

$$\eta = 5.83 \times 10^{-2} \frac{Q_p}{P_{in}}$$

Displacement of variable-displacement piston pump:

$$X = 2r \tan \propto$$

where:

b = Clearance between centered spool and sleeve, in.
F = Flow force, lb.
d = Diameter of sleeve, in.
d_1 = Diameter of fixed orifice, in.
d_2 = Diameter of nozzle, in.
L = Length of leakage path, in.
P_{in} = Input power, hp
p = Pressure differential or pressure, psi
p_c = Back pressure of flapper-nozzle, psi
p_l = Pressure drop across load, psi
p_s = Supply pressure, psi
Q = Flow rate, gpm
Q_c = Control flow, gpm
Q_r = Rated flow, gpm
r = Radius, in.
x = Length, in.
α = Tilt angle, deg
η = Overall efficiency, percent
ν = Centistoke, cSt
ρ = Density, lb/in.^{3}

REFERENCES

1. Blackburn, J. F., et al. (eds.). *Fluid Power Control.* Cambridge, MA: The Technology Press of MIT, 1960.
2. Atchley, R. D. A more reliable electrohydraulic servovalve. *Robots VI Conference Proc.,* Detroit, MI, 1982.
3. Mockler, J. Preventing hydraulic leakage: A simple solution. *Robots 8 Conference Proc.,* Detroit, MI, 1984.

Chapter 6
PNEUMATICS

AIR CHARACTERISTICS

Pneumatics may be used in two ways in robotic systems: (1) to drive all axes of smaller robots, including some robots that are specifically designed for educational purposes; and (2) merely to drive the actuators for a great number of end effectors.

Air is a readily compressible fluid. Hence, an air motor or cylinder does not have the stiffness of an electric or hydraulic actuator. Its dynamic response thus is inferior, and the power that can be developed is usually limited. But, otherwise, air is clean, expendable, and—if shop air can be used, as it usually is—readily available. Actuators are generally of lower cost, primarily because of the lower pressure that is involved with pneumatics. Best of all, air is intrinsically explosion-proof.

What are the characteristics of compressed air as compared with hydraulic fluids? In many respects its behavior is identical to that of hydraulic fluids; for example, in both cases force is equal to the area to which the fluid is applied multiplied by the pressure. Some factors, however, that are negligible in hydraulics can become significant in pneumatics because the gas is so much more compressible than liquids. Some of the peculiarities of air will be discussed in the following paragraphs.

Gas Laws for Air

An ideal gas is one that follows the gas laws discussed below. Air is so nearly "ideal" that it can be considered an ideal gas for all purposes of robotic pneumatics.

For an ideal gas, the absolute pressure varies inversely as the volume, provided the temperature remains constant. This relationship is known as Boyle's law. According to this law, doubling the absolute pressure of a given weight of gas halves its volume; tripling the absolute pressure reduces the original volume by two-thirds; and so on.

196

This law can be expressed mathematically by:

$$\frac{p_{a,1}}{p_{a,2}} = \frac{V_2}{V_1}$$

(6.1)

from which it follows that for a given weight (or mass) of air, $p_{a,1} V_1 = p_{a,2} V_2 = p_{a,3} V_3 = $ constant. The numerical subscripts refer to the various conditions. Furthermore, p_a is the absolute pressure in psia, which means gage pressure (psig or simply psi) plus 14.7, as explained on page 114, and V is the volume, which is usually expressed in cubic feet. Actually, it is irrelevant in which units either absolute pressure or volume is expressed in Eq. 6.1, as long as this is consistently done. Usage of pounds per square inch (psi) for pressure and cubic feet for volume is customary in dealing with gases.

Equation 6.1 applies only for isothermal conditions. This means that while the pressure changes, enough heat is added or removed to maintain a constant temperature.

By contrast, adiabatic conditions exist when everything is well insulated and no heat is removed or added while pressure and volume change. For adiabatic conditions, the relationships are:

$$\frac{p_{a,1}}{p_{a,2}} = \left(\frac{V_2}{V_1}\right)^{1.41}$$

(6.2)

which may be written $V_1/V_2 = (p_{a,2}/p_{a,1})^{0.71}$.

Suppose that 40 ft³ of air at atmospheric pressure (14.7 psia) is compressed to 4 ft³. What pressure will result? If the temperature is kept constant (i.e., if conditions are isothermal), Eq. 6.1 applies, and the final pressure is 147 psia or about 132 psig.

However, if no heat is added or removed in the process (i.e., if conditions are adiabatic), Eq. 6.2 applies, and the final pressure is 378 psia or 363 psig.

In the practical world of pneumatic power, conditions are between isothermal and adiabatic. Thus, in the preceding example, much but not all of the heat would have been dissipated through the cylinder walls and other thermal passages. Hence, the final pressure would be somewhere between 142 and 363 psi. In other words, on the basis of these data it would be rather unpredictable.

This shows the strong influence of temperature on air pressure. If the volume of a given weight of gas is kept constant (i.e., if isometric conditions prevail), the absolute pressure will vary directly as the absolute temperature of the gas. This is expressed by:

$$\frac{p_{a,1}}{p_{a,2}} = \frac{T_1}{T_2} \tag{6.3}$$

where T is absolute temperature, which, in U.S. customary units, is usually expressed in degrees Rankine (°R). Adding 459.67 or, in round numbers, 460 to a reading in degrees Fahrenheit gives the absolute reading in degrees Rankine. For example, 32°F equals 492°R. (Note, however, that a temperature *change* of 100°F is equal to a change of 100°R.)

A further gas law must be added. It assumes isobaric conditions, which means that the absolute pressure of a given weight of gas is kept constant. In this case, the volume changes directly as the absolute temperature. That is:

$$\frac{V_1}{V_2} = \frac{T_1}{T_2} \tag{6.4}$$

It should be noted that Eq. 6.1 applies for isothermal or constant-temperature conditions; Eq. 6.3, for isometric or constant-volume conditions; and Eq. 6.4, for isobaric or constant-pressure conditions. The three equations can be combined, and the result is:

$$\frac{p_{a,1}V_1}{T_1} = \frac{p_{a,2}V_2}{T_2} \tag{6.5}$$

To determine pressure of air that has an initial volume of 400 ft³ and a pressure of 30 psi (44.7 psia), and is compressed to a volume of 150 ft³ while its temperature increases from 80°F (540°R) to 120°F (580°R), Eq. 6.5 can be used. The result is 128 psia, or a gage pressure of about 113 psi.

It is obvious from Eq. 6.5 that for a given weight of gas, the quotient of the product of its absolute pressure and volume, divided by its absolute temperature, is a constant. This means that for a given weight of gas, $p_a V/T$ = constant. Values of this constant for one pound of gas have been determined experimentally for practically all of the known gases. They are called gas constants. For air, the gas constant is 53.34. A general gas law can thus be established, which for air reads $p_a V/WT$ = 53.34. However, in this case p_a would have to be expressed in pounds per square foot, rather than per square inch. To be able to express p_a in pounds per square inch requires dividing the gas constant for air by 144, making it 0.37. In this form, the general gas law as applied to air is expressed by:

$$\frac{p_a V}{W T} = 0.37 \tag{6.6}$$

Here, p_a is the absolute pressure in pounds per square inch, V is the volume of air in cubic feet, W is the weight of air under consideration, in pounds, and T is the absolute temperature in degrees Rankine.

Consider, for example, a volume of 23 ft³ of air, weighing 5 lb and having a pressure of 30 psi, which is equal to 44.7 psia. According to Eq. 6.6, its temperature would be 556°R, or 96°F.

The preceding equations are the practical working formulas for the engineer who has to consider relations between pressure, volume, and temperature in pneumatic circuits.

Flow of Air

Air changes in density. However, by reducing the volumetric flow to so-called standard conditions, it can be manipulated as a constant-density fluid for the purposes of calculations. According to Standard B93.2 issued by the American National Standards Institute, "standard air" is air at a temperature of 68°F, a pressure of 14.7 psia, and a relative humidity of 36%. Density of air under these conditions is 0.0750 lb/ft³.

Density of *dry* air under the same conditions would be 0.07524 lb/ft³, and under saturated conditions, 0.07104 lb/ft³. This is a relatively small change for practical considerations; hence, the influence of humidity on the change in density generally can be neglected in pneumatics. Its influence on the reliable operation of pneumatic components, however, is a different matter and will be treated later.

The problem now is to convert from flow rates under standard conditions to those of actual conditions.

The flow rate of air, like that of any other fluid, is expressed by the product of flow velocity and the inside area of the pipe through which it flows. Hence:

$$Q = 0.007 \, vA \tag{6.7}$$

where 0.007 is a conversion factor to permit using inconsistent units, namely, cubic feet per minute (cfm) for the flow rate, Q; feet per minute for the velocity, v; and square inches (rather than square feet) for the cross-sectional area, A, of the conduit through which the flow takes place.

Flow rate is volume per unit time; that is, $Q = V/t$, or $V = Qt$. Equation 6.5 stated that $p_{a,1} V_1 / T_1 = p_{a,2} V_2 / T_2$. Using the equality $V = Qt$, this can be written:

$$\frac{p_{a,1} Q_1}{T_1} = \frac{p_{a,2} Q_2}{T_2} \tag{6.8}$$

This equation is the one needed to convert air flow at standard conditions to any other condition. As stated above, standard air in pneumatic systems means 14.7 psia and 68°F (528°R). Substituting these conditions in Eq. 6.8 gives:

$$Q = 0.028 \, \frac{Q_s T}{p_a} \tag{6.9}$$

where Q_s is the flow rate under standard conditions in standard cubic feet per minute (scfm), and Q, T, and p_a refer to the flow rate for any other combination of absolute temperature and absolute pressure. Thus, if $Q_s = 80$ scfm, then the flow for an absolute temperature of 580°R and an absolute pressure of 115 psia is 11.3 cfm.

Flow through Restrictions

Flow through conduits is of minor importance for robotic pneumatics. The air is taken from a shop air tap, just as electric power is taken from a power line. Losses in the transmission lines usually would not concern the user of electric power, nor would they influence robotic pneumatics, if the required minimum power were available. Within the pneumatic system of the robot, conduits usually are short and losses negligible.

This is not the case with restrictions that occur in valves and other devices. In order to evaluate their flow rates, it is necessary to determine their pressure losses.

In hydraulic flow, the volumetric flow rate for a given density is constant as long as incompressibility is assumed. In fact, since volume and mass (or weight) can be considered constant, it is hardly necessary to differentiate between volumetric and mass flow. In pneumatics, it is different: the weight of a given volume of air depends on pressure and temperature, as expressed in Eq. 6.6, where all four factors are combined. Thus, it must be stressed that in discussing flow rate in the following text, it is the volumetric flow rate (cfm) that is considered.

The rate of flow through a valve or an orifice at a given upstream pressure depends on the pressure differential across the restriction. This is basically true for liquids and gases. However, for gases there is also a maximum differential pressure that corresponds to the maximum flow rate that can pass through the restriction. It is given by the critical ratio, which is different for each type of gas. For air it is 0.528, or simply 0.53. When the ratio of the absolute upstream pressure to the absolute downstream pressure is 0.53, the critical ratio is reached. Thus, if the upstream pressure of a restriction is 85 psi, or 99.7 psia, then the maximum flow rate is reached

when the downstream pressure is 52.6 psia. Further reduction of downstream pressure will not increase the flow rate.

Within these limitations, air flow through a restriction can be calculated
by the equation:

$$Q = 45.6\ C_d A\ \sqrt{T_1\left[\left(\frac{p_2}{p_1}\right)^{1.43} - \left(\frac{p_2}{p_1}\right)^{1.71}\right]} \qquad (6.10)$$

Here, Q is the flow in cubic feet per minute, C_d is the coefficient of discharge, A is the cross-sectional area of the restriction in square inches, T_1
is the absolute temperature in degrees Rankine of the upstream gas, p_1 is
the absolute pressure upstream in psia, and p_2 is the same downstream.

To facilitate calculations, one can write

$$k = \sqrt{(p_2/p_1)^{1.43} - (p_2/p_1)^{1.71}}$$

in which case Eq. 6.10 becomes:

$$Q = 45.6\ k\ C_d\ A\ \sqrt{T_1} \qquad (6.11)$$

and the value of k for various ratios of p_2/p_1 can be read from Fig. 6-1.

Coefficients of discharge, C_d, were discussed on page 123 for hydraulic
systems. They are the same for pneumatics; so for valves it may be assumed
that $C_d = 0.625$. Consequently, for an absolute pressure upstream of 100
psia and downstream of 80 psia, that is, for $p_2/p_1 = 0.8$, together with a
cross-sectional area of the valve port opening of 0.5 in.^{2}, and an absolute
temperature of 580°R, the flow rate would be about 72 cim (cubic inch
per minute). Here, the value of k is read to be 0.21 from Fig. 6-1.

Furthermore, at an upstream pressure of 100 psia, the maximum flow
rate would be attained at 53 psia. Using the value of $k = 0.256$, according
to the graph, this means that the maximum flow through the valve at this
port area, air temperature, and supply pressure is about 88 cim.

The capacity of pneumatic valves is frequently expressed by the flow
coefficient or C_v factor, where:

$$C_v = 0.06\ Q_s\ \sqrt{\frac{T_1}{p_1^2 - p_2^2}} \qquad (6.12)$$

This permits determination of the flow rate of the valve for any specified
condition if the C_v factor of the valve is known. Thus, a valve with a flow
coefficient of 1.5 that is to operate at an upstream pressure of 100 psia with
a pressure drop of 20 psi and an upstream air temperature of 530°R, will

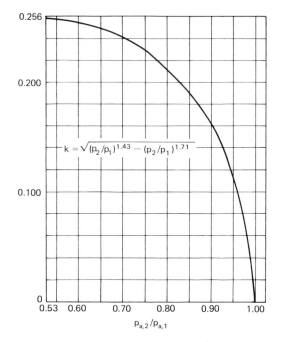

Fig. 6–1. Values of k for Eq. 6.11

have a flow rate of 65 scfm. For conditions other than standard, this value can readily be converted by means of Eq. 6.9.

PNEUMATIC CONTROL SYSTEMS

The M50 robot made by International Robomation/Intelligence (IRI) is shown in Fig. 6-2. This is a robot with five degrees of freedom, all of them air-operated. The payload is 50 lb, and the reach 20 to 80 in. The robot is controlled by a hierarchy of microcomputers with a Motorola 68000 as the systems manager. Each of the five axes has its own Motorola 680X microprocessor receiving and passing information to the manager. Air requirements are 20 to 130 scfm air at 100 psi. Repeatability of programmed positions is ±0.040 in.

The schematic in Fig. 6-3 shows the air distribution for the pneumatic system, in a typical configuration for one axis. The branch-off points for the other four axes are indicated in the schematic. Each one of those circuits is identical to the one shown here.

The actuator is an air motor with eight vanes. By switching pressure and

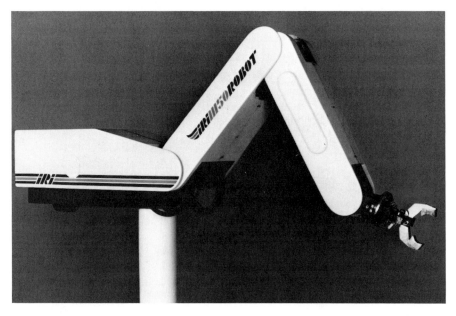

Fig. 6-2. M50 pneumatic robot. (*Courtesy of International Robomation/Intelligence*)

exhaust connections, the direction of rotation can be reversed. The switching is effected by the spool valve which is an integral part of the motor.

The air motor rotates at 6000 rpm. This requires a reduction of 300:1 to arrive at the pitch speed of 120°/sec. To obtain this, a multistage set of chain drives is used. Four successive sets are required. The chains are tensioned to eliminate the effect of their elasticity on dead band and dynamic behavior.

The motor is combined with a four-way, three-position valve, which is centered in its neutral position by springs. A solenoid controls the two pilot pressures, which are directed against the end faces of the spool. In the center position, air pressure is applied to both sides of the rotor. By switching the valve from one side to the other, the rotation of the motor is changed.

The electrohydraulic servovalve described in the previous chapter normally would receive analog signals in proportion to which the flow would be regulated. Because of the compressibility of air, accurate flow regulation by analog methods is almost impossible with an air valve. Other methods must be used. IRI developed a digital control scheme for this purpose. A two-position pressure-control valve is made to dither by switching its solenoid on and off in rapid succession. Changing the size and frequency of

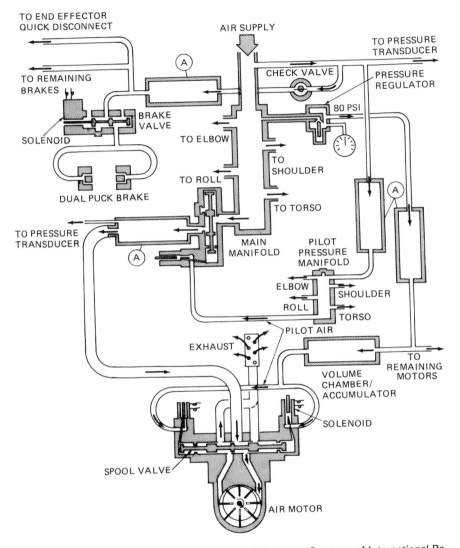

Fig. 6-3. Schematic of pneumatic system of M50 robot. (Courtesy of International Ro-bomation/Intelligence)

the pulses produces various rates of acceleration or sustained velocity for the air motor. The "dithered" air flows into a volume chamber/accumulator that smoothes out the pulses. The valve is pilot-operated and alternates between two positions. The pilot pressure is derived directly from the air supply line.

The pilot pressure for the spool valve of the air motor is kept at a constant 80 psi by means of a pressure regulator. The spool valve is also dithered to overcome frictional effects. The chain drives of the speed reducer smooth out the pulses, and the movement does not reflect the pulsating action of the valve.

To lock the robot arm in position, a solenoid-operated two-position air valve is provided to operate a brake. This is a single-stage valve without pilot action. When the solenoid is activated, air pressure is applied to a dual set of brakes. They engage the first-stage sprocket in the multistage chain reducer.

The exhaust from the air motor passes through a silencer to diminish the noise generated. Also, this air carries lubricating oil for the motor and is used to provide lubrication for the chain. A number of additional volume chamber/accumulators are provided. Their purpose is to smooth out the system against transient pressure changes. Because of their large surface, they can also serve as an additional cooling means.

In order to close the loop of the control system, an optical encoder sits on the end of the motor shaft and feeds back a positioning signal, assuring a repeatability of the programmed position of ± 0.040 in.

The system described above incorporates essential elements of pneumatic robotic systems. These elements will be discussed in somewhat greater detail in the following section.

COMPONENTS

Pressure Regulators

Pneumatic systems for robots usually depend on shop air. An exception may be minirobots for educational purposes, which have their own small compressors. In production, shop air is the usual energy source. However, it has its hazards, since its pressure rarely is constant, and it may need to have oil added for lubrication of air motors, valve spools, and so on; or it may contain too much humidity, tramp oil, and/or dirt, which must be removed. These problems have led to several accessories. The first one is the pressure regulator, whose function is to transform a fluctuating air pressure supply to a constant one by lowering the pressure output. The output is usually referred to as regulated or secondary pressure.

A regulator without its own exhaust can only be used in a system with a constant bleed downstream of the regulator. It would not work in a dead-ended line, since secondary air would build up to supply pressure due to leakage across the regulator. Hence, only the so-called pressure-relieving regulators are considered here. They are provided with exhaust or relief ports.

A particular characteristic of air pressure regulators is droop, that is, the deviation between secondary pressure at no flow and at some given flow.

Suppose the pressure regulator is set for a secondary pressure of 70 psi at no flow. A downstream valve is opened, and a flow of 50 cfm passes through the regulator. This leads to an additional pressure drop across the regulator, bringing its secondary pressure to 64 psi. The 70 − 64 = 6 psi is the droop of the regulator, also referred to as the flow characteristic.

The regulation characteristic, on the other hand, is the change in regulated or secondary pressure as a result of a change in supply pressure. For example, a regulator passing constant flow may change its regulated pressure by 0.5 psi for a supply pressure change of 10 psi. This would be its regulation characteristic.

The pressure regulator may be either a single-stage or a pilot-operated two-stage design. An example of a single-stage pressure regulator is the Prep-Air regulator made by the Schrader Bellows Division, which is shown in Fig. 6–4.

Flow passes from the supply pressure side over the poppet valve to the secondary pressure side. The necessary pressure drop is provided by the

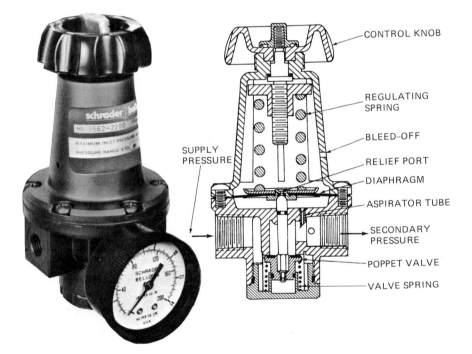

Fig. 6–4. Single-stage pressure regulator. (*Courtesy of Schrader Bellows Division*)

opening of this valve. By regulating its opening by means of the external control knob, which changes the compression of the regulating spring, the secondary pressure can be controlled at the desired setpoint. In the condition shown, the regulator is nonoperative. The regulating spring is relaxed, and the valve spring pushes the poppet against the seat, closing it completely.

Connecting the regulator to the supply pressure will not change this situation until the control knob increases the compression of the regulating spring sufficiently to push down the stem of the poppet valve, thereby opening the valve port. The secondary pressure is fed back through the aspirator tube to the area underneath the diaphragm. The narrow passage of the aspirator stabilizes what would otherwise be an underdamped system.

The secondary pressure under the diaphragm, together with the valve spring, counteracts the force of the regulating spring. This tends to close the poppet valve until a balance of the forces involved is obtained. A secondary pressure results that is proportional to the compression of the regulating spring produced by the control knob.

Should the secondary pressure become excessive at any time, it would lift the diaphragm sufficiently to open the relief port and relieve the pressure through the bleed-off opening provided.

The larger the relative size of the diaphragm of a single-stage pressure regulator is, the better its regulation characteristics will be. Obviously, the optimum size is not always needed, and miniature or midget regulators with minimum space requirements may be used to advantage.

On the other hand, when the highest accuracy is desired, the pilot-operated pressure regulator is in its element. Its principle is illustrated in Fig. 6-5. Two stages are involved: the pilot stage and the main stage.

The pilot stage operates on the basis of a small amount of air that flows from the supply side through a fixed orifice and the pilot valve nozzle to the secondary pressure side. The pressure drop of this pilot flow that takes place between the supply and the secondary pressure is produced in the fixed orifice and the variable nozzle opening. Opening of the nozzle is controlled by a baffle, which has the same function as the flapper in the previously described flapper-nozzles. The baffle is suspended between flat springs and positioned by a balancing diaphragm. The latter is balanced between the forces of the regulating spring and the secondary pressure.

The slightest drop in secondary pressure throttles the pilot valve over a very narrow band. This increases the back pressure between the nozzle and the fixed orifice, which pushes down on the main stage diaphragm assembly. The assembly consists of two diaphragms moving in unison. In moving up, the supply valve will open the port that connects secondary pressure to the exhaust. In moving down, the supply valve will open the port that con-

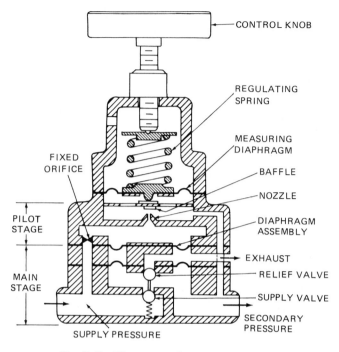

Fig. 6-5. Pilot-operated pressure regulator.

nects supply pressure to secondary pressure. In the nonoperative condition, as shown in Fig. 6-5, both ports are closed.

When the back pressure increases, the supply valve opens until the secondary pressure has increased sufficiently to push the baffle back to the balance point. This is obtained when the back pressure has dropped sufficiently to open the supply valve port just enough to supply the necessary air and maintain the desired pressure.

In case of increasing secondary pressure, the supply valve will close, and, if necessary, the port to the exhaust will open to vent air through the exhaust until the desired pressure is again established.

Very minute changes of secondary pressure will produce the action described. The great accuracy is due to the high gain of the control system: a very small motion of the pilot valve baffle generates enough force to control the main stage.

Pressure regulators with even greater accuracy are available, but are hardly necessary for the specific conditions of robotic pneumatics.

Filters

Abrasives and suspended liquid droplets carried by the air, as well as tramp oil, cause most of the downtime and lost production in pneumatic systems. The filters immediately after the compressor are generally inadequate. Too many contaminants can be picked up as the compressed air is distributed. Hence it is advisable to provide an extra filter for the pneumatic system of a robot.

The most widely used type is the so-called general-purpose filter, which usually operates in two stages as shown in Fig. 6-6. This is a product of the Wilkerson Corp. First, the air is swirled by deflectors to remove gross particles and liquid drops. They are thrown by centrifugal force against the inside of the filter bowl, from where they reach the sump area, which is protected from the turbulent flow by a baffle that also supports the filter

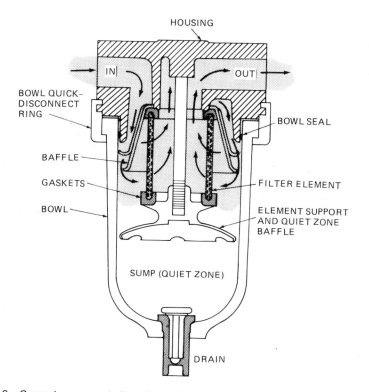

Fig. 6-6. General-purpose air filter from Wilkerson Corp. (*Courtesy of MACHINE DE-SIGN, a Penton/IPC Publication*)

element. The partially cleaned air then passes through the filter element, where a portion of the remaining contaminants is removed. The amount removed depends on the rating of the element.

Various materials are used for the filter housing, with aluminum and zinc being most common. The bowls can be transparent plastic or metal. Transparent bowls are generally preferred because they allow easy monitoring of liquid levels, and do not require a sight glass. Plastics may be incompatible with certain chemicals, however, as discussed below under lubricators. For added protection, slotted or perforated bowl guards in metal or plastic are available. Provision for drainage should be provided at the bottom of the bowl.

There are both surface filters and depth filters. The most common surface filters consist of one-wrap wire or cloth mesh as the filter medium. They offer little resistance to air flow because the air need not negotiate a tortuous path through the screen. They can be cleaned and reused many times.

Surface filter elements come with ratings from 5 to 70 microns. The pressure drop across the elements, unlike that with depth filters (see below), is nearly independent of rating, since the total free-flow area is little affected by pore size.

Disadvantages of surface filters are that particles larger than the rating can pass through, depending on their shape and orientation, and that the filter element has no liquid-holding capacity. Suspended liquid droplets in the air (aerosols) may not have been thrown out by the centrifugal action of the swirling motion. They collect on the screen and are promptly blown downstream. One other disadvantage of such filters is that they are fragile and readily torn or distorted.

Because of these disadvantages, depth filters are usually preferred. Here, the air must pass through a labyrinth of paths and channels within the thick layer of the filtering medium. Sintered elements composed of tiny beads of glass, plastic, or metal that are joined on the surfaces by pressure and/or heat are widely used. The micron rating is determined by the bead size, which in turn determines the average size of the pathways between beads. The smaller the beads, the lower the micron rating, and the higher the pressure drop. Therefore, to obtain the desired flow rate at a reasonable pressure drop and filter size, micron ratings are relatively high.

Several filter manufacturers supply sintered elements rated for 40 to 50 microns as the standard, and offer 5 to 10 microns as an option. However, the increase of pressure drops for such options should be carefully checked before they are chosen. Sintered elements are rigid, structurally strong, and relatively low in cost. Although the depth elements can be cleaned, cleaning is rarely worthwhile because particles lodging inside the filter are difficult and expensive to remove.

Depth filters may also use felt and multiwound cloth or string elements. Felt elements generally use cellulose fibers impregnated with a resin to provide rigidity. Because of their very porous nature, they can provide a 5-micron rating without loss of flow or pressure drop rating. Felt elements must be handled carefully. Momentary high pressure differentials may also damage them. When they are installed, care should be taken during the first few minutes of operation, since small amounts of loose fiber can dislodge from the inner surface and migrate downstream. The behavior of multiwound cloth or string elements is similar to that of felt. They are particularly well suited for absorbing aerosols. However, once they become saturated, they must be discarded. Their pressure drop as compared with felt is high, and is usually compensated for by reducing flow.

All general-purpose filters require the same frequency and kind of maintenance. The pressure drop across a filter should be checked at regular intervals and logged, unless automatic methods are used as were described for Fig. 5-2. Logging provides a guideline for replacement of the filter element. Periodic draining of liquid from the bowl is another requirement. Reliable operation of a pneumatic robot depends on well-filtered air.

Lubricators

While it is necessary to remove tramp oil from the air supply, lubrication is required to minimize the wear of many components, including most packings of air cylinders, and so forth. Lubrication of such components is most efficient when oil is injected into the air that feeds them. The air will then spread a lubricating film over all exposed parts.

The principle of the Watts Air Lubricator, which is made by Watts Fluidair Inc., is illustrated in Fig. 6-7. The bowl contains the lubricating oil to be added. These bowls are available in polycarbonate and metal. Transparent polycarbonate simplifies inspection of the oil level and checking for dirt and liquid condensate in the oil. However, caution must be exercised when using polycarbonate bowls in any area where certain chemicals are used. Even if the compressor is lubricated with synthetic oils or with oils containing phosphate esters or chlorinated hydrocarbons, the chemicals can attack and weaken the bowl.

Air enters the lubricator housing and is channeled in either of two directions, depending on flow rate. At low flow rates, all the air passes through the venturi which creates a low pressure area to draw oil from the reservoir through a capillary tube to the point of injection into the air-

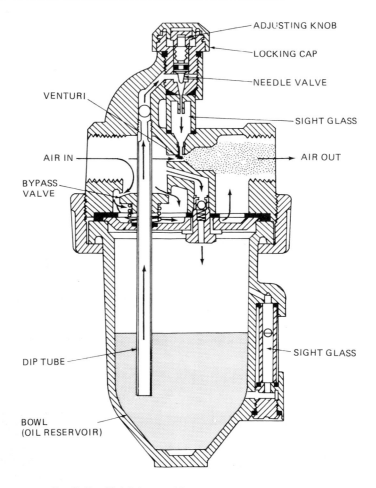

Fig. 6-7. Air lubricator. (*Courtesy of Watts Fluidair Inc.*)

stream. A manually adjusted needle valve in the housing sets the oil drip-rate so that an exact, metered quantity of droplets can be added. A sight glass allows this rate to be observed.

Under higher flow conditions, the spring-loaded bypass valve opens. The excess air then flows through the valve, bypassing the venturi section, and joins the oil-carrying air before the outlet. This method results in a wide operating range at a very low pressure drop.

In selecting a lubricator, minimum and maximum flow rates and pressure requirements should be considered. Pressure drop curves, available from the manufacturer, should be reviewed to make certain the selected model will not create a higher pressure drop than the system can tolerate.

Directional Control Valves

Valves with shear action, such as the valve in Fig. 5–5, will differ in construction depending on whether they are used for hydraulic or pneumatic service. The reasons are that: (a) pneumatic valves generally operate with lower pressure; and (b) since air instead of hydraulic fluid is the controlled medium, lubrication must either come from a lubricator, or the valve must be so designed that it operates without lubrication. Air valves are almost always on–off valves with or without a neutral center position.

Figure 6–8 shows the Alpha valve made by the Aro Corp. It is a four-way valve with either a two- or a three-position spool. The illustration pictures the two-position design. In this case, the spool is spring-loaded on one end face, so that with pilot pressure vented to the atmosphere, it is pushed

Fig. 6–8. Directional control valve. (*Courtesy of The Aro Corp.*)

by the spring into a hard-over position. When pilot pressure is applied to the opposite end of the spool, the spring is compressed, and the spool shifts over to the other end.

Details of the design (except for the spring) are shown in Fig. 6-9. The spool is made of aluminum, and a urethane seal is bonded over the midsection of the spool and provided with a grease lubricant (Key Lube) that eliminates the need for a lubricator. The valve body is of die-cast aluminum, and the bore in which the spool moves is honed to a 16-microinch finish.

The solenoid is easily slipped on and off after removal of the knurled nut on top of it. It can be supplied with various voltage ratings and also with a connector that has a built-in micro relay for the transistor-transistor logic (TTL) of microprocessors. The plunger of the solenoid is pulled up when the solenoid is energized. This opens the orifice of the pilot valve,

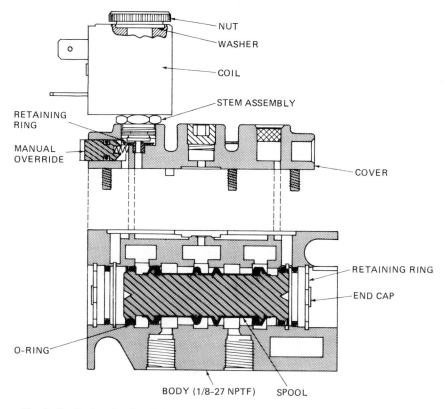

Fig. 6-9. Design details of directional control valve. (*Courtesy of The Aro Corp.*)

pressurizing the pilot chamber and actuating the valve spool. When the solenoid is de-energized, the plunger is held down by a loading spring, and pilot pressure is vented to atmosphere. The button on the left side of the coverplate is a manual override. When it is pushed in from the outside, it lifts the plunger from its seated position, thus allowing pilot air to shift the spool.

The valves are rated from a maximum pressure of 150 psi to vacuum. Pilot pressures can be as low as 25 psi for the spring return models and 10 psi for double pilot models. The valves come in two sizes. One is for $\frac{1}{8}$-in. N.P.T. connection, the other for $\frac{1}{4}$-in. N.P.T. They have C_v ratings of 0.9 and 1.5, respectively. This means that at a supply pressure of 80 psi with a 15% pressure drop across the valve, the flow rates will be 30 and 50 scfm, respectively. There are aslo stackable body models that are mounted on stacking pins and held together by end plates. Their ratings are 1.3 for the $\frac{1}{8}$-in. size and 1.9 for the $\frac{1}{4}$-in size.

Mufflers

Air exhausts tend to be noisy because of the sudden expansion of air, and mufflers are used to combat the noise. The two most frequently used mufflers are exhaust silencers and muffler-filters. They are both screwed into the exhaust ports of pneumatic components.

Exhaust silencers decrease the noise by tuned resonant control of the expansion of the exhausted air. They produce a minimum of back pressure, which is important for air motors and cylinders, where back pressure may slow down the motion.

Muffler-filters decrease noise by back pressure control of the expansion of the exhausted air. They usually consist of sintered material with a relatively large area to let the air escape. They also protect against outside dirt that could migrate into the system through an unprotected exhaust port, although exhaust air may blow dirt out during operation.

Pneumatic Actuators

Pneumatic cylinders are often used with pneumatic systems. Their construction is similar to that of the hydraulic cylinders described in the preceding chapter. However, because they lack lubrication, design modifications are necessary, and air lubricators are usually required or the cylinders are provided with a built-in system for feeding lubricant into the airstream. Also, the usually lower pressure of pneumatics calls for a different design approach.

The same applies to fluid motors that are used with pneumatics. With high-speed vane motors, such as the one used in Fig. 6-3, friction of the vane edges that are thrown with centrifugal force against the inner wall of the surrounding ring is of particular concern. It is diminished by reducing the diameter and making the vanes long rather than high. Figure 6-10 shows an exploded view of a small vane motor produced by Cooper Air Tools, as well as its operating principle.

Whereas the vanes are thrown out by centrifugal force when the motor is running, they must be pressed outward to start the motor up. This is usually done either by springs or by applying air pressure under the vanes.

In general, pneumatic motors chosen for robotic applications are vane motors. They may have anywhere from three to ten vanes. The volumetric efficiency increases significantly with the number of vanes, as does the starting torque.

By using high-speed air motors, such as the 8000-rpm unit in IRI's M50 robot, dimensions, and thus weight, can be reduced in a small, compact unit. This is very desirable for robots. However, the penalty is in speed

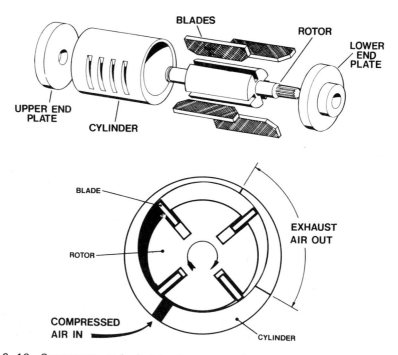

Fig. 6-10. Components and principle of vane-type air motor. (*Courtesy of Cooper Air Tools*)

reducers. The extent to which their bulk, weight, and maintenance offset the advantage of the high-speed motor must be weighed from case to case.

To combat noise that originates with air motors, they may be provided with double walls. Exhaust air passes through the interspace, which may be provided with muffler-like baffles and slotted openings. The air passing through the interspace simultaneously acts as a coolant for the motor.

An unconventional concept of a robot with pneumatic actuators stems from a joint development by Hitachi, Ltd. and Bridgestone Corp. It utilizes rubber tube elements that expand and contract to produce the different degrees of freedom. The steel skeleton of the robot is controlled by muscle-like rubber tubes with braided-fabric coverings. Double sets of rubber tube elements are used. The principle is shown in Fig. 6–11. Inflation produces a shortening of the tube length, and deflation an extension. By applying pressure to one tube and releasing it from the other, the pulley is rotated and the robot axis moved.

The arms of such a design would behave as springs, providing flexibility in case of an unforeseen impact and thereby increasing the safety of operation. The concept is primarily intended for assembly robots. At this time, only a prototype has been produced. It has a weight of 12 lb and a load capacity of 4 lb, making a ratio of weight to load capacity that is much smaller than that of any hydraulic or electrical robots.

CONTROL OF AIR MOTORS

Air motors are usually controlled by open-loop systems such as that described with reference to the M50 robot from International Robomation/ Intelligence. It uses a spool valve operating in the on–off mode to control flow to the air motor. To reduce friction, dither is provided.

A closed-loop system for air motors with a positive positioning force for

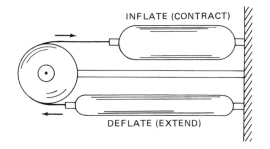

Fig. 6–11. Principle of rubber-tube actuators.

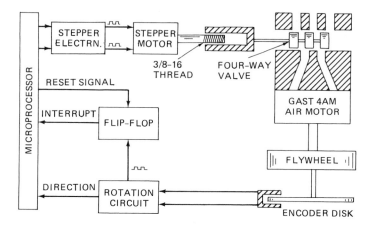

Fig. 6-12. Closed-loop control system. (*Courtesy of University of Michigan*)

the spool valve was developed at the University of Michigan by Dr. R. B. Keller and a former student, W. N. Verge. The system is illustrated in Fig. 6-12. A rotary vane air motor, Model 4AM from Gast Manufacturing Corp., was used. A flywheel was added to simulate the load in the test set-up. A stepper motor drives a screw to position the valve spool. Each command pulse changes the open port area of the valve by 2%. Thus, 50 pulses are needed to open the valve completely. For a maximum stepper frequency of 500 Hz, this requires 0.1 sec.

An encoder (see pages 267-271) feeds back the position of the air motor shaft. It uses two photo-interrupter modules and a disk with 20 equally spaced holes. Electronic circuitry produces 80 pulses for each shaft revolution, plus a direction-of-rotation signal. Each pulse locks into a flip-flop, with a reset signal required after each rotation pulse. Computer simulation of the test setup showed that the response was twice as fast as that of an open-loop system without feedback.

SUMMARY OF GAS LAWS

Isothermal conditions (constant temperature):

$$\frac{p_{a,1}}{p_{a,2}} = \frac{V_2}{V_1}$$

Adiabatic conditions (no heat removed or added):

$$\frac{p_{a,1}}{p_{a,2}} = \left(\frac{V_2}{V_1}\right)^{1.41}$$

Isometric conditions (constant volume):

$$\frac{p_{a,1}}{p_{a,2}} = \frac{T_1}{T_2}$$

Isobaric conditions (constant pressure):

$$\frac{V_1}{V_2} = \frac{T_1}{T_2}$$

General gas law for air:

$$\frac{p_a V}{W T} = 0.37$$

Flow rate of air:

$$Q = 0.007 \ vA$$

Conversion of air flow rate at standard conditions to air flow rate at any other condition:

$$Q = 0.028 \ \frac{Q_s T}{P_a}$$

Air flow through restriction when $p_2 \geq 0.53 \ p_1$:

$$Q = 45.6 \ C_d A \ \sqrt{T_1 \left[\left(\frac{p_2}{p_1} \right)^{1.43} - \left(\frac{p_2}{p_1} \right)^{1.71} \right]}$$

or:

$$Q = 45.6 \ k \ C_d A \ \sqrt{T_1}$$

Flow coefficient of pneumatic valves:

$$C_v = 0.06 \ Q_s \ \sqrt{\frac{T_1}{p_1^2 - p_2^2}}$$

where:

A = cross-sectional area, in.^{2}
C_d = coefficient of discharge

C_v = flow coefficient
k = constant from Fig. 6-1
p = pressure, psi (psig)
p_a = pressure, psia
Q = flow rate, cfm
Q_s = flow rate under standard conditions, scfm
T = Absolute temperature in degrees Rankine, °R
v = flow velocity, ft/in.
V = volume, ft³
W = weight, lb

Chapter 7

SERVOMOTORS

It was not until NC machine tools and robots were seen to constitute a market of history-making proportions that the need for compact lightweight industrial servomotors was recognized. Supported by rapid developments in electronics, transducers, and permanent magnets, a number of new concepts emerged, based on various traditional motor designs and their modification for the purpose of specific servosystems.

Servomotors are part of the system that controls the axis motions of a robot. An error signal (cf. Fig. 1–15) is fed into the motor control system, and feedback devices are used to close the loop. The electric power supplied to an ac motor may be rectified, regulated, and converted back into some form of alternating current, shaping amplitude, frequency, and wave pattern.

Add to this the overall control system that coordinates the motions of three or more servomotors in the planned action of the robot, and the complexity of the network of many interacting components becomes apparent.

The most common basic concepts that are used for servomotors are the following:

- ac induction motors, primarily the two-phase version.
- Synchronous reluctance-type ac motors.
- Commutator-type dc motors.
- Brushless dc motors.
- Pancake motors.
- Torque motors.
- Stepper motors.

Principles of these various actuators are described in the following pages, albeit only in some basic concepts. It should be understood here that all these motors are building blocks of drive systems, and thus can operate only with suitable support components, which will be covered separately.

AC INDUCTION MOTORS

The ac motor, in the typical fashion of other motors, consists of a stator and a rotor, both made of iron. Slots in the stator carry coils, and by interaction of the electromagnetic effects of the stator and rotor currents, the force to create rotation is produced.

The rotor consists of a laminated cylindrical iron core with slots in which conductors are embedded. The most common type of rotor carries copper or aluminum conductors. The ends of the conducts are connected by rings on both sides of the rotor forming the "squirrel cage." Rotation is produced by the moving magnetic field produced by the alternating current in the coils of the stator, which induces a current in the shorted conductors of the rotor.

For the sake of simplicity, Fig. 7–1 shows only two of an otherwise multiplicity of stator coils. Current flowing through a coil in a direction that goes into the paper in the illustration (indicated by a "cross") goes around the coil and returns in the opposite direction (indicated by a "dot").

A current flowing through a wire produces an electromagnetic field that passes through the iron of the stator as well as of the rotor; and where there is a magnetic field, there are a north pole and a south pole.

As seen in Fig. 7–1a, current flowing "into the paper" produces a clock-

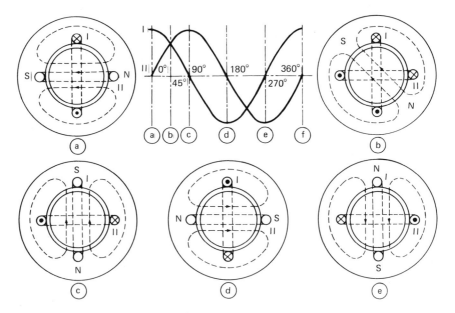

Fig. 7–1. Rotating field produced by two-phase control.

wise electromagnetic field, while current flowing "out of the paper" creates a counterclockwise field. The result is that the right side of the stator becomes a north pole (N), and the left side a south pole (S).

The illustration refers to a two-phase induction motor, a type often used as a servomotor. The ac supply enters the motor in two separate phases, I and II, each connected to a different set of coils. The phases are 90° apart, as illustrated by the sine waves of phases I and II, shown between diagrams (a) and (b) of Fig. 7–1. Likewise, the corresponding coil windings in the five diagrams are 90° apart.

The condition of Fig. 7–1a corresponds to point *a* on the two-phase sine wave diagram. This is the instant when no current flows through coil II, but current is at its maximum in the indicated direction through coil I. Gradually, as one goes from Fig. 7–1a through Fig. 7–1e, this changes, and the north and south poles rotate in the iron of the stator. Thus, a rotating electromagnetic field has been produced, but as yet a mechanical motion has not. This requires consideration of two other phenomena.

One is that electromotive force (emf) is induced in any current that cuts across or is cut by a magnetic flux; and emf is the pressure that causes electricity to flow. The other is that if a current-carrying conductor is placed in a magnetic field, a force will act upon it.

Both phenomena are illustrated in Fig. 7–2.

- If the conductor is moved at right angles to the magnetic field, an emf will be induced in it that will make current flow through the conductor.
- If current flows through the conductor while it is in the magnetic field, a force will result that will tend to push the conductor out of the magnetic field.

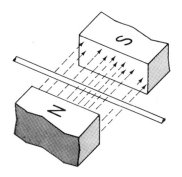

Fig. 7–2. Conductor in magnetic field.

The force that acts on the conductor is given by the equation:

$$F = 5.7 \times 10^{-7}BLI \tag{7.1}$$

where F is the force in pounds, B is the flux density in gauss, L is the length of the conductor in the magnetic flux in inches, and I is the current in amperes.

As soon as current flows through the coils of the stator and the rotating field is produced, magnetic flux cuts through the conductors of the rotor. The direction of the induced currents is such that they oppose the direction of the rotating field. The result is that the rotor starts to move in the direction of the field.

This may appear to be a paradox. It can best be illustrated by visualizing a person who is trying to stop a moving merry-go-round. Instead of stopping it, he or she is being pulled with it. Since, in similar fashion, the rotor attempts to rotate with the speed of the rotating field, it runs faster and faster. As long as there is no load, it almost attains synchronism, that is, the speed of the rotating field.

If full synchronism were achieved, then the conductor would no longer cut through magnetic flux lines; but as long as it does, the necessary emf is produced. Thus, there is a balance of slip or asynchronism between rpm of the field and rpm of the rotor that increases with the load. The greater the slip is, the more flux lines are cut; and, according to Eq. 7.1, the greater is the force the rotor can develop.

The slip or asynchronism of the rotor increases the emf induced in the rotor. This must come from the power supply of the stator coils. Consequently, an increased load means not only an increase in slip but also an increase in power drawn from the supply network by the motor.

When the induction motor is used as a servomotor, the same two windings as in Fig. 7–1 are used, both being 90 electrical degrees apart. There are, however, some modifications. One of these windings is the fixed or reference winding, the other the control winding. Operating from a single-phase supply, the 90° phase shift for the control winding can be produced by a phase-shifting circuit in the amplifier. Torque is obtained by varying the magnitude of the voltage in the control winding. Directional control is obtained by phase reversal in the control winding.

Figure 7–3 shows the characteristics. At zero torque, the speed approaches synchronous speed, independent of the voltage V_c in the control winding. Starting torque, when the speed is zero, is almost proportional to V_c.

Servomotors of this type have small-diameter, high-resistance rotors. The small diameter reduces the inertia.* The high resistance can produce a nearly

---

*It should be noted that inertia as used here and in the following is really "moment of inertia." The word "inertia" is used instead in accordance with general practice. (see page 12.)

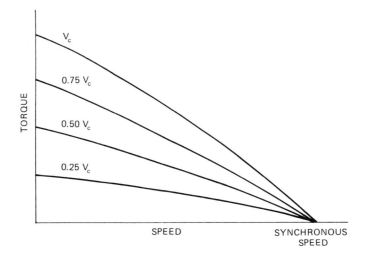

Fig. 7–3. Induction servomotor characteristics.

linear speed–torque relationship for accurate control. Induction servomotors usually are designed for horsepower ratings that exceed robot requirements, and thus are preferred for other industrial applications. However, with the development of suitable electronic control systems, the induction motor can be expected to be seriously considered in robotic servo-control systems.

SYNCHRONOUS RELUCTANCE-TYPE AC MOTOR

Contrary to the asynchronism characteristic of the induction motor, a reluctance-type synchronous motor runs at synchronous speed. The two motors have basically different designs. In the synchronous reluctance motor its salient poles are part of the rotor, whose laminations are stamped in a shape that includes the pole formations.

The principle is illustrated in Fig. 7–4. At a given instant, alternating current flows through the stator winding 1–2 in the direction indicated by the "cross" and the "dot." The rotor may at first be considered to be at rest at this instant. Its salient north and south poles are produced by dc windings that are connected to the outside by slip rings. They are facing the slots of the stator coil, as shown.

Suppose the alternating current in the stator coil has reached its maximum in the given direction at this instant. The electromagnetic field generated by this current produces an interaction between the salient rotor poles and the stator that tends to turn the rotor in the direction indicated by the arrow.

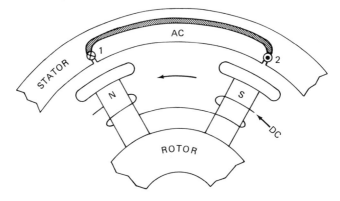

Fig. 7-4. Principle of synchronous reluctance motors.

One of the characteristics of the synchronous reluctance motor is that the more poles there are, the more torque there is developed and the slower the speed. The reason is that because of the increased number of poles, the rotor moves through a smaller angle each time a winding is energized. However, the power remains the same, and therefore the torque increases correspondingly.

A problem with the usual synchronous reluctance motor is that before the mechanical inertia of the rotor has been overcome and the rotor gets a chance to turn, the alternating current has changed its direction, and the force pushes in the opposite direction. The best one can obtain under these conditions is oscillation, not rotation.

To overcome this, it is necessary first to put the rotor into synchronous motion. This may be done by an auxiliary squirrel cage as part of the rotor. Once synchronism is obtained, the rotor will continue running on the basis of the above described principle, and will always run in absolute synchronism with the supply frequency.

Another approach is to use suitable electronic starting circuits. This applies particularly to motors of, say, under 5 hp. An example is the RA10 Series servomotors that Toshiba Corporation developed specifically for robotic application. In this design, the electromagnets of the rotor are replaced by permanent magnets of rare-earth materials. Rated outputs range from 0.02 to 2.7 hp, depending on motor size. Rated speed for all but the smallest motors (0.07 hp and below) is 3000 rpm. Maximum speed is 4000 rpm. A 2-hp motor, for example, which weighs only 24 lb, has a rated torque of 7.2 lb-ft and a maximum torque of 28.8 lb-ft.

A synchronous servomotor that deserves special mention is the Megatorque motor by Motornetics Corporation. It makes use of a multitude of poles to obtain a high-torque, low-speed motor, and by a special magnetic design it avoids the need for an auxiliary squirrel cage for starting purposes.[1] A cutaway diagram is shown in Fig. 7–5. The design combines two fixed concentric stators with an annular rotor mounted on bearings so that it rotates between them. All three are constructed from laminations of low silicon steel which has high saturation flux densities.

To obtain a large number of poles, the stator laminations are shaped with

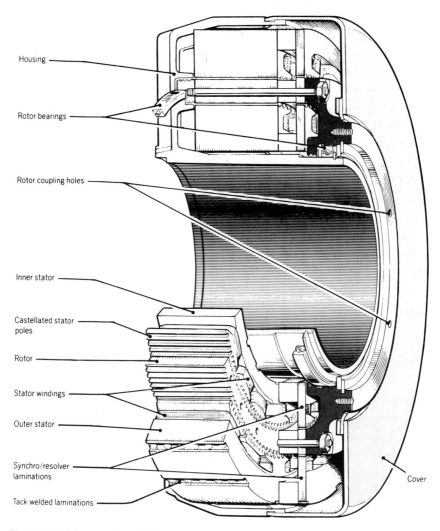

Fig. 7–5. Cutaway view of Megatorque motor. (*Courtesy of Motornetics Corporation*)

salient (i.e., castellated) polepieces on their inner or outer peripheries, depending on their location. Around them are placed the windings. The rotor is made of similar castellations on both inner and outer edges, but without the polepieces. The clearance between the stator and the rotor does not exceed 0.008 in. in the larger model and 0.005 in. in the smaller one.

The Megatorque is a three-phase synchronous reluctance motor with the castellations in the laminations forming salient poles. It is used with a power amplifier generating the waveform currents required for rotation. Sequential and proportional energizing of the phase windings produces a rotary magnetic field that generates torque in the rotor by magnetic interaction of the teeth on the stator and the rotor. The tooth pitch is such that it produces one revolution of the rotor for every 150 cycles of ac input. In the units offered, this represents a maximum no-load speed of one revolution per second (60 rpm).

In addition, because the magnetic flux runs parallel to the lamination direction, eddy current losses are low. Rotor eddy current losses are also lower than those of conventional synchronous reluctance motors, which require eddy currents for starting and stability.

Input frequency and voltage are changed to control the motor speed, and maximum continuous torque is rated at 35 ft-lb for the smaller model and 262 ft-lb for the larger model of the Megatorque. The overall diameter of the smaller model is 6.94 in., and its weight without frame is 6.3 lb, while the larger model measures 14.75 in. and weighs 87 lb without frame. Figure 7-6 shows the speed-torque characteristic of the larger model.

Since the horsepower, P, at a given rpm, N, is:

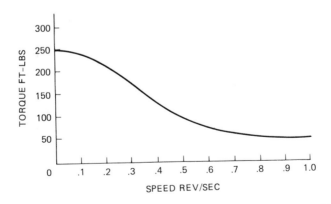

Fig. 7-6. Speed-torque characteristic of Megatorque motor. (*Courtesy of Motornetics Corporation*)

$$P = \frac{T_i N}{63,025} \tag{7.2}$$

where torque, T_i, is expressed in lb-in., or:

$$P = \frac{T_f N}{5252} \tag{7.3}$$

where torque, T_f, is expressed in lb-ft, it follows that based on the data of Fig. 7–6, the approximate horsepower output of the larger model of the Megatorque servomotor is 0.5 hp at 0.5 rev/sec or 30 rpm, and also at 1.0 rev/sec or 60 rpm. The smaller motor has horsepower ratings of about 0.1 hp at either speed.

Because of its low-speed, high-torque characteristics the motor could not only drive conventional transmissions but could also be made an integral part of a robot joint. However, its weight, particularly for the larger size, could become a detriment if it were used for the elbow joint, which not only moves by itself but also has to be moved by the shoulder joint.

On the other hand, the required speed reducers would have to be minimal with a Megatorque, and inertia, backlash, and ultimately positioning accuracy could be considerably improved.

Servomotors, as already stated, need feedback devices. They are discussed separately. Here, it merely should be understood that they are a necessary adjunct of the motor. Thus, Fig. 7–5 shows the laminations of the synchro/resolver, which is the feedback of the Megatorque. It is constructed of same laminations that are used for the stators and mounted a short distance away from them. The signals of the synchro/resolver are used to control position and speed of the servomotor.

COMMUTATOR-TYPE DC MOTORS

The basic construction of a dc motor consists of an armature and a magnetic field. The armature is a cylinder of soft steel laminated to minimize eddy current losses and mounted on a shaft so that it can rotate about its axis and simultaneously drive the load. Embedded in longitudinal slots in the surface of the armature are a number of copper conductors. They are bars or bundled copper conductors formed into winding loops or coils, with two free ends.

The commutator of the dc motor is an extension of the cylindrical armature. It carries a multitude of copper bars that are isolated from each other. Each free end of the winding loops or coils that are embedded in the armature slots is connected to one end of the copper bars of the commutator. A total of 48 coils would require 96 bars.

Figure 7-7 shows the principle on the basis of only four commutator bars and two coils. The magnetic field is set up by two permanent magnets that are part of the stator. The interaction between the magnetic field and the long sides of the coils that cut at right angles through it results in a torque on the armature, making it rotate.

The direction of the current changes in either coil as its commutator bars pass under one or the other set of brushes. As the armature rotates through 180°, the polarity is reversed. Thus, the current alternates its polarity as the armature rotates. This means that the essential purpose of the commutator is to convert direct current into alternating current. In this respect, every dc motor becomes an ac motor.

However, the dc motor, as described here, requires commutators and brushes, which are said to present a maintenance problem. Much of this claim may apply to wound-field dc motors, which are hardly used as servos. With proper design, sizing, and control of the motor, the maintenance does not exceed a minimum corresponding to that of other maintenance procedures. Thus, PMI Motors can claim for their ServoDisc motor, which is described further below, that a brush life "of 20,000 hours is not uncommon for these motors, particularly in stop–start incremental motion operation." And 20,000 hours is the life expectancy of many motors. As E. F. Kohn [2] has pointed out:

It must be said that the currently available transistor drives, when sized properly with the appropriate permanent magnet motor have little or no brushwear, and the brushes do not create commutation or wear problems.

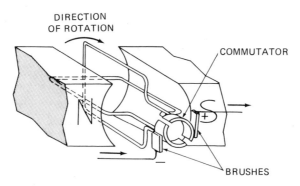

Fig. 7-7. Principle of dc motor.

Wound-Field and Permanent-Magnet Motors

Commutator-type dc motors, as originally conceived, were provided with electromagnets rather than permanent magnets. These field-wound motors are usually connected to the same power supply as the brushes for the armature windings. They are classified by the three different ways of connecting the field, as the series motor, the shunt motor, and the compound motor.

With the series motor, the field windings are connected in series with the armature windings; with the shunt motor, they are connected in parallel; whereas with the compound motor, part of the field windings are connected in series and part in parallel with the armature windings. Each one has its own speed–torque characteristics and its own requirements for speed control.

With the advent of stronger magnetic materials, it became advantageous for servomotors to produce the magnetic field by permanent magnets rather than by electromagnets. This simplifies power supply connections and eliminates thermal heating due to field windings. The torque is higher, and the torque vs. speed characteristics are more linear. For a given output power, the permanent-magnet motor can be made smaller and lighter than the equivalent wound-field motor.

Permanent-magnet motors have a speed range from zero to maximum. For a given size, they allow high torque and acceleration, particularly when magnets with high-energy products are used.

As shown by Eq. 7.1, the force exerted on a conductor is proportional to the flux density of the magnetizing field. But the flux density is proportional to (a) the magnetic field intensity and (b) the length of the magnet, and it is also inversely proportional to the length of the air gap that the flux has to traverse. Thus, the product of flux density and magnetic field intensity (i.e., the energy product) represents the maximum energy per volume of magnet.

Flux density is commonly expressed in gauss or megagauss (10^6 gauss), symbolized by Gs or MGs, respectively. Magnetic field intensity is commonly expressed in oersteds, symbolized by Oe. The energy product becomes megagauss-oersted, or MGsOe.

The three magnet materials most commonly used are ceramic, alnico (*al*uminum/*ni*ckel/*co*balt), and samarium-cobalt (SmCo). Magnets produced of SmCo belong to the generic group of rare-earth magnets. So do the neodymium-iron-boron (Nd-Fe-B) magnets, since samarium as well as neodymium belongs to the class of rare-earth elements. Nd-Fe-B magnets are a serious competitor of SmCo magnets, as will be seen.

Energy products and costs of typical magnetic materials are given in Table 7-1. The figures that are shown are only approximate values. Furthermore, it should be noted that while the costs per pound of SmCo and Nd-Fe-B magnets are about the same, the latter are about 10% less dense, so that they have a 10% advantage over SmCo.

Besides the SmCo magnet listed, there is also an 18 MGsOe grade of SmCo that is widely used.

For the user, an even more important figure often is the cost per unit energy product. Figures on cost per energy basis have been presented by M. A. Bohlmann.[3]

The ceramic materials, mostly strontium and barium, are the lowest in cost, but their energy product is also low. Alnico has been the most popular material for years and continues to be the favorite, although samarium-cobalt, because of its high energy product, is finding increasing acceptance in spite of its price.

Neodymium-iron-boron is the most recently developed of the magnetic materials. It offers the great advantage that it does not require cobalt, a strategic material of ever increasing cost. Nd-Fe-B has a Curie point between 250°F and 300°F. (If a magnet reaches the Curie point, it loses its magnetism immediately.) It is expected that the Curie point of Nd-Fe-B can be further improved, but for ceramic magnets the Curie point is about 860°F, and alnico materials can be operated at temperatures of 1020°F and slightly above without loss of flux. This is the highest operating temperature of any currently available permanent magnet material. It should be noted, however, that it is not only the Curie point that has to be observed with high-temperature operation. The temperature effects are more complex than that, particularly since a number of metallurgical effects are also involved.

IG Technologies, Inc. produces two grades of Nd-Fe-B permanent magnets, NeIGT 27 and NeIGT 35, which provide energy products of 27 MGsOe and 35 MGsOe, respectively. The energy products are dependent on the magnetic orientation obtained during manufacturing. IG Technologies uses a powder metallurgy technique to fabricate the magnets. The Curie point for these materials is actually 590°F, but IG suggests a maximum practical operating temperature of 300°F. The value is dependent on operating con-

Table 7-1. Energy products and costs of magnetic materials.

MAGNET MATERIAL	ENERGY PRODUCT (MGsOe)	COST ($/LB)
Ceramic	4	<10
Alnico 5	5.5	15
Alnico 8	6	30
SmCo	22–26	90 to 100
Nd-Fe-B	27–35	90 to 100

ditions and demagnetizing influences, but metallurgical changes occur at about 350°F to 390°F.

Other magnetic materials are being developed in various places, and undisclosed compositions have been reported to exhibit energy products of 60 MGsOe.

BRUSHLESS MAGNETS

These are "inside-out" motors: the rotor carries the permanent magnets, and the stator the windings. Looking at Fig. 7-4, it would only be necessary to replace the electromagnets of the rotor with permanent motors, and the basis for a brushless dc motor would be obvious. But, to quote P. M. Bartlett and H. Shankwitz,[4] "The brushless permanent magnet motor is a true dc motor even though its construction follows in all essentials that of an ac synchronous machine."

The brushless motor does have commutation, but this is done not by commutator bars and brushes but electronically, by either a resolver or a Hall-effect sensor (see Chapter 8). These devices rotate with the rotor shaft and transmit the rotor position at every instant in order to supply current to the appropriate stator winding. Brushes and commutators are eliminated, so that motor operation at higher speeds and higher peak torques is possible.

However, since brushless motors are similar to permanent-magnet ac synchronous motors, they usually can also be used with variable-frequency ac servo systems.

A Pacific Scientific brushless motor is shown in Fig. 7-8. It operates with a Y-connected three-phase stator winding and a four-pole permanent-magnet rotor. The rotor can be either ceramic or rare earth (samarium-cobalt), depending on the model chosen. In addition to the main rotor magnet, there is a four-pole sensor magnet. The sensor magnet plus three Hall-effect sensors, which are located 60° apart, constitute an encoder that provides the necessary shaft position information for electronic commutation by the controller.

Figure 7-8 also shows the motor winding configuration, consisting of phases R, S, and T. The controller turns on the appropriate solid-state switches to sequentially excite two of the three windings at a time. The net effect is that of a rotating magnetic field. The permanent magnet rotor follows the rotating field, and torque is developed, its value being a function of the product of the motor winding current and the torque constant of the motor.

Figure 7-9 shows the performance characteristic of the Pacific Scientific BL 3232 brushless servomotor, which is equipped with ceramic magnets. It

Fig. 7-8. Cutaway view of brushless motor and winding configuration. (*Courtesy of Pacific Scientific, Motor & Control Division*)

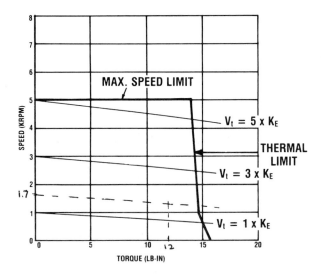

Fig. 7-9 Performance characteristic of Pacific Scientific brushless servomotor. (*Courtesy of Pacific Scientific, Motor & Control Division*)

is available with two different windings: X and Y. They are for two different torque constants. Suppose winding Y is chosen for the high torque constant. In this case, the "voltage constant" KE is, according to catalog data, 72.0. Let it be desirable to run this motor at 1700 rpm with a torque of 10 lb-in. plus 20% for the form factor (see page 241) of 1.2 (i.e., 12 lb-in.). Using Fig. 7-9, the intersection of these two parameters can be used to run the dotted line that intersects the krpm ($= 10^3$ rpm) axis at 1.7, which means 1700 rpm. Hence, the necessary voltage would be 1.7 × KE = 122 V.

One could also choose, of the same design, a motor with samarium-cobalt magnets, such as the BL 3412-21. Disregarding the fact that this motor allows an 11% higher continuous stall torque, its weight is only 9.7 lb instead of 13.8 lb, and its length only 5.5 in. instead of 8 in. Its diameter would be 4.25 in. instead of 4 in.

The inertia of the ceramic-magnet design BL 3232-1 is 0.00036 lb-ft-sec². That of the samarium-cobalt design is only 0.00010 lb-ft-sec².

It is always dangerous to compare servomotors of different manufacturers, unless the entire spectrum for a specific application is carefully analyzed. Having this in mind, however, it may be interesting that a typical wound-rotor dc servomotor of a different manufacturer, which has about the same power and also uses samarium-cobalt magnets in the stator, has an inertia including tachometer of 0.012 lb-ft-sec². Since the tachometer has an inertia of probably not more than 1% of that of the rotor, its influence in this respect can be safely neglected.

The above comparisons clearly show not only the difference in inertia between ceramic and samarium-cobalt motors but the even more pronounced difference between brushless and wound-rotor motors.

PANCAKE MOTORS

Unlike conventional dc servomotors with cylindrical rotors, the rotor of a pancake motor is made up of two, four, or six mechanically stamped copper or aluminum windings separated and bonded together by insulating layers made of epoxy/fiberglass composites. The entire assembly, which has a thickness of not more than a fraction of an inch (typically 0.06 in.), is rigidly bonded to the shaft.

Figure 7-10 shows the elements of a pancake motor. It is the ServoDisc motor made by PMI Motors, which is used among others in ASEA's welding robots. Permanent magnets are located on both sides of the armature. The magnetic flux passes through the narrow air gap of the armature and the nonmagnetic rotor to the iron of the stator, and then passes through

Fig. 7-10. Elements of a ServoDisc pancake motor. (*Courtesy of PMI Motors*)

the iron and turns around, returning the way it came to the magnets of opposite polarity.

The inductance of a motor armature is one henry when a current variation of one ampere per second induces one volt of counter emf. Since there is no iron in the armature, the motor has extremely low inductance, typically less than 85 microhenries. Conventional iron-core servomotors generally have armature inductances 20 times higher than the ServoDisc motors. As a result of this low inductance, virtually no arcing occurs at the brushes during commutation, even when high drive currents are used. Consequently, brush life is three or more times that obtainable with iron-core servomotors.

Besides the eight to ten magnetic poles, which is a relatively large number, ServoDisc motors have some 117 to 196 commutator bars. The result is a very constant and smooth torque over the entire speed range, while low-speed torque ripple is an occasional problem with servomotors with fewer magnetic poles and commutator bars.

A typical ServoDisc motor, the JR16M4C from PMI, which is specially designed for industrial automation and robotics applications and uses Alnico 5 magnets, would have specifications such as the following (the significance of some of the characteristics listed here is discussed further below):

Rated speed, 3000 rpm. No-load speed at rated voltage 3425.5 rpm. Armature resistance incl. brushes, 0.740 ohm. Continuous torque at stall, 312.9 oz-in. Peak torque, 3370.8 oz-in. Torque sensitivity, 33.56 oz-in./A. Average friction torque, 11 oz-in. Damping constant, 5.06 oz-in. per 1000 rpm. Moment of inertia, 0.084 oz-in.-sec^2. Electrical time constant, <0.14 msec. Mechanical time constant without load, 7.79 msec. Open-loop speed regulation at constant voltage, 0.89 rpm/oz-in. Maximum acceleration at no load, 10,064 rpm/sec^2.

It should be noted that:

- The speed rating of 3000 rpm is for reference purposes only. The motor can be run at higher speeds, as seen in Fig. 7–11, although continuous operation above 4000 rpm is not recommended.
- Peak torque, peak current, and peak acceleration are calculated for a maximum pulse duration of 50 msec and a 1% duty cycle.
- Motor performance data are based upon 150°C and with the motor mounted on an aluminum plate with no forced cooling.

Figure 7–11 shows the speed vs. torque characteristic of the motor. For the uncooled motor with 86.93 V (i.e., about 87 V) supply voltage, it can be seen that at 3000 rpm the maximum continuous torque is 320 oz-in. If

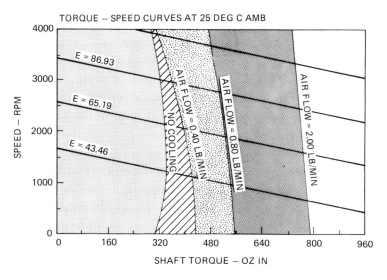

Fig. 7–11. Speed–torque characteristic of JR16M4C motor. *(Courtesy of PMI Motors)*

the motor were cooled with 0.80 lb/min of air flow, the continuous torque could be increased to about 500 oz-in., while the speed would drop to about 2700 rpm.

DC SERVOMOTOR CHARACTERISTICS

Continuous torque at stall has already been mentioned. It is also known as *continuous stall torque* and is often referred to simply as *continuous torque.* It is usually designated T_c. It is the maximum torque the motor can sustain when the motor is stalled and the dc current has a form factor of one, which, as will be defined below, means that the dc current possesses a pure straight-line characteristic. Continuous torque is based on a temperature in the armature winding that does not exceed the limitation of its insulation class.

Torque sensitivity is also known as *torque constant* and usually designated by K_T. It is the continuous torque at stall the motor can develop per one ampere of armature current. Hence, $K_T = T_c/I_c$, where T_c is the continuous torque for which the motor is rated and I_c the current that the motor absorbs at this torque.

Continuous torque at stall as well as torque sensitivity should be determined at a specified temperature. For continuous torque, it is the ambient temperature of the room where the motor is being tested. For torque sensitivity, it is the armature temperature. At Kollmorgen Industrial Drives, for example, continuous torque is based on an ambient temperature of 40°C (104°F), and torque sensitivity on an armature temperature of 170°C (338°F). Obviously, ratings given at 25°C (77°F), as may be done, give better readings—apparently. Kollmorgen recommends multiplying such ratings by a factor of 0.94 for continuous stall torque and by a factor of 0.88 for torque sensitivity.

Peak torque, T_p, is usually defined as five to six times the continuous torque. Servomotors are tested at this peak torque level for 10 sec. Peak torque is available from standstill to speed at rated horsepower, which is generally 200 to 600 rpm. It is primarily of interest for the acceleration period of a servomotor.

Different manufacturers may have different interpretations of peak torque ratings. Some consider peak torque as the torque that can be produced by a current pulse. The maximum current pulse is usually determined by the magnitude that can be applied without demagnetizing the motor. Peak torque determined by this method can be twice the peak torque as defined in the preceding paragraph.

Armature resistance either may include the resistance of the brushes, R_m, or may only refer to the armature, in which case the symbol R_a generally applies. In case of R_a, brush resistance may be determined by the fact that the voltage drop across the brushes is approximately 2 V at rated continuous current. The armature resistance, R_a, is usually rated at 25°C (77°F) and has a tolerance of +12.5%. This resistance increases by about 40% at the maximum armature temperature.

Insulation class of electrical windings gives the maximum permissible temperature limits as determined by an increase of their electrical resistance in accordance with the Standards of the National Electrical Manufacturers Association (NEMA). The basis is an ambient temperature of 40°C (104°F). There are two classes of interest here:

- Class F, allowing a temperature rise of 130°C (234°F) for armature windings of totally enclosed nonventilated and totally enclosed fan-cooled dc integral-horsepower motors, of 105°C (189°F) for all totally enclosed fan-cooled integral-horsepower single-phase and polyphase induction motors, and of 110°C (198°F) for the same ac motors non-ventilated.
- Class H, allowing a temperature rise of 155°C (279°F) for armature windings of totally enclosed nonventilated and totally enclosed fan-cooled dc integral-horsepower motors, of 125°C (225°F) for all totally enclosed fan-cooled integral-horsepower single-phase and poly-phase induction motors, and of 135°C (243°F) for the same ac motors non-ventilated.

Thermal time constant is the time, t_c, in minutes that it takes the motor temperature to reach 63.2% of the maximum temperature it eventually obtains (cf. page 17). The termal time constant is usually measured with the motor mounted on a heat sink. It should not be confused with the electrical and mechanical time constants of the motor response.

Duty cycle is a ratio expressed by:

$$R = \frac{t_{on}}{t_{on} + t_{off}} \tag{7.4}$$

where R is the duty cycle, t_{on} is the time the motor runs at full load or torque, and t_{off} is the time the motor runs at no load.

The motor may run at a torque in excess of its continuous stall torque rating for a certain time, t_{on}. However, this requires that this time be followed by a cooling-off period, t_{off}. In other words, it is possible to run a motor at excess torque, provided the corresponding duty cycle is observed. The excess torque permissible in relation to a corresponding duty cycle can be calculated by the equation:

$$\frac{T_e}{T_c} = \sqrt{\frac{1 - e^{-(t_{on}/Rt_c)}}{1 - e^{-(t_{on}/t_c)}}} \qquad (7.5)$$

where T_e/T_c is the ratio of the excess torque relative to the continuous stall torque for which the motor is rated, t_{on} is the time the motor runs at full load in minutes, R is the duty cycle, and t_c is the thermal time constant in minutes. Equation 7.5 has been used to plot the curves of Fig. 7-12.

Thus, for an "on" time of 25 minutes and a thermal time constant of 62.5 minutes, the corresponding point on the ordinate of Fig. 7-12 is at $t_{on}/t_c = 0.4$. Suppose, excess torque is to be 45% above rated continuous torque; then $T_e/T_c = 1.45$. According to Fig. 7-12, this permits a duty cycle of about $R = 0.34$, as shown by the dashed line. Inserting these values in Eq. 7.4 gives $t_{off} = 48.5$ minutes. In other words, if the motor is to be run in periods of 25 minutes at a torque that exceeds the continuous stall torque by 45%, then the total cycle should last 73.5 minutes, of which the motor runs for 25 minutes followed by 48.5 minutes at no load, after which it may run again for 25 minutes at 45% excess load, and so on.

Speed regulation is the variation in actual speed expressed as a percentage of set speed. There are two forms of speed regulation. One refers to the

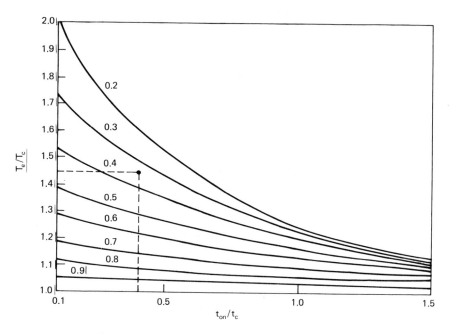

Fig. 7-12. Duty-cycle characteristics.

open velocity loop, which means the motor without velocity feedback; the other to the closed velocity loop, which means the motor as part of a control system with velocity feedback.

In the case of an open velocity loop, the motor speed will be a function of torque only—provided that no change occurs in supply voltage, armature current and resistance, voltage drop across brushes, and counter emf. The last quantity is the voltage developed in the armature windings by the continuously reversing currents, which is at every instant directed against the applied emf.

In the case of a closed velocity loop, the motor speed is independent of torque, provided the amplifier gain is sufficiently high. Since this is the condition under which a robotic servomotor runs, speed regulation for a closed velocity loop is of primary importance. The feedback produces a high degree of regulation from zero to maximum speed. Factors that can affect speed regulation, however, are inertia of the system, precision of the motor and gears, range of the velocity controller and power amplifier, and accuracy of the feedback device. Long-term regulation is typically 0.1% of the set speed of the motor down to 100 rpm. Below 100 rpm, regulation degrades rapidly.

Dynamics of a motor are determined by the usual characteristics of dynamic systems, such as inertia, viscous damping, static friction, and so on. Of particular interest in servomotors are the following:

- Mechanical time constant (cf. page 17), which is defined as the time required by the motor to attain 63.2% of its final speed when a step voltage is applied and the motor temperature is 25°C (77°F). It is proportional to the rotor inertia and the armature resistance including brushes, and inversely proportional to the torque sensitivity and the counter emf per rpm.
- Electrical time constant, which is measured with the motor shaft locked. The time required for the armature under these conditions to reach 63.2% of its steady state value is the electrical time constant. It is a function of the ratio of armature inductance to armature resistance.

Form factor is not so much a characteristic *of* the motor as it is an effect *on* the motor. It is a figure of merit indicating how much the current departs from pure dc, which is represented by unity. Values greater than one indicate an increasing departure from pure dc. Since supply voltage is usually rectified ac, some remnant of the ac may survive, causing a ripple effect in the dc current. The form factor is given by the ratio I_{rms}/I_{avg}, where I_{rms} is the root mean square of the ripple (see below) and I_{avg} is the ideally pure, straight-line dc current.

Root mean square (rms) means the average square root of the sum of the squares of the magnitudes of a changing current within a given period. For sinusoidal variations, the rms value of an ac current is 0.707 times its maximum value. Thus, if the amplitude of the sinus wave of the ac current is 3 A, then its rms value is 2.12 A.

The rms ripple current produces the same heating effect in the windings as would be produced by an equal value of dc, but it provides no torque. Thus, it increases the heating, reduces brush life, and may excite mechanical resonance of machines producing audible noise.

Form factors can be relatively high. They depend primarily on the type of motor control being used. Typical values are:

1.4 for SCR control, one-phase, full wave.
1.2 for SCR control, three-phase, half wave.
1.05 for SCR control, one-phase, full wave.
1.01 for pulse-width-modulation (PWM) control.

To avoid difficulties caused by form factors, torque and current ratings of motors must be reduced correspondingly.

SELECTION OF A DC SERVOMOTOR

Consider the selection of a servomotor for the shoulder axis of a robot. The following data may be used for example purposes:

Maximum load to be lifted: 130 lb
Maximum reach of arm: 6.7 ft
Maximum speed of load: 5.0 ft/sec
Effective mass of arm: 132 lb at 3.3 ft
Inertia of arm: 48 lb-ft-sec^2
Inertia of gear reducer: 0.004 lb-ft-sec^2
Ratio of gear reducer (input to output rpm): 200:1
Frictional load torque: 3.8 lb-ft

In these data, the speed of the load is specified as 5.0 ft/sec at 6.7 ft. Since 6.7 ft is equivalent to the radius of a circle, the circumference of the circle is 42.1 ft. Hence, a speed of 5.0 ft/sec is equivalent to 7.13 rpm. With a gear ratio of 200:1, this calls for a motor speed of 1426 rpm. (This is actually low for the usual rating of dc servomotors, where speeds of at least 2000 rpm are more common. By increasing the gear reducer ratio, such a faster motor could well be employed, but this detail may be disregarded for the purposes of this discussion.)

The torque acting on the shoulder joint is the effective mass of the arm plus the torque of the maximum load to be lifted. This means 132 × 3.3 + 130 × 6.7 = 1306.6 lb-ft. This torque translated to the motor must be divided by the gear ratio, which gives 6.53 ft-lb. To obtain the total torque at the motor, the frictional load torque of the given data must be added. Since it is 3.8 lb-ft, the result is 10.33 lb-ft.

Equation 7.3 established that $P = T_f N/5252$. Since in the above case the torque is $T_f = 10.33$ lb-ft, and the motor speed is $N = 1426$, the power, P, is 2.8 hp.

A dc motor is chosen that has a performance curve as shown in Fig. 7-13. Other data may be assumed as follows: The peak torque is 66 lb-ft. At a speed of 1426 rpm, it can deliver about 11 lb-ft of torque. Since the total torque requirement (above) is 10.33 lb-ft, this would appear to be an adequate motor. However, the form factor needs to be considered, as described below.

The next step is the selection of the motor control system. It may be either a PWM (pulse width modulation) or an SCR (silicon controlled rectifier) system. For the purposes of this example, an SCR system with a form factor of 1.2 is chosen.

The total torque requirement (above) is 10.33 lb-ft, which would have to be increased by 20% to 12.4 lb-ft, a figure slightly above the torque rating of the motor. Where such marginal conditions are involved, the manufacturer should be consulted. In this case, it probably would be thought that

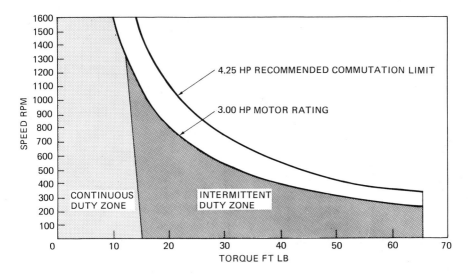

Fig. 7-13. Performance curve of typical dc servomotor.

the slight excess of torque could be tolerated, since form factors improve (i.e., get closer to unity) at higher speeds.

To determine acceleration, the components of the total inertia of the mass as reflected on the motor have to be determined. There are four separate inertias that are involved: arm, load, speed reducer, and motor. The inertia of the arm was listed as 48 lb-ft-sec^2. To translate it to the motor, it must be divided by the square of the gear ratio. The result is 0.0012 lb-ft-sec^2.

The load inertia must be calculated on the basis of the standard equation $I = r^2w/g = (6.7)^2 \times 130/32.174 = 181,38$ lb-ft-sec^2, which in turn must be divided by the square of the speed reducer ratio. The result is 0.0045 lb-ft-sec^2.

The inertia of the speed reducer was given as 0.004 lb-ft-sec^2, and the motor inertia, according to catalog data, is 0.012 lb-ft-sec^2.

Adding all four inertias results in a total inertia at the motor of 0.0217 lb-ft-sec^2.

If we know the different inertias, it is also of interest to know to what extent the speed reducer approaches the optimum reducer ratio. This is given by $J_l/(J_m - J_g)$. Here, J_l is the total load inertia, J_m is the motor inertia, and J_g is the motor pinion inertia.

There are several ways of accelerating a motor. Only two need to be mentioned here. One is initially to apply a maximum constant torque up to a certain velocity level, and then to slow down to a constant power acceleration that smoothes over into the steady-state velocity. The corresponding motion profile is shown in Fig. 7–14. This two-step acceleration has the advantage that at light loads the acceleration is faster, but the disadvantage that it is slower at high loads than the other method, which uses only constant torque as the basis for the acceleration.

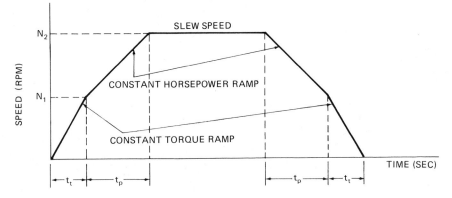

Fig. 7–14. Motion profile of two-step acceleration and deceleration.

Assuming that more often than not this robot has to carry only a relatively light load, the two-step acceleration (and deceleration) profile is used. The question then is: At what point should the acceleration be switched from constant torque to constant power?

According to Fig. 7–14, the commutation limit of the motor is 4.25 hp, the maximum torque is 66 ft-lb, and the maximum speed under these conditions is 338 rpm. (The 338 rpm is read with difficulty from the graph, but the peak torque of 66 lb-ft is also listed in the catalog data of the motor.) From these data the rpm can be calculated, since, according to Eq. 7.3, $P = T_f N/5252$, and, hence, $N = 4.25 \times 5252/66 = 338$ rpm.

Thus, the constant-torque slope can be used up to 338 rpm. From there on, acceleration is at constant power.

There are two equations that give the acceleration time for (a) the constant-torque phase and (b) the constant-horsepower phase, respectively, both phases or ramps being shown in Fig. 7–14. The time applied for constant-torque acceleration is:

$$t_t = \frac{J_t \, (N_i/9.55)}{T_p - T_m} \qquad (7.6)$$

and for the constant-horsepower acceleration phase, the time applied is:

$$t_p = \frac{J_t}{T_m} \left\{ \frac{550 \, P_L}{T_m} \, \ln\left[\frac{5252 \, (P_L/T_m) - N_1}{5252 \, (P_L/T_m) - N_2} \right] - \frac{N_2 - N_1}{9.55} \right\} \qquad (7.7)$$

The symbols and their quantities are as follows:

- J_t is the total inertia at the motor, which was shown to be 0.0217 lb-ft-sec^2.
- N_1 is the maximum motor speed to be obtained with constant torque acceleration, which was shown to be 338 rpm.
- N_2 is the final motor speed (slew rate), which is the motor rpm corresponding to the desired load velocity and which was shown to be 1426 rpm.
- T_p is the peak torque, which was shown to be 66 lb-ft.
- T_m is the total torque at the motor, which was shown to be 10.33 lb-ft.
- P_L is the recommended commutation limit, which, according to Fig. 7–13, is 4.25 hp.
- t_t is the acceleration time required for the constant torque phase, in seconds.

- t_p is the acceleration time required for the constant power phase, in seconds.

To calculate the acceleration time for the constant torque phase, the corresponding values are inserted in Eq. 7.6. The result is $t_t = 0.014$ sec.

Equation 7.7 is similarly used to obtain the acceleration time for the constant power phase and gives:

$$t_p = 0.0021\ (205.736 - 113.927) = 0.193\ \text{sec}$$

Thus, the total time required for acceleration is $0.014 + 0.193 = 0.207$ sec.

If we decide to use a constant-torque acceleration with a one-step profile, Eq. 7.6 again applies, but with two differences: the velocity, N_1, is now equal to the slew speed of 1426 rpm, and the peak torque must be replaced by the continuous torque for which the motor is rated, which is 15.5 lb-ft. The result is 0.627 sec. This is an acceleration that takes three times longer to arrive at the slew speed of the robot than the two-step profile described before.

The next question concerns the distance the robot load has moved during this acceleration.

The first part of the acceleration brought the motor to 338 rpm in 0.014 sec. This means an average velocity of 169 rpm during 0.014 sec, which is equivalent to a total motor rotation of 14.2°.

During the second part of the acceleration, the motor went from 338 to 1426 rpm in 0.193 sec. This means an average speed of $(1426 - 338)/2 + 388 = 932$ rpm during 0.193 sec, which is equivalent to a motor rotation of 1079.3°.

Thus, the motor turned a total of $1079.3 + 14.2 = 1093.5°$ or 3.0375 revolutions in 0.207 sec; so the average motor speed during acceleration was 880 rpm. Since it was established before that a speed of the load of 5 ft/sec was equivalent to 1426 rpm, the average speed of the load during acceleration was 3.086 ft/sec with a final speed or slew rate, according to the data initially given, of 5 ft/sec. During the 0.207 sec of acceleration, the load moved through 0.639 ft.

Regarding deceleration, Eq. 7.6 can be used as before. For constant-horsepower deceleration, however, Eq. 7.7 must be modified as follows:

$$t_p = \frac{J_t}{T_m}\left\{\frac{550\ P_L}{T_m}\ \ln\left[\frac{5252\ (P_L/T_m) + N_1}{5252\ (P_L/T_m) + N_2}\right] - \frac{N_1 - N_2}{9.55}\right\} \qquad (7.8)$$

As small as the times lost in acceleration and deceleration may seem, they can be of great importance in calculating the productivity of a robot that may go through the same motion thousands of time during a workday, particularly when the distances moved are short enough to make acceleration and deceleration an essential part of the total working cycle.

TORQUE MOTORS

There are various kinds of torque motors. In Chapter 5, torque motors for hydraulic servovalves were discussed. Their motion is limited to about 0.5°. This limitation and their particular application require a very specialized design. The torque motor considered here is entirely different. It is a legitimate motor capable of continuous rotary motion. It is called a torque motor because it can maintain its maximum torque in a stalled condition for prolonged periods.

The Robotics Institute of the Carnegie-Mellon University[5] developed a robot based on the direct-drive method using samarium-cobalt dc torque motors. Figure 7-15 shows the concept. The motor has become part of the robot joint. The stator is part of one connecting link of the joint; the rotor,

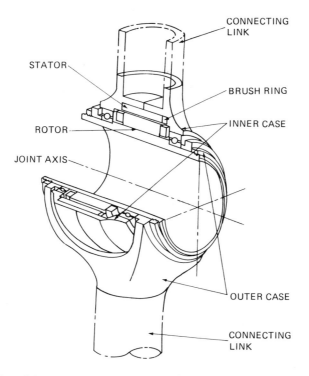

Fig. 7-15. Direct joint drive with torque motor. (*From Asada and Kanade, Ref. 5*)

of the other. Maximum speeds at the arm tip of 13 ft/sec were obtained, which is as good as most commercially available electrically powered robots can give.

This concept eliminates all transmission mechanisms and with it backlash and friction. However, the developmennt of a practical robot that incorporates this concept would require powerful active joints with light weight and good heat dissipation.

STEPPER MOTORS

Stepper motors are characterized by discrete angular motions of essentially uniform magnitude rather than continuous rotation. They have advantages primarily because no feedback devices are required, such as the encoders described in Chapter 8, although those devices still may be used and are compatible. Error in positioning a stepper is nonaccumulative as long as a fixed relation between the pulse and the step is maintained.

There are several design concepts for stepper motors. One is the variable-reluctance stepper. It operates on the principle of the synchronous reluctance motor (see Fig. 7-4). The number of teeth on the rotor and stator, as well as the number of winding phases, determines the step angle. The soft-iron rotor is made to step by sequential excitation of stator coils. Electronic circuitry furnishes the sequencing pulses.

The hybrid concept is a combination of a variable-reluctance rotor with a permanent magnet in its magnetic path. This is the most common design of stepper motors. Oriental Motor, for example, uses it for steppers of 100, 200, and 400 steps per revolution.

The concept of a hybrid stepper[6] is shown in Fig. 7-16. The rotor shows one set of toothed cylindrical magnets—actually it may have up to three

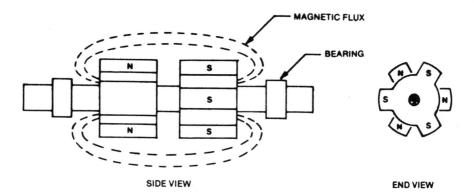

Fig. 7-16. Concept of hybrid stepper. (*From Slingland, Ref. 6*)

such sets—and a toothed stator, which is wound in such a way that alternate poles are driven by two separate phase currents. Each set of toothed magnets on the rotor consists of two magnets offset against each other in the manner shown.

Figure 7-17 shows the schematic end view of a motor with 12 steps per revolution. In Fig. 7-17a, phase 1 of the stator current is energized. The north pole of the permanent magnet assumes a position that is in alignment with the south pole generated in the stator by phase 1, while the south pole of the permanent magnet aligns with the north pole of the stator.

To step the motor, phase 1 is de-energized, and phase 2 becomes energized as shown in Fig. 7-17b. The rotor lines up with the new magnetic poles of the stator and, in doing so, advances 30°. Thus, the motor can be stepped in increments of 30°.

The rotor may also be moved one-half of the step (i.e., 15°) by energizing both phases together. This is illustrated in Fig. 7-17c. Here, the rotor is held in the middle of one full step. In other words, it moves in increments of 15°. By adding additional teeth on stators and rotors, the number of steps per revolution can be further increased.

In order to make smooth acceleration and motion possible with a stepper motor, steps have to be minimal. This has led to microsteppers, as represented by the stepper motor developed and produced by Compumotor Corp. The design allows expansion of 200 steps per revolution to a total of 50,000 steps, by electronically producing 250 intermediate positions within each step.

Microstepping is accomplished[6] in a way similar to half-stepping except that currents in phase 1 and phase 2 are changed in small increments relative to each other by electronic means. If current flows through phase 1 and phase 2 simultaneously, but produces a larger flux in one phase than in the other, then the rotor will assume some intermediate position. The rotation of the stepper motor assumes almost the continuity of a rotary motor, but it can be controlled in discrete steps.

Indexers provide some form of microprocessor-based pulse generation logic to control acceleration, deceleration, velocity, and position of stepper motion.[7] A block diagram is shown in Fig. 7-18. Control information is received by a microprocessor through an interface. The microprocessor uses this information to generate pulse rates from zero to 500,000 steps/sec. When optical encoders are used, the microprocessor interprets encoder output and produces the pulses needed to correct positional errors.

Each pulse generated by the indexer will cause the driver to move the motor one microstep. A string of pulses of a given frequency will cause the motor to rotate at a velocity proportional to that frequency. Thus, a 25 kHz pulse train will make a 25,000 steps/rev Compumotor microstepper rotate at exactly 60 rpm.

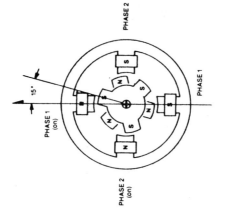

PHASE 1
(on)

0°

ROTOR

PHASE 2

STATOR

PHASE 2
(off)

PHASE 1

(a)

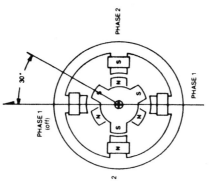

PHASE 1
(off)

30°

PHASE 2

PHASE 2
(on)

PHASE 1

(b)

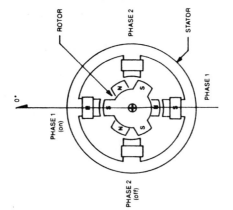

PHASE 1
(on)

15°

PHASE 2

PHASE 2
(on)

PHASE 1

(c)

Fig. 7–17. Stepper action incl. half-step action. *(From Slingland, Ref. 6)*

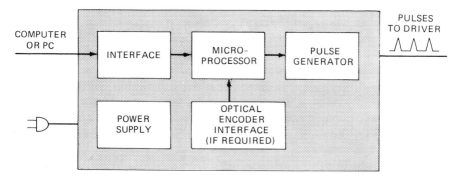

Fig. 7–18. Block diagram of indexer. (*From Ording, Ref. 7*)

The typical microstepping motion profile is trapezoidal, where acceleration and deceleration follow a smooth, linear ramp, ascending and descending, respectively. The rate of acceleration/deceleration and the preset velocity depend on the load, and determine the overall move time. The motion profile can become triangular where the preset distance is too short or the acceleration rate too slow to allow the motor to reach the preset velocity. In this case, one-half of the move is spent in acceleration, the other half in deceleration.

STEPPING CONTROL FOR DC MOTORS

Stepper motors are often criticized because they are susceptible to resonances that are said to lead to long settling times and to result in loss of synchronization. On the other hand the closed-loop control system of a dc motor is more complex and expensive than that of the open-loop stepper motor.

However, the stepping-servo controller chip GL–1200, which was developed by Galil Motion Control, accepts input commands in the same pulse-train format as those traditionally given to move a stepping motor[8]. It accepts the feedback signal from an incremental encoder and compares the position feedback pulses with input command pulses to derive a position error.

Outputs include a pulse-width-modulated command signal and two TTL-level signals (see chapter 9) required for damping. These three signals are externally added, and the sum is a single analog signal that controls motor speed, position, and direction via an external amplifier. This controller mechanism lets the motor follow the input command very accurately.

Thus, the GL–1200 offers a low cost alternative to stepper motors and permits at the same time the use of a conventional dc servomotor.

SUMMARY

Equations of this chapter are repeated here for ready reference.

Force acting on a conductor in a magnetic field:

$$F = 5.7 \times 10^{-7} \, BLI$$

Motor horsepower required for a given torque and rpm:

$$P = \frac{T_i N}{63,025} = \frac{T_f N}{5252}$$

Duty cycle:

$$R = \frac{t_{\text{on}}}{t_{\text{on}} + t_{\text{off}}}$$

Permissible excess torque for electric motor:

$$\frac{T_e}{T_c} = \sqrt{\frac{1 - e^{-(t_{\text{on}}/Rt_c)}}{1 - e^{-(t_{\text{on}}/t_c)}}}$$

Constant-torque acceleration:

$$t_t = \frac{J_t \, (N_1/9.55)}{T_p - T_m}$$

Constant-horsepower acceleration:

$$t_p = \frac{J_t}{T_m} \left\{ \frac{550 \, P_L}{T_m} \, \ln \left[\frac{5252 \, (P_L/T_m) - N_1}{5252 \, (P_L/T_m) - N_2} \right] - \frac{N_2 - N_1}{9.55} \right\}$$

Constant-horsepower deceleration:

$$t_p = \frac{J_t}{T_m} \left\{ \frac{550 \, P_L}{T_m} \, \ln \left[\frac{5252 \, (P_L/T_m) + N_1}{5252 \, (P_L/T_m) + N_2} \right] - \frac{N_1 - N_2}{9.55} \right\}$$

where:

$B =$ flux density in gauss, Gs
$F =$ force, lb

I = current generated in conductor, A
J = total friction inertia, lb-ft-sec^2
L = length of conductor, in.
N = speed of motor, rpm
N_1 = motor speed attained on constant-torque acceleration (see Fig. 7–14)
N_2 = motor speed attained on constant-horsepower acceleration (see Fig. 7–14)
P = horsepower, hp
P_L = commutation limit of motor, hp
R = duty cycle
T_c = continuous stall torque, lb-ft
T_e = excess torque, lb-ft
T_f = friction torque, lb-ft
T_i = friction torque, lb-in.
T_m = total torque as seen by motor, lb-ft
T_p = peak torque, lb-ft
t_c = time constant, min
t_{on} = full-load running time, min
t_{off} = no-load runnning time, min
t_p = time during constant-horsepower acceleration or deceleration, sec
t_t = time during constant-torque acceleration or deceleration, sec

REFERENCES

1. Robot motor puts the drive where the motor is. *Engineering Materials and Design* (England), February, 1984, pp. 28–30.
2. Kohn, E. F. AC–DC drive system; Past, present, and future trends. *Powerconversion International,* September, 1983, pp. 4,6.
3. Bohlmann, M. A. Permanent magnet materials for electric machines. *1984 Coil Winding Proc.,* Chicago, 1984.
4. Bartlett, P. M. and Shankwitz, H. The specification of brushless dc motors as servomotors. *Powerconversion International,* September, 1983, pp. 10–16.
5. Asada, H., and Kanade, T. *Design of Direct-Drive Mechanical Arms.* Pittsburgh, PA: Carnegie-Mellon University, April 28, 1981.
6. Slingland, E. Small steps turn into big improvements. *Powerconversion International,* October, 1983, pp. 20–25.
7. Ording, J. J., II. Servo-like performance comes to step motors. *Power Transmission Design,* August, 1984, pp. 37–40.
8. Tal, J., and Baron, W. Stepping servo - low cost alternative to step motors. *Powerconversion International,* March, 1985, pp. 42–49.

Chapter 8
FEEDBACK SENSORS FOR VELOCITY
AND POSITION

Feedback sensors are needed on a robot for feeding back the position and velocity of motion of the various joints. The information is necessary to maximize the dynamic accuracy, to accelerate and decelerate the motion, to stop it at a programmed position, to signal the commutation sequence of a brushless dc servomotor, and so on. In the last case, the sensors are usually referred to as commutation sensors. Hall generators, which are described below, are frequently used for this purpose. Other feedback sensors are potentiometers, tachometers, resolvers, and optical encoders. The following gives an overview of these different types.

COMMUTATION SENSORS

What is known today as the Hall effect was discovered in 1879 by E. H. Hall. It is a complex theory that leads to a basically simple device. A small square-shaped semiconductor chip of, say, indium arsenide or indium antimonide is used. The chip is not more than 0.004 in. thick. It is usually enclosed in a plastic envelope or the equivalent. Two of its opposing edges are connected to a constant dc source, as shown in Fig. 8-1. The other two edges are connected to an amplifier to transmit the Hall voltage, which is generated as soon as the chip is exposed to a magnetic field.

There are numerous applications for this phenomenon. One is with servomotors, where Hall generators are used to sense rotating shaft speed and/or angular position because they respond to the magnitude and position of magnetic fields.

Most frequently, they are used as commutation sensors for brushless permanent magnet dc motors. By sensing the approach of the rotating magnets of the motor, they feed back to the motor controller the instants for reversal of armature current in rapid succession. The action is similar to that of

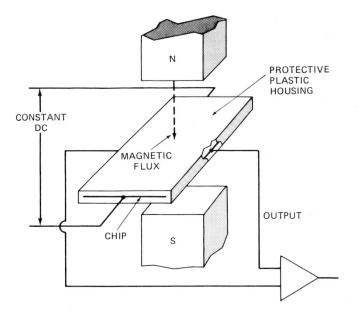

Fig. 8-1. Principle of Hall generator.

commutator bars and brushes, except that here the logic of the controller prevails, and not mechanical contact with bars and brushes.

This method can also provide some compensation for armature reaction effects which may occur in some motor designs.[1] Armature reaction produces a magnetic field shifted by 90 electrical degrees with respect to the direction of the stator magnetic field. The armature in effect becomes an electromagnet that tends to oppose the main stator magnetic flux. This produces a demagnetization effect that in most cases is reversible. That is, when the current goes to zero, the magnets return to full strength. However, there is a definite value of current that will cause a permanent demagnetization, shutting the motor down. When located in the stator, the Hall generator helps to prevent this, although temperatures in the stator may reach 320 to 350° F, and such temperatures can in turn affect the performance of the Hall generator.

It is quite possible to locate the Hall generator away from the active stator and use a separate magnet on the motor shaft for sensing. However, then the compensation for armature reaction effects is lost.

Another commutation sensor is an electro-optical switch, consisting generally of a light-emitting diode (LED) and a phototransistor as detector.

The LED transmits infrared light, to which the phototransistor is more sensitive than it is to light in the visible region. The LED and the detector are often combined in a single U-shaped interrupter module. A disc containing optical slits is mounted on the shaft and moves through the light beam, changing light intensity and inducing output voltage pulses. The resolver, described below under a separate heading, is still another device that finds frequent use as a commutation sensor.

POTENTIOMETERS

The potentiometer consists of a resistance element and a wiper. It is provided with three terminals: one on each end of the resistance element and one for the wiper. The wiper either moves longitudinally on a linear resistance element, or is connected to a shaft and rotates with it while sliding on a circular resistance element. It is shown in Fig. 8–2 and is the type generally used as a robotic sensor.

The output signals of the potentiometer are the resistances between either end of the resistance element and the wiper. While the resistance between the wiper and one end increases, it decreases between the wiper and the other end. The ratio of these two resistances is a function of the shaft position.

According to their resistance elements, potentiometers can be divided into three different classes: wirewound, cermet, and plastic potentiometers.

The wire used in wirewound potentiometers is usually nickel-chromium (75% Ni, 25% Cr). The alloy has a temperature coefficient of less than ±5 ppm/°C. This means that if the temperature of the wire changes from 25°C (77°F) to 125°C (257°F), its resistance changes not more than 0.05%. The independent linearity of wirewound potentiometers is usually within ±0.003%.

Wirewound potentiometers are available in single-turn and multi-turn

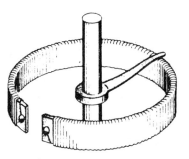

Fig. 8–2. Single-turn potentiometer.

versions. With the single-turn potentiometer of Fig. 8–2, the wire is wound over a mandrel, and the wiper moves from winding to winding changing the resistance between the wiper and either terminal in steps. Although the steps are small, the resolution of the single-turn potentiometer is nevertheless limited.

The situation is different with the multi-turn potentiometer of Fig. 8–3. Here, the wire is wound in a helix, and the wiper tracks the helical path of the wire not only by rotation but also up and down with the three-dimensional helix. This requires a somewhat more complex structure but assures practically infinite resolution.

Cermet is an acronym for *ceramic metal*. It consists of very fine particles of a powdered glass binder (glass frit) with almost equally fine particles of precious metals. An organic resin mixture is added to make a paste out of these particles. The paste, which is often referred to as "ink" is applied to a flat ceramic substrate, such as alumina or steatite. Baking in a kiln burns off the organic resin and fuses the glass particles and ceramic substrate together. The metallic particles provide a film with the characteristics of an electrical resistor. A wide range in resistance values is achieved by variations in:

- Compostion of the paste
- Time and temperature in the kiln
- Physical size of the element

Fig. 8–3. Cutaway-view of multi-turn potentiometer. (*Courtesy of Bourns, Inc.*)

The temperature coefficient of cermet materials averages about ±100 ppm/°C. This is essentially higher than that of nickel-chromium wirewound potentiometers, which was stated as ±5 ppm/°C. However, their resolution, even as single-turn units, is infinite.

The life of wirewound as well as of cermet potentiometers is generally rated at one million cycles. Since in a wirewound potentiometer the failure mode is usually a wire defect, this means if a wiper moves over the same spot 1500 times a day, it would likely have to be replaced after about 2 years.

The failure mode of the cermet potentiometer is different. It is usually wear of the wiper rather than the resistance element. The reason for this is a certain microscopic roughness in the surface of the resistance element due to the particles used. This is avoided with the third type, the plastic potentiometer, whose life expectancy thus is considerably longer.

Potentiometers made of conductive plastic film have a life expectancy of 10 to 20 million cycles. That is 10 to 20 times more than the life of either wirewound or cermet potentiometers. G. J. Gormley[2] described life tests by dithering the plastic potentiometers at a high frequency that showed almost no deterioration after one billion cycles.

Conductive plastics are made of epoxy, polyester, and other resins, which are blended with carbon powder to make them conductive. Their thermal coefficients are relatively high. They usually range between 200 and 400 ppm/°C. This means they are the most temperature-sensitive of the three types. Humidity changes may also affect their resistance, although improved sealing methods have minimized this effect. Linearity can be held to ±0.025% with laser or mechanical trimming.

It is possible[3] to combine a wirewound element with a conductive coating to obtain a hybrid element that has the temperature coefficient and resistance stability of wirewound potentiometers, combined with the long operational life, low resolution, and low noise of the conductive plastic element.

In spray-painting robots, for example, where accuracy and repeatability requirements are not so severe as they may be with assembly robots, potentiometer feedback may well be used. The potentiometer can be driven by the robot axis or by reducing gears directly from the actuator.

Thus, Thermwood Corp. uses for its robots a custom-designed potentiometer driven by the servomotor. The external command is provided by a second potentiometer. Potentiometers for such purposes are generally connected as in Fig. 8-4. Here, the potentiometer becomes a voltage divider, with output voltage changing relative to the wiper position. In the arrangement used by Thermwood, both potentiometers act as voltage di-

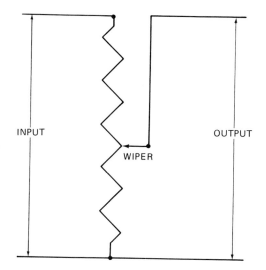

Fig. 8-4. Voltage divider.

viders, and the difference between the two voltages provides the input to a comparator circuit.

The comparator starts the motor whenever there is a difference between the two voltages, with the rotating direction of the motor dependent on which is the greater voltage. This assures that the voltage difference will be eliminated by the action of the motor. When this is done, the motor stops, and the system is in balance.

TACHOMETERS

The tachometer, or more precisely the tachometer generator, is a small dc or ac generator coupled to a rotating shaft. It is mounted on the shaft of the servomotor or coupled to it, and generates a voltage that is a function of the rpm of the motor.

Generally, these tachometers use permanent magnets, either in the stator or —in brushless types—in the rotor. They are often called generators, since they act like them, producing a voltage in proportion to speed. The power they have to provide is extremely small; so the contention that brushes require special maintenance and that therefore brushless tachometers should be used, becomes even more meaningless than it is with servomotors.

The voltage a tachometer produces is supposed to be proportional to the shaft speed, but there are some unavoidable minor deviations. These lead to errors in the motor speed. However, since deviations from absolute linearity in the voltage vs. speed response are rarely more than 0.5%, this is hardly a serious problem. It merely means that a motor that should run at, say, 2000 rpm may run at 1990 rpm.

Another point is that because of the use of permanent magnets, the voltage output of a tachometer may change with temperature. The temperature coefficient should not be higher than 0.02%/°C. This would mean that for a temperature change of 100°F or 56°C, the voltage output, and hence the motor speed, would change by 1.1% unless some temperature compensation were provided.

Tachometers are generally mounted rigidly to the servomotor shaft. A unique solution to the motor–tachometer combination is represented by Electro–Craft's patented Motomatic motor–tachometer. Figure 8–5 shows the motor commutator on the left and the tachometer commutator on the right. The windings for both are wound on the same armature. However, the device has some limitations in its ability to respond rapidly in high-performance servosystems. Since the latter are generally used in robots, it is preferred that the motor and tachometer be separate, though mounted on the same shaft, as is the case with the Electro-Craft E-576 in Fig. 8-6, to obtain an improved response. Such a design combines the rigidity of direct coupling, which is necessary for stability of the system, with magnetic separation of the motor and tachometer, which is desirable for system response.

The purpose of the tachometer is to continuously sense and feed back the speed of the motor. This can be done either for controlling the speed of the servomotor, as in the arrangement of Fig. 8-7, or for improving the stability of a position servosystem, as in the arrangement of Fig. 8-8.

The speed servo of Fig. 8-7 feeds the voltage generated by the tachometer into the feedback compensation network of the motor control system. The

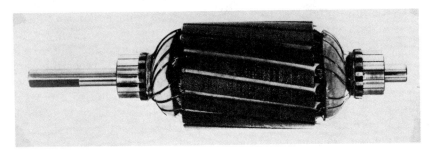

Fig. 8–5. Motor and tachometer windings on same armature. (*Courtesy of Electro-Craft Corp.*)

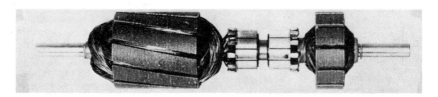

Fig. 8-6. Motor and tachometer mounted on same shaft. (*Courtesy of Electro-Craft Corp.*)

signal is conditioned there to be comparable with the input signal which calls for a specific motor speed. The feedback signal and the input signal come together in the summing point. The difference between the two is the error signal which feeds into the amplifier. From the amplifier, the signal passes to the power amplifier which controls the motor current.

There is another loop shown which feeds back the motor current to the amplifier and stabilizes it. Sometimes, this feedback loop takes off between the amplifier and the power amplifier as in Fig. 8-8. Its function would be the same.

Figure 8-8 includes a position sensor. This concept applies to a robot much more often than the speed servo. The input signal, here, comes from the robot control system and commands the speed with which the motor

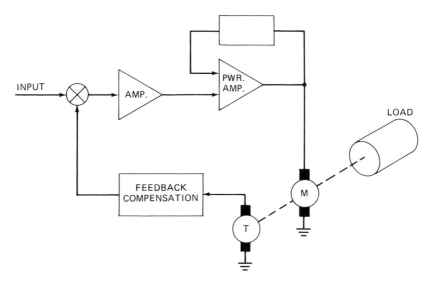

Fig. 8-7. Speed servo with tachometer feedback. (*From DC Motors, etc., Ref. 1*)

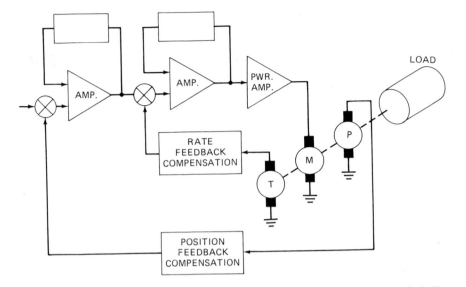

Fig. 8-8. Position servo with tachometer damping. (*From DC Motors, etc., Ref. 1*)

must move the load, and the final position it has to assume. In this case, the main purpose of the tachometer is to provide damping in a dynamic system.

RESOLVERS

The resolver is driven by the motor shaft or by a robot axis. It develops an output voltage that for a given input voltage is a function of the sine (and cosine) of the shaft angle. Electronic circuitry converts the trigonometric information into angular position and/or speed. Thus, it is suitable for either speed or position measurement.

In the case, for example, of the ac servodrive system from Gould, Inc., the resolver furnishes the velocity feedback from the motor. It also has the optional feature of a resolver for position feedback. The general advantage of resolvers is that the electronics and the oscillator (see below) can readily be installed at distances 200 ft or more away from the resolver itself.

A resolver looks like a motor. In its electrical structure, it operates like a rotary transformer, as illustrated in Fig. 8-9. A transformer usually has two separate windings. The input voltage in the primary winding produces a changing magnetic field with the frequency of the applied voltage. This induces a voltage in the secondary winding. The resolver has two secondary windings, wound 90° apart on the laminated structure of the stator. The

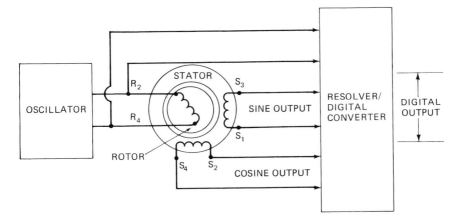

Fig. 8–9. Components of a resolver system.

primary is wound on the rotor. This, at least is the usual arrangement, although the reverse may also be true: the primary on the stator, the secondaries on the rotor. The relative position of the three windings in the normal arrangement is shown in Fig. 8–9. As the rotor moves, the relative position changes, and the output voltages of the two stator windings change with the sine and cosine, respectively, of the shaft positon.

Consider the instantaneous value of an ac input. Let it be 100 V. Let the angle of the rotor be $\theta = 30°$. The output voltage between S_1 and S_3 in Fig. 8–9 is then 100 sin $\theta = 50$ V and between S_2 and S_4, 100 cos $\theta = 43.3$ V.

Connection to the rotor is made by brushes and slip rings or through inductive coupling. Resolvers using the inductive method are referred to as brushless types, and are the type generally used in robot applications.

Another component of a resolver system is the oscillator (see Fig. 8–9), which is needed to supply a stable reference frequency to the rotor winding of the resolver. Its frequency ranges from 400 Hz to several KHz.

The resolver/digital (R/D) converter is an essential element of the resolver system. It converts the resolver signal to a digital output, in order to synchronize it with the robot control system.

Robots may drive resolvers directly from the robot axes without gearing. This is done, for example, with the T^3746 robot by Cincinnati Milacron Corp. To achieve good resolution in such applications, a 16-bit resolver/digital converter rather than the usual 12-bit R/D converter may be needed. Such a 16-bit R/D converter is the 168H200 made by Control Sciences Incorporated (CSI), which was developed specifically for robotic requirements.

The robotic industry generally requires that an R/D converter has an absolute accuracy of at least ±6 minutes of arc (i.e., 0.1°), and a repeatability of ±2 minutes of arc or better, even under the worst conditions of temperature, direction of approach, and hysteresis.

CSI improved on these requirements by achieving in the 168H200 converter an accuracy of ±3.5 minutes of arc and a repeatability that varies from ±0.7 minute of arc at room temperature to ±1.8 minutes at other points of the temperature range from 0°C to 50°C (32–122°F).

Figure 8–10 shows the block diagram of an R/D converter. The reference signal to the resolver is supplied by the oscillator, which is not shown. The output signals from the resolver are applied to an isolation transformer in the R/D converter. Since the reading of the resolver repeats itself every 90° of rotation, it is necessary to determine to which quadrant of a revolution the shaft or axis angle refers. Hence, a special feedback system is required that utilizes sine and cosine multipliers feeding into an amplifier. The sine and cosine multipliers are connected to the digital feedback from an up–down counter that keeps track of the preceding cycles, adding and subtracting the various quadrants. The output provided incorporates the count that is in the up–down counter, thus supplying the absolute position of the shaft. Besides the error output that feeds the signal to the position control system, a tap is also provided to get a direct signal of the output velocity.

J. R. Currie and R. R. Kissel of NASA's Marshall Space Flight Center[4] developed a somewhat different approach. In Fig. 8–11, the same type of resolver is used, but instead of the oscillator, Currie and Kissel used a pulse technique that leads to a different way of calculating the angle position. The microprocessor triggers the pulse generator, causing it to energize the rotor winding of the resolver. The microprocessor then quickly reads the resolver sine and cosine windings with the aid of the data-acquisition system. The sine winding is read first, then the cosine winding; then the cosine winding is read a second time. Both sine values are used in the electronic calculations to eliminate the effect of a slope in the winding-output voltages. The microprocessor calculates the ratio of sine to cosine to obtain the tangent, then calculates the arctangent to obtain the principal value of the angle. From the polarities of the sine and cosine voltages, the quadrant is determined, and the complete value of the angle is then known.

Each pulse-generator output pulse has two rising slopes. Slope 1 is the steeper of the two, and is adjusted to prevent the resolver from ringing. Slope 2 is somewhat flatter, and is selected so that sine and cosine outputs are as nearly constant as possible while they are being read.

The data-acquisition system sequentially reads the sine and cosine windings. From the readings, the microprocessor determines the angle θ through which the resolver shaft has turned in relation to the reference (zero) angle.

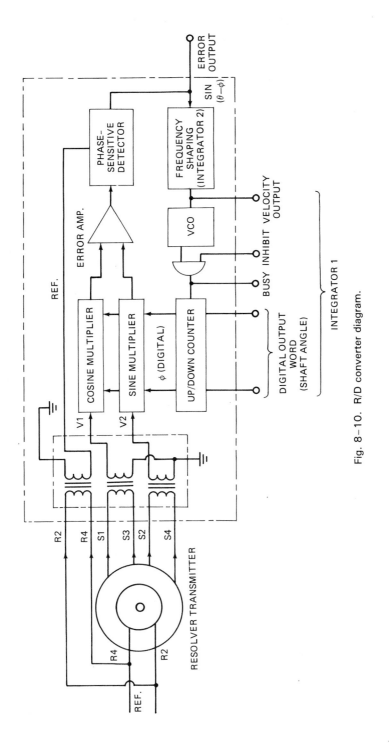

Fig. 8–10. R/D converter diagram.

265

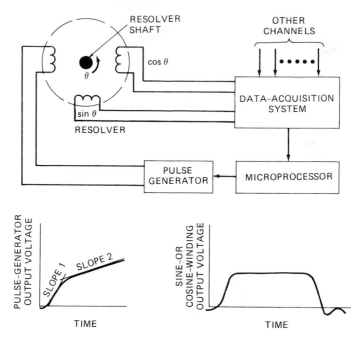

Fig. 8-11. Resolver circuit. (*From NASA Tech Briefs, Ref. 4*)

A single microprocessor and data-acquisiton equipment can be used in a system that combines a number of resolvers.

COMPARISON BETWEEN RESOLVERS AND ENCODERS

Resolvers as described above are rugged, with a wide temperature range and high reliability. Like encoders they can operate at slow speeds all the way down to zero. They have a high immunity to electrical disturbances or spurious signals from the outside.

A disadvantage of resolvers, however, is that they require an oscillator and an R/D converter. In spite of these accessories, according to ILC Data Device Corp.,[5] a 14-bit optical absolute encoder costs four times what a resolver system costs. The encoder, however, has one absolute digitized value for every angle of its shaft.

Resolvers as well as encoders are used with hydraulic servo systems and can give hydraulically actuated robots an accuracy and repeatability usually associated with electrically driven robots.

Encoders need much simpler electronics than resolvers. On the other hand they require as many as 14 wire connections for a 12-bit resolution, while an equivalent resolver needs only 6.

OPTICAL ABSOLUTE ENCODERS

The optical encoder, or the encoder for short, is fundamentally oriented to digital rather than analog techniques. It responds to the velocity on its shaft. However, when the output is differentiated, the device is also an accurate speed sensor. It consists of a disc that rotates between a light source and a photodetector. The disc has clear and opaque segments that either let light pass to the photodetector or block it out. Light pulses are generated as the angular position of the disc changes.

There are two basic types: the incremental encoder and the absolute encoder. The incremental encoder generates a pulse for every angular increment of shaft rotation. The output is a pulse train that is read as the number of increments traversed on the code disc. The output pulses are counted with an external digital counter that permits reading speed and position.

The absolute encoder discussed here is the type mostly used in robotics. Its great advantage is that its output always represents actual shaft position, even if the power fails and comes back on, or if a non-reoccurring erroneous pulse is received.

The light source of the absolute encoder is either a light emitting diode (LED) or an incandescent lamp. The incandescent lamp can tolerate higher temperatures than the LED, while the latter can operate under higher vibration levels and has a longer average life within its temperature rating. But even the life of an incandescent lamp can be greatly prolonged by slightly decreasing the applied voltage, since its life is inversely proportional to the 12th power of the voltage.

Usually, the light is concentrated on its target by first passing it through a collimating lens and then a focusing lens. Since there are a number of concentric rings or tracks on the code disc, each ring with its alternate opaque and clear segments, photosensors and light sources for each track must be provided. Usually a single light source suffices, and the light beams are distributed by a parabolic mirror. The light sensors are phototransistors or photovoltaic cells.

The output from the light sensors is transmitted to the electronics that interpret, analyze, and convert the signals for further use. They also diagnose faulty signals.

The code disc converts each position into a specific code, which is generated by a row of light signals transmitted through the clear segments of the tracks in the code disc. The parabolic mirror in Fig. 8-12 reflects six light beams upward to the six tracks of the code disc. Since only three tracks

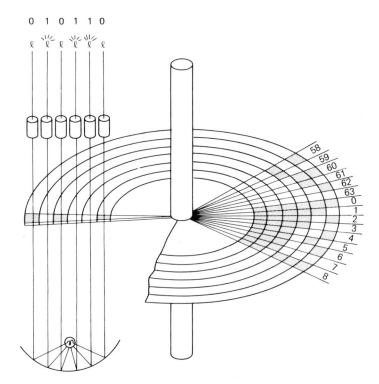

Fig. 8-12. Diagram of code disc function.

have clear segments in the position shown, only they are passing light beams to the photosensors on top. No light means "0," while "1" means that light is passing. Consequently, in this case the signal received is 010110. This is a six-bit binary number.

Code discs are usually made from glass. The opaque segments can be produced by photo emulsion or by metallization. Metallized glass has a harder surface than a photo emulsion, and is more resistant to harsh environments.

Since the encoder can only distinguish between two conditions—light and dark (on and off)—counting requires the binary system, which has only two digits, 0 and 1. This also corresponds to the switching technique of electronic circuits.

The basis of the so-called straight or natural binary code is conversion from digital to binary numbers, shown in Table 8-1. It is based on six digits, but could be continued in the same pattern for any number of digits. Addition of these binary numbers is the same as with digital numbers. Thus,

Table 8–1. Conversion from digital to straight binary numbers.

DIGITAL NUMBER	BINARY NUMBER
0	000000
$2^0 = 1$	000001
$2^1 = 2$	000010
$2^2 = 4$	000100
$2^3 = 8$	001000
$2^4 = 16$	010000
$2^5 = 32$	100000

adding $2 + 4 + 8 = 14$ in binary numbers gives 001110, and the largest six-bit binary number is 111111, or 63 in the decimal system. Each of these binary numbers is called a word. A code disc with decimal 0 to decimal 63 thus has 64 words.

The preceding discussion, as stated, refers to a straight binary code. There are numerous other codes, and the one chosen depends on the particular conditions of the case. For optical encoders, the so-called gray code is usually preferred. This requires a relatively simple translation in the electronics from gray code to straight code, but avoids ambiguities in the response and facilitates diagnostics, as will be shown.

Table 8–2 compares the straight with the gray code. In the gray binary

Table 8–2. Comparison of digital, straight binary, and gray binary numbers.

DIGITAL	STRAIGHT BINARY	GRAY BINARY
0	000000	000000
1	000001	000001
2	000010	000011
3	000011	000010
4	000100	000110
5	000101	000111
6	000110	000101
7	000111	000100
8	001000	001100
9	001001	001101
63	111111	100000
0	000000	000000

code, only one bit changes from one word to the next. For example, changing from decimal 7 to decimal 8 requires one bit to change from 0 to 1 in the gray code. In the straight binary code, however, three bits change from 1 to 0, and one changes from 0 to 1.

The fact that with the gray code only one bit changes from one word to the next makes it easy to discover a change of more than one bit as the code disc moves from word to word. If this were to happen, it would mean a spurious signal or a malfunction, and could be shown as such on the screen of the control system or in some other fashion.

The other reason for preferring the gray binary code is the problem of ambiguity. Suppose a code disc with a straight binary code is aligned for the digital 63. This means that all photosensors receive light, and the binary number is 111111. As the disc proceeds to its next position, which is the initial digital 0, all lights become blocked. The binary number becomes 000000. Now, in between there is a transition where each photosensor receives 50% of the light. In fact, since perfect alignment and perfectly equal detector sensitivity do not exist, one photosensor will receive somewhat more light than the other. Reading a word under these conditions can become very ambiguous and can produce errors.

A similar situation exists every time that the next word produces a change of more than one bit. The only way to get clear, unambiguous readings is to avoid having more than one bit to change when the disc proceeds from one word to the next. This is exactly what the gray code does.

Robotic encoders actually have more than six tracks or bits. Thus, the absolute position encoder, model R-2533m, made by Renco Corporation is available with up to 12 bits in either gray or straight binary code. One has a choice of the following:

$$8 \text{ bits } = 256 \text{ words per turn}$$
$$9 \text{ bits } = 512 \text{ words per turn}$$
$$10 \text{ bits } = 1024 \text{ words per turn}$$
$$11 \text{ bits } = 2048 \text{ words per turn}$$
$$12 \text{ bits } = 4096 \text{ words per turn}$$

The maximum diameter of this encoder is 2.65 in. Since, for the 12-bit code disc, there are 4096 words, this means that the resolution would be 0.088° or 5.27 minutes of arc. One could go further than this. With a 16-bit code disc, for example, there would be 65,536 words, and the resolution would be 0.005° or 0.33 minute of arc.

Renco's R-2553 encoder provides a "whole word" output that assures an exact reading of shaft position in those applications where power to the encoder or robot has been interrupted and then reinstated. The operating temperature of the encoder is 32 to 122°F. The accuracy is ± ½ bit.

REFERENCES

1. *DC Motors—Speed Controls—Servo Systems,* 5th ed. Hopkins, MN: Electro-Craft Corporation, 1980.
2. Gormley, G. J. Conductive plastic film precision potentiometers. *Electronic Engineering Times,* March 26, 1984, pp. 65, 66, 68, 73, and 95.
3. Todd, C. D. *The Potentiometer Handbook.* New York: McGraw-Hill Book Co., 1975.
4. Circuitry for angle measurements. *NASA Tech Briefs,* Summer 1983, p. 383.
5. Morris, H. M. Robotic control systems need accurate positional feedback inputs. *Control Engineering,* January, 1984, pp. 90–93.

Chapter 9
ELECTRONIC HARDWARE

The electronic hardware of robotic technology consists of circuits necessary to control the actions of the robot. This refers specifically to power supply, control of servosystems, treatment of feedback signals, and storage and execution of a program according to which the robot is made to respond.

To a large extent, these controls use digital concepts. This implies that electronic switching circuits are used that operate with discrete increments or decrements. Basic components for this purpose know only two conditions: "on" and "off." This does not always mean a complete cutting off of the current in the "off" state, but it does mean a significant decrease of it. The two conditions or two discrete levels correspond to the 1 and 0 of binary systems, which were discussed under "Encoders" in Chapter 8.

Sometimes, as in stepper motors, even the actuators are digital. However, other actuators and feedback devices, such as servovalves, tachometers, and so on, are analog devices. Their action is either an analog function of the digital input signal, or they produce a signal that is an analog function of the measured variable. This requires digital to analog or analog to digital conversion.

The following discussion deals with the basic elements and logic of digital switching required for these purposes in robotic control systems.

BITS, WORDS, BYTES, AND CLOCKS

As stated, robotic control systems consist largely of on–off (i.e., binary) switches. Myriads of them may be involved, but still they are nothing but microscopic binary switches. They are put together into logic networks that, in the end, control servomotors and so forth, and execute programs.

Only the miniaturization of electronics made such an agglomeration of switching components practical. Without this technology, the robot would never have left the realm of fantasy and become a practical universal tool.

It is based on the particular characteristics of semiconductors, as will be described in this chapter.

Bit means binary digit. It is either a "one" (also written logic 1 or binary 1) or a "zero" (also written logic 0 or binary 0). Instead of "one" and "zero," we may also refer to "yes" and "no," or "high" and "low," or "on" and "off," and so on. A combination of bits in parallel, such as 0110, is called a word.

There are both serial operation and parallel operation. In serial operation, information is transmitted such that the bits are handled sequentially, rather than simultaneously as they are in parallel operation. Serial operation is slower than parallel operation, but it requires less complex circuitry.

In parallel operation, a single wire can transmit only one bit at a time with $2^1 = 2$ different words. Four wires can transmit $2^4 = 16$ different words. These words represent all possible combinations of 0 and 1 in groups of four, such as 0000, 0001, 0010, 0011 . . . 1111. With 8 wires, there are $2^8 = 256$ different words. Thus, the number of words that can be transmitted over a cable with n number of wires is equal to 2^n. Each wire represents a bit of either 0 or 1.

A byte is a word consisting of a specific number of bits treated as a single entity and processed together in parallel. Usually, a byte is considered to consist of eight bits. Sometimes, long words are divided into a high byte and a low byte. This makes it possible, for example, to transmit a 16-bit word over an 8-bit bus, which has only eight wires. In such cases, low bytes are stored at even-numbered memory locations, and high bytes are stored in odd-numbered memory locations. The information is then transferred into a register of 16 flip-flops for processing.

The myriad of switches mentioned above that manipulate the bits, can only be controlled if they are synchronized to operate simultaneously in fixed time intervals. These intervals are extremely short, generally less than 10^{-6} second. The time reference is established by a digital master clock, usually a crystal-controlled oscillator, that enables or disables the switches to proceed in synchronized steps. Between clock signals, the switches will have had time to assume their new states, and transients will have passed.

CIRCUIT ELEMENTS

Semiconductors

Semiconductors as discussed here, are crystals, such as silicon, germanium, and so on. They have an electrical conductivity that lies about midway between that of metals and that of insulators. The flow of electricity through these semiconductors can be controlled by a number of specific methods that cannot be applied with full conductors, such as copper.

Electricity is a flow of electrons. But electricity flows from " + " to " − " while electrons flow in the opposite direction. The reason for this apparent contradiction is that the naming of polarities was a matter of arbitrary convention long before the relationship between current and electrons was realized.

The flow of electrons through a semiconductor is different from the flow through a copper wire. In the latter case, the electrons that flow into one side of the wire come out on the other. With semiconductors, the electrons that are put into the material do *not* necessarily come out at the other end.

There are two types of semiconductors: the p-type and the n-type. A p-type semiconductor has a deficiency of electrons. This is brought about by doping, which means that certain impurities, called acceptors, are added to the crystal. Each atom of the acceptor has the property of accepting an electron from a neighboring atom and thus creating a hole in the lattice structure of the crystal. Under the influence of an electric field, other electrons will try to fill the hole, thereby creating another hole. Thus, the flow of electricity through a p-type semiconductor is often referred to as a flow of holes rather than electrons.

Adding another sort of impurities (donors) to such crystals creates an n-type semiconductor, which has an excess of electrons. If a p-type and an n-type semiconductor are joined back to back as in Fig. 9-1, current can be made to flow from the p-side to the n-side, since the excess electrons from the n-side tend to fill the holes in the p-side. Very little, in fact a negligible current flow would pass in the opposite direction.

This phenomenon is used in the diode rectifier. It permits flow in one direction, but essentially blocks it in the other. In the blocked condition, very little flow passes. However, it does increase slightly as the reverse voltage increases. Finally, a breakdown voltage is reached, at which the blocking action of the diode completely disappears. Thus, the diode as rectifier

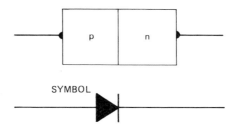

Fig. 9-1. Rectifier diode.

has to operate considerably below the breakdown voltage to be effective. On the other hand, the breakdown characteristic permits another use of diodes.

Breakdown produces what is called an avalanche action, where suddenly the blocking action of the diode breaks down, and an avalanche of electrons supplied by the external voltage penetrates the barrier. Diodes used for this purpose are specifically tailored to such application and are called avalanche, breakdown, or Zener diodes.

In the curve of Fig. 9–2, rectifier action and avalanche action are combined. In the lower left quadrant, the reverse voltage increases with little current passing up to the breakdown point, where almost instantaneously the blocking action breaks down. Zener diodes can be designed for a wide range of breakdown voltages.

In Fig. 9–3, a Zener diode is connected across a load that is to be protected from excessive voltage. At a predetermined point, which lies slightly below the excessive voltage, the breakdown voltage will be reached, and the Zener diode will bypass the current, reducing the voltage to the desired limit. Many variations of this concept are possible and are being used.

Another diode variation is used in optoelectronic diodes, which are either triggered by light or emit light. The first type is called a photodiode, the other a light-emitting diode, or LED for short. In the case of the photodiode, light is directed to the junction between the p- and n-regions. This greatly increases the reverse leakage current. The consequence is that the

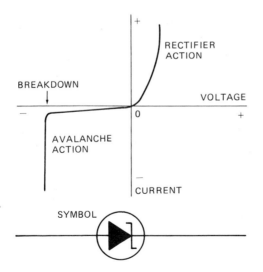

Fig. 9–2. Avalanche or Zener diode.

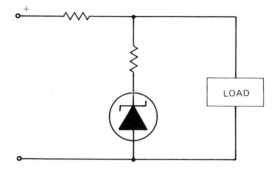

Fig. 9-3. Zener diode as voltage protector.

conductivity of the diode becomes a function of the intensity of the impinging light.

LEDs have a p-n junction like all other diodes. Here, the design is such that as the electrons travel from one side to the other, they give up energy in the form of electromagnetic radiation which is converted into light.

If a third element is added to a p-n structure, either another n-type or another p-type of semiconductor, and they are joined together as in Fig. 9-4, a bipolar transistor results. The center region is the base (*B*), and the two outer regions are the emitter (*E*) and the collector (*C*), respectively. In the pnp transistor, the negative region is in the center, with the emitter connected to the positive side of the power supply and the collector to the negative side. Under these conditions, the control current, which controls the action of the transistor, flows from the emitter to the base. The working current would flow from the emitter to the collector.

In the npn transistor, the positive region is in the center, and the polarities

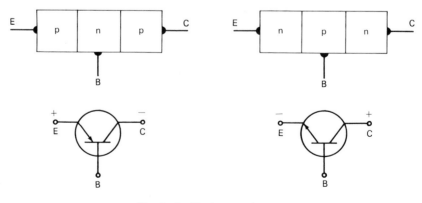

Fig. 9-4. Bipolar transistors.

between the emitter and the collector are reversed. The control current would flow from the base to the emitter, and the working current from the collector to the emitter. The symbols for both types of transistors indicate the direction of the control current by an arrow.

Either type of transistor, npn or pnp, has two pn junctions. The base is kept extremely thin. Sometimes it is only a 20-millionth of an inch thick, while a human hair is at least 100 times thicker. Furthermore, the base is only slightly doped. In the case of an npn transistor, this means the number of holes in the positive base is kept to a minimum. Also, the collector is less doped than the emitter.

In Fig. 9-5, a small control current flows through the base into the emitter. Because of the narrow base and the few holes in it, only a few electrons pass from the emitter into the base; most of them flow right into the collector. The result is that the collector current is much larger than the base current. The collector current, I_c, is the working current that actuates the next element. As long as no base or control current, I_b, is flowing, the transistor is off. When control current flows, it is on. Thus, amplifier action is produced. The smaller control current triggers the larger working current.

Obviously, a circuit such as that in Fig. 9-5 requires two different voltage sources: V_{be} and V_{ce}. In a typical case, the voltage V_{ce} is 8 V, and the base current I_b is 0.7 mA. For a widely used transistor, the 2N706, this would produce a collector current of about 22 mA. In other words, the amplification factor or current gain would be 31 times the control current. One could also increase the base current to 1.4 mA, in which case the collector current would jump to 50 mA, and the gain to about 36. Furthermore, since the base current would thus be increased by 0.7 mA, resulting in an increase of collector current of 28 mA, the gain in this respect would be 40.

The npn transistor is much more widely used than the pnp transistor, but the operation of the pnp is very similar to the npn, as is illustrated in Fig. 9-6, the difference being that now I_e becomes the control current flowing into the base. Increasing I_e by a small amount will increase I_c by a much larger amount, thus producing the desired amplification.

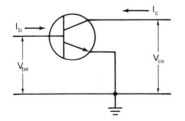

Fig. 9-5. Circuit with npn transistor.

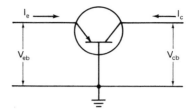

Fig. 9-6. Circuit with pnp transistor.

The bipolar transistor may well be used as a linear amplifier, which means that it amplifies an input signal at a more or less proportional ratio. However, for the purposes considered here, it is a switch that responds to a relatively minute signal. It is part of a binary system that involves certain logic elements, some of the more essential of which are discussed below.

Gates

The various theories, methods, and systems that deal with binary switching systems are called logic. This involves Boolean algebra, truth tables, and so forth. Certain basic concepts, however, suffice to interpret the inherent action of many such circuits.

The two basic logic operations are AND and OR. They may be compared to conventional mathematical concepts such as addition and multiplication. Any of the corresponding circuits is called a gate. The gate can be represented either as a schematic or as a logic diagram, as will be shown. There are other gates derived from these basic two, such as the NAND, NOR, and exclusive OR. The NAND and NOR gates involve an inverter.

The basic devices that perform arithmetic and logic operations are built from a multitude of gates. For example, to switch the output of a register from on to off, and vice versa, by means of two independent signals requires two gates in a common circuit, called a flip-flop.

All such gates are described here. No attempt is made to present a complete description of the vast variety of such elements; the purpose is merely to discuss some basic concepts on which logic elements are built.

The AND gate has several inputs. An output will be produced only when all of these inputs are "on," that is, on binary 1. This is equivalent to an electrical conductor with several switches in series. Only when all switches are closed will current flow. In an electronic circuit, an arrangement such as the one in Fig. 9-7 may be used. The schematic shows three input connections, *A, B,* and *C,* and one output connection, *D.* As long as all inputs are on logic 0, no current flows from the collectors to the emitters of tran-

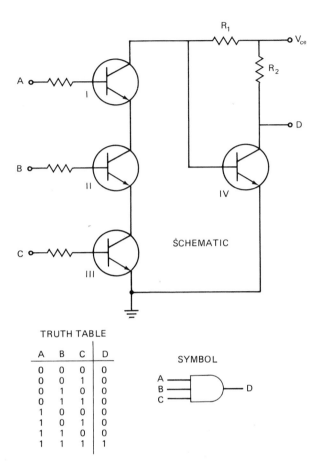

TRUTH TABLE

A	B	C	D
0	0	0	0
0	0	1	0
0	1	0	0
0	1	1	0
1	0	0	0
1	0	1	0
1	1	0	0
1	1	1	1

Fig. 9-7. AND gate.

sistors I, II, and III. However, a control current can flow from V_{ce} (the voltage applied across the collector and emitter) over resistor R_1 through the base and emitter of transistor IV to ground. This activates transistor IV, making current flow through resistor R_2 and the collector and emitter of IV to ground. Disregarding voltage drop in the transistor, output voltage will be at logic 0. (Even though this voltage may not be entirely zero, any level can be logic 0, as long as it is significantly less than logic 1.)

If input signals A, B, and C are binary 1, current will flow from V_{ce} through resistor R_1 and transistors I, II, and III to ground. This short-circuits the path to the base of IV, and flow through that transistor stops. The result is that voltage at D assumes the level of V_{ce}, which means that the output is logic 1. Note that all three inputs must be logic 1 to make the output logic 1. If any one remains logic 0, transistor IV remains activated, and output *D* stays logic 0.

Instead of the more elaborate details of the schematic, the AND gate may also be expressed by the generic symbol given in Fig. 9–7. On the flat end are the various inputs—from two to any number—and on the rounded end is the output.

A third way of expressing the action of a gate is the "truth table." For each combination of inputs the corresponding output is tabulated in the way shown.

Figure 9–8 shows some of the more frequently used gates, their functions,

FUNCTION	SYMBOL	TRUTH TABLE		
		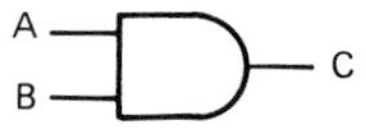 A	B	C
AND	A ⟶ ⟩ ⟶ C B ⟶	0	0	0
		0	1	0
		1	0	0
		1	1	1
		 A	B	C
OR	A ⟶ ⟩ ⟶ C B ⟶	0	0	0
		0	1	1
		1	0	1
		1	1	1
		 A	B	C
NAND	A ⟶ ⟩o⟶ C B ⟶	0	0	1
		0	1	1
		1	0	1
		1	1	0
		 A	B	C
NOR	A ⟶ ⟩o⟶ C B ⟶	0	0	1
		0	1	0
		1	0	0
		1	1	0
		 A	B	C
EXCLUSIVE-OR	A ⟶ ⟩⟩ ⟶ C B ⟶	0	0	0
		0	1	1
		1	0	1
		1	1	0

Fig. 9–8. Basic gate configurations.

symbols, and truth tables. Here, all truth tables are limited to two inputs, since this is sufficient to recognize the basic actions of these gates.

The OR gate requires that if any one of a number of inputs applies a signal, an output results. As shown in the truth table, even when more than one input is logic 1, the output will still be logic 1. The only way to make the output logic 0, is to make *all* inputs logic 0. The OR gate is the equivalent of an electrical circuit with all the switches in parallel.

The NAND gate simply produces the reverse of the AND gate. The output always shows logic 1, except when *all* inputs are logic 1. In this case it turns logic 0.

The NOR gate, in the same way, is the reverse of the OR gate, as can be seen by comparing the two truth tables.

The exclusive-OR gate is another modification of the OR gate. Its action is the same as the OR, with the exception of the situation where all inputs are logic 1. In this case, the output goes to logic 0, instead of logic 1 as is the case with the OR gate.

NAND and NOR gates are actually AND and OR gates, respectively, combined with an inverter. An inverter, such as the RTL inverter in Fig. 9–9, may consist of one npn transistor and two resistors. RTL means resistor-transistor-logic; that is, a logic circuit built from resistor and bipolar transistor elements.

With logic 0 input, no current will flow from the base to the emitter. Consequently, there is no current flow from the collector to the emitter either. Under this condition, the output voltage equals the supply voltage of 3.5 V, which, in this case, is equivalent to logic 1.

On the other hand, if the input is at logic 1, which may be, say, 3 V, then current flows from the collector to the emitter, dropping the voltage in the 640-ohm resistor, and the output will be logic 0. Thus, the input/output relation of the inverter reverses a 0-input into a 1-output, and a 1-input into a 0-output.

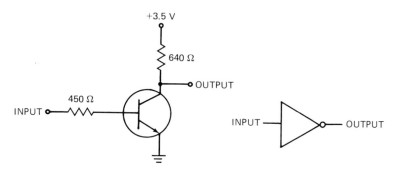

Fig. 9–9. RTL inverter.

Flip-flops

An element that is used extensively in electronics is the flip-flop. The name implies that it can be at either one of two stable states. The outputs are Q and its inverse $\overline{Q}$ in one stable state, and $\overline{Q}$ and Q in the other. Therefore, the flip-flop can be used as a device that stores one bit of information. Flip-flops can form storage registers, counters, shift registers, and other functional elements.

The flip-flop is controlled by two independent input signals. At any time, there is at least one of the two input signals on logic 1.

A simple flip-flop may be made by cross-connecting two NAND gates. Figure 9-10 labels one of the two inputs S, for set, and the other R, for reset. This arrangement, therefore, is frequently called an RS flip-flop.

One condition of the RS flip-flop is that the situation where $R = S = 0$ does not occur. This means either the R- or the S-input has to be on logic 1, as already mentioned. If the R-input is on logic 1, it has to stay on logic 1 until the S-input also becomes logic 1. Only then can it be made logic 0 without affecting the function of the device.

According to the truth table for NAND gates in Fig. 9-8, the conditions of Table 9-1 apply for Fig. 9-10. Considering the case where $S = 1$ and $R = 0$, it follows that this must result in $S = 1$, $R = 0$, $Q = 0$, and $\overline{Q} = 1$.

If input R goes on logic 1, there is no change, since the inputs to NAND II are now $R = 1$ and $Q = 0$, which leaves $\overline{Q}$ at logic 1 as before. However, as soon as input S goes to logic 0, while R remains on logic 1, the situation changes.

The input signals to NAND I are now $S = 0$ and $\overline{Q} = 1$. This makes $Q = 1$. Consequently, the inputs to NAND II are now $R = 1$ and $Q = 1$, which makes $\overline{Q} = 0$. The fact that in this transaction $\overline{Q}$ has changed from $\overline{Q} = 1$ to $\overline{Q} = 0$ does not change the output from NAND. The condition is now $S = 0$, $R = 1$, $Q = 1$, and $\overline{Q} = 0$. The flip-flop is complete until again S becomes logic 1, and R returns to logic 0. Either Q or $\overline{Q}$, or both, can be used as an output signal.

When entire circuits are shown, functional groups of gates, such as the

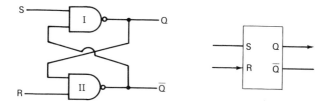

Fig. 9-10. Flip-flop.

Table 9–1. Truth Table of RS Flip-Flop

NAND I			NAND II		
S	$\overline{Q}$	Q	R	Q	$\overline{Q}$
0	0	1	0	0	1
0	1	1	0	1	1
1	0	1	1	0	1
1	1	0	1	1	0

flip-flop, may be shown in block-diagram form together with the condition in which the entire circuit is initially assumed. Arrowheads are used for the latter purpose. The block diagram in Fig. 9–10, for example, shows the flip-flop in the condition where $S = 0$, $R = 1$, $Q = 1$, and $\overline{Q} = 0$.

FETS and MOSFETS

MOSFET is short for *metal-oxide-semiconductor field-effect transistor*. It started out as the FET, the *field-effect transistor*, which operates by using an electric field to control current flow in a semiconductor channel.

In Fig. 9–11, three basic structures are compared: bipolar transistors,

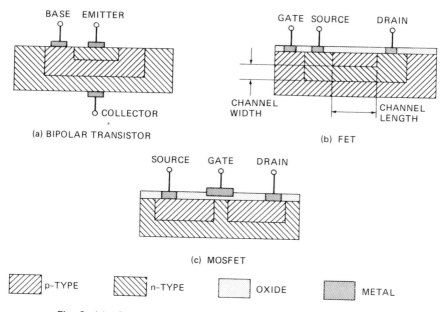

Fig. 9–11. Structures of bipolar transistors, FETs, and MOSFETs.

FETs, and MOSFETs. All three transistor types operate on the basis of silicon semiconductors, which are doped either for an excess of holes (p-type) or an excess of electrons (n-type). In the bipolar transistor, the same working current passes through p- and n-type material in succession; hence, it is called bipolar. With FETs (and hence with MOSFETs), the working current flows only through either p- or n-type material; this makes them unipolar transistors. While the three connnections in bipolar transistors are referred to as base, emitter, and collector, they become gate, source, and drain in the unipolar FETs and MOSFETs. Putting an electrical voltage on the gate terminal controls current flowing between the source and the drain. Note that the signal *current* applied to the base of the bipolar transistor becomes a signal *voltage* applied to the gate in the FET and MOSFET.

FETs that are controlled by negative gate voltage are n-channel FETs; those that are controlled by positive gate voltage are p-channel FETs. The magnitude of the input voltage, either negative or positive, controls the changes of the current flow passing from the source to the drain. The resistance between the source and the drain is determined by the channel width and length of the n-type material, as shown in Fig. 9–11b.

MOSFETs are FETs of a particular structure. The main difference is that the metal of the gate connection in the MOSFET is not in direct contact with the semiconductor, but is separated by a thin layer of silicon oxide. Oxides are good insulators. There is thus a sequence of layers at the gate: metal, oxide, and silicon (which led to the abbreviation MOS). The oxide layer is extremely thin. The L²FET, for example, which RCA developed, has an oxide layer of a 2-millionth of an inch. Using the same comparison as before, a human hair is at least 1000 times thicker.

The action of the MOSFET can be described as follows: The metal area of the gate connection with the n-type substrate in Fig. 9-11c, and the insulating oxide layer in between form a capacitor. If a negative potential is applied to the gate, a corresponding positive charge is produced in the n-type substrate. At a certain magnitude of the negative potential, the positive charge induced in the channel is increased to a point where it causes the n-type material beneath the oxide temporarily to assume the characteristics of p-type material. This transformation causes current to flow between the source and the drain.

PMOS and NMOS differ in the way the current flow through the channel is produced. With PMOS devices, n-type source and drain regions are diffused into a p-type substrate to create an n-channel for conduction; that is, current is produced by holes in the silicon semiconductor. With NMOS devices, current is produced by excess electrons that flow between the n-type drain and source terminals.

There are also depletion-mode and enhancement-mode MOSFETs. With the depletion-mode MOSFET, a positive gate voltage decreases channel resistance, and a negative gate voltage increases channel resistance. With the enhancement-mode MOSFET, which is the one shown in Fig. 9–11c, a positive gate voltage decreases channel resistance, but negative gate voltages have no effect. It is the enhancement mode that is used in most applications.

Comparison of MOSFETs with bipolar transistors[1] shows that in general MOSFETs:

1. Operate at much higher speeds than bipolar transistors.
2. Have a positive temperature coefficient, which means that their resistance increases with temperature, so that the MOSFET is inherently stable in response to temperature changes and protected against thermal run-away.
3. Can be operated in parallel with other MOSFETs without "current hogging"; that is, if any device overheats, its resistance increases, and the current is rerouted to the cooler chips.

Conventional MOSFETs require a gate voltage of 10 V or more for maximum output current. That is more than can be obtained directly from CMOS logic, which is discussed below, or from logic circuits that are constructed with bipolar transistors, the so-called TTLs (*t*ransistor-*t*ransistor-*l*ogic). To operate with CMOS or TTL, special MOSFETs are required, such as the above-mentioned L²FETs from RCA, which are capable of interfacing directly with 5-V TTL or CMOS logic because the thin gate-oxide layer allows the device to switch at just one-half the gate voltage required by conventional MOSFETs.

There are two different symbols for MOSFETS in general use. One is for NMOS devices, the other for PMOS devices. Both are shown in Fig. 9–12, where the drain is D, the source is S, and the gate is G. The center tap with the arrow is the only difference between the two symbols. It refers to the active flow of carriers—electrons or holes—in the substrate.

Symbols for MOSFETs are not uniform, however. Figure 9–13 shows two different methods for comparison. Both diagrams depict the same inverter circuit. MOSFET I is of very high resistance, say, 50,000 ohms. A

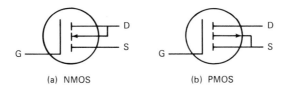

(a) NMOS   (b) PMOS

Fig. 9–12. Symbols for MOSFETs.

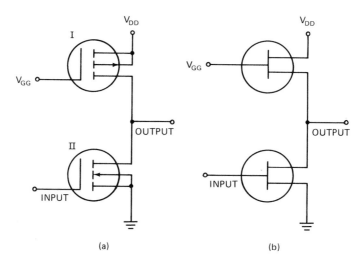

Fig. 9–13. Alternate methods to show the same inverter.

fixed negative voltage is applied at V_{GG}. Another fixed voltage, which is slightly less negative than V_{GG}, is applied at V_{DD}. A typical value for V_{DD} is -30 V, which suffices to keep MOSFET I active. However, as long as the input signal is zero, no current will flow through MOSFET II, and, hence, none through MOSFET I. Consequently, the output will be at -30 V, corresponding to a logic 1. If a negative voltage is applied at the input, current will pass through MOSFET I, and output voltage will go to logic 0. All this is quite similar to the RTL inverter of Fig. 9–9, except that a MOSFET is being used instead of a resistor. This makes the package smaller. It is also easier to insert a transistor than a resistor in an integrated circuit. All the other logic gates that were discussed above may also be constructed with MOSFETs as well as with other transistors.

A great advantage of MOS logic is its high noise immunity, due to the rather large voltage steps from binary 0 to binary 1, which make for inherent filtering out of extraneous noise. As a result, nonregulated power supplies can be used rather than the more expensive regulated supplies.

The most important criterion for the performance of MOSFETs is probably the resistance, R_{DS}, between the drain and the source. It may well be considered a criterion for the performance of different MOSFETs. Obviously, the current-handling capabilities are increased by a lower R_{DS}, since the heating effect of the current passing through the MOSFET is reduced.

Modern production techniques have led to modifications that have reduced the R_{DS} to about 5% of that of traditional MOSFETs. The General Electric devices that accomplished this are called insulated-gate transistors

(IGTs); the RCA devices are conductivity-modulated FETs (COMFETs); and the Motorola versions are gain-enhanced MOSFETs (GEMFETs). Ben Carlisle[2] uses the generic term "modified FETs" for all these devices.

The primary difference between MOSFETs and modified FETs is that the substrate of the latter is composed of p-type rather than n-type silicon. Modified FETs are usually slower than conventional MOSFETs. This is a serious drawback of modified FETs in general, and particularly in their use for control of servomotors by pulse-width modulation, and it is the reason why bipolar transistors or MOSFETs are more frequently used for this purpose.

HMOS transistors are n-channel MOS that permit speeds as short as those obtained from costly TTL circuits. The HMOS process also produces denser circuitry. The entire 16-bit data and microprogrammed control structure of the typical microprocessor uses 29,000 transistors integrated on a chip 0.225 × 0.225 in.

Intel Corp. developed the CHMOS process, which improves the speed–power product of its earlier HMOS II process—which was already a high-speed variant of the n-channel MOS by a factor of 10. Here, the improvement comes entirely from power reduction. CHMOS transistors operate at HMOS speeds but at CMOS power levels.

Integrated Circuits

Integrated circuits (ICs) combine a number of gates on and in a single crystal of semiconductor material (a die or a chip) by etching, doping, diffusion, and other techniques. The components are connected to each other and are capable of performing complete circuit functions. If there are fewer than 100 gates combined on a chip, one usually refers to medium-scale integration. Large-scale integration (LSI) refers to 100 to 1000 gates and very-large-scale integration (VLSI) to 1,000 to 80,000 gates. The latter may embody a whole microprocessor. Anything beyond this falls into the realm of ultra-large-scale integration (ULSI) and even covers entire microcomputers on a single chip.

The chips may not be larger than a square of about 0.1 × 0.1 in. They require packaging, primarily because outside connections must be provided. In the case of a VLSI device, 64 input/output connections may be needed. There are various means of packaging. A common form for industrial usage is the "plastic dual-in-line package," or plastic DIP, which is a molded plastic container with two rows of closely spaced connecting leads, which give it the name "dual-in-line." DIPs are sometimes called "bugs."

Production of an IC begins with growing crystals, either by pulling a tiny silicon seed through molten silicon or by placing it at the bottom of the melt under very closely controlled conditions. A large crystal will grow around the seed and will have a molecular structure that is exactly the same as that of the seed, but will be large enough to weigh as much as 30 lb. The finished crystal is cut into wafers that are usually 0.01 in. thick and up to 5 in. in diameter. One wafer can readily hold 60 identical ICs, which are obtained when the wafer with all circuitry finished is eventually cut into that many pieces or chips.

The above-mentioned HMOS circuit fits on a chip that was 0.225×0.225 in. Thus some 380 identical chips would fit on the face of one wafer. The ICs are reproduced by ultraviolet light imprinting the 380 (or whatever the number may be) circuit patterns side by side on the wafer.

"Monolithic" is a term that pertains to an IC that has been built on a single chip. A circuit that can be produced as a monolithic chip is usually cheaper and more reliable than that obtainable by any other method of fabrication.

CMOS

The complementary metal oxide semiconductor, or CMOS, is an arrangement that uses both PMOS and NMOS devices on the same silicon substrate. In the basic concept, only one of the two is turned on at a time, keeping power dissipation low. Memory systems use CMOS because of the low standby power, wide operating range, and relatively high speed.

Inverters and NAND gates in CMOS technology could be constructed as shown in Fig. 9-14. The inverter consists of an NMOS in series with a PMOS, that is, a single CMOS. When the input is logic 0, the NMOS is off, and the PMOS is logic 1. This puts voltage +V on the "out" terminal, disregarding the voltage drop in the PMOS, which is kept very low. On the other hand, if positive voltage is applied as the input signal, the NMOS is on and the PMOS is off, dropping the output voltage to zero.

The NAND gate uses two CMOS devices. The two NMOS are connected in series, and the two PMOS in parallel. If either input, A or B, goes on logic 1, the corresponding PMOS is turned on, and the series path to ground is interrupted because the NMOS turns off. This makes the output logic 1. When both inputs are logic 1, both PMOS are turned on, and the output is connected to ground, thus making it logic 0. Without any input signal, the output remains logic 1. Comparison with the truth table of Fig. 9-8 will show that the requirements of the NAND gate are thus satisfied.

CMOS transistors are primarily used in ICs. Originally, their usage was limited because they were slow. This has changed. For example, Integrated Device Technology (ITD) developed the "CEMOS" technology. Among other ICs, they offer a RAM storage device (see note on p. 311) which

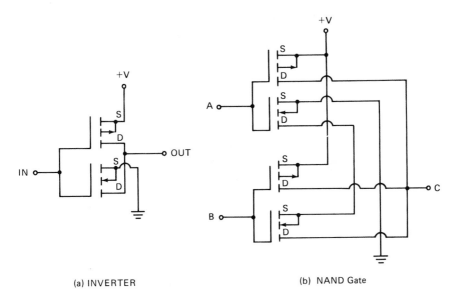

(a) INVERTER   (b) NAND Gate

Fig. 9-14. CMOS gates.

provides immediate access to any storage location point, with the data stored in flip-flops without the need for refreshing (i.e., a static RAM). It has a capacity of 65,536 storage units (64K bytes) using an equal number of CMOS transistors. The time to recover any specific data from this memory device is 0.000000055 sec (55 nsec).

Another example is the G655C02, an eight-bit CMOS microprocessor made by GTE Microcircuits Division. This unit, shown in Fig. 9–15, is capable of controlling multiple processes in remote data acquisition and control systems. Such devices are suitable for robots to read sensor outputs, transmitting data, and controlling servosystems.

Silicon Controlled Rectifiers

The silicon controlled rectifier, or SCR, is a semiconductor that consists of four alternate layers of p- and n-material. This arrangment is shown in Fig. 9–16. Any such pnpn structure falls under the generic name "thyristor." The SCR is only one type of this general category.

Another thyristor, for example, is the Triac, which is essentially two SCRs in parallel. This arrangement enables the Triac to conduct in two directions, while the SCR can only conduct in one direction. Being a unidirectional conductor makes the SCR similar to a rectifying diode. Unlike the diode, however, the SCR starts conducting working current only at the so-called

Fig. 9–15. Eight-bit CMOS microprocessor. (*Courtesy of GTE Microcircuits Division*)

breakover voltage, which can be controlled by a signal applied to its gate terminal. It remains conductant as long as the working current stays above zero or changes in direction.

Besides the gate terminal, there are two other terminals: the anode and the cathode. With a voltage applied across the anode and the cathode, the SCR does not conduct current, provided the voltage does not exceed a certain maximum value. However, by applying a gate control signal, the SCR fires and becomes a rectifier, capable of conducting current but in one direction only. The higher the gate current, the lower the breakover voltage at which the transition occurs.

This characteristic of the SCR leads to many applications. The most important one for robots is to control the speed of dc servomotors while simultaneously converting the ac supply to dc. This application is covered under a separate heading.

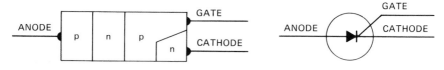

Fig. 9–16. Silicon controlled rectifier (SCR).

RC NETWORKS

Resistance–capacitance networks, or RC networks for short, are frequently used in electronic systems. The capacitor consists of two metal plates or two electrically conducting foils with a di-electric (i.e., insulating) material between them.

If dc is applied to a capacitor, current will flow into it for a short time, producing an excess of electrons on one of the plates and a deficiency on the other. In this condition, the capacitor is charged. Once it is fully charged, it behaves like an insulator, preventing current flow from passing into it. The extent to which a capacitor can absorb and store electric energy is its capacitance.

If a resistor is connected in series with the capacitor, as in Fig. 9-17, and the power is disconnected by the switch as shown, so that the resistor and the capacitor are now connected in a closed circuit, the capacitor discharges its stored energy through the resistor. The direction of current is now reversed.

The larger the resistance or capacitance is, the longer is the time required to charge the capacitor, and equally, to discharge it. The process is expressed by:

$$q = Q\,(1 - e^{-t/RC}) \qquad\qquad (9.1)$$

Here, q is the charge at any given time, and Q is the final charge (in coulombs), t is the time (in seconds), R is the resistance (in ohms), and C is the capacitance (in farads). The result of this equation is the exponential curve in Fig. 9–18, which, in accordance with the equation, approaches the final charge only gradually, (i.e., asymptotically). In case of a discharging capacitator, the characteristic is precisely the reverse, as shown in the illustration.

The time required to reach 63.2% of the final charge is the time constant (see page 17), also called in this instance the RC constant. Expressed in seconds, it is equal to the product of resistance in ohms and capacitance in farads.

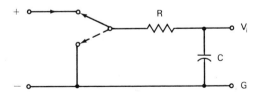

Fig. 9–17. RC network.

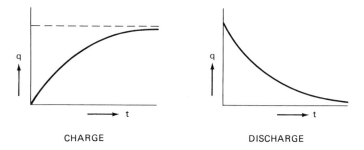

CHARGE   DISCHARGE

Fig. 9-18. Charge and discharge characteristics of capacitor.

Since capacitance in farads is usually a very small number, it is generally expressed in microfarads (μf) or even in picofarads (pf). As an example, if $C = 1\ \mu f = 10^{-6}$ farad and the resistance is 10^6 ohms, then the RC constant is $RC = 10^6 \times 10^{-6} = 1$ sec.

When the switch in Fig. 9-17 is connected to the power supply, voltage at V_i increases gradually. Used this way, the RC network becomes an integrator, where V_i increases as a function of time until it reaches the supply voltage.

The integrator circuit is also the principle of a simple capacitive filter used to smooth out the pulses of rectified ac (cf. Fig. 9-22). The capacitor is charged to its peak value with the positive half cycles of the rectifier. As the ac pulse starts to decrease after an increase of 90°, the capacitor starts to discharge into a load resistance across V_i and G in Fig. 9-17, thus maintaining the load current more or less constant and reducing the ripple of the pulsations.

NETWORK INSULATION AND OTHER PRECAUTIONS

Electrical noise consists of disturbances originating in the supply power due to sudden changes—such as the opening and closing of electromagnetic relays—in the condition of connected machinery and apparatuses, that are not related to the robot system. Such transient phenomena may cause voltage surges of considerable magnitude in the power line and thus produce an electromagnetic field. If an electronic network is nearby, the noise can be transmitted to it by capacitive or inductive coupling and produce spurious signals in the servosystem.

Electrical noise can be counteracted in a number of ways. It is of primary importance that all control wiring of a robotic system be twisted and shielded, and that the shield be grounded at terminals provided by the manufacturer as part of the equipment, and nowhere else.

Noise filters may also be provided, or, to avoid grounding to the common ground wire of the power supply, an isolation transformer may be used. This is a 1:1 transformer designed to prevent any significant conductive or electrostatic coupling between primary and secondary windings.

Another way to minimize noise problems is to use fiber-optic cables, since they are immune from picking up electromagnetic interference. But it is also possible to use an optoelectronic transistor, as shown in the optical coupler of Fig. 9-19. On the input side is a light-emitting diode (LED). The light activates the output transistor which reproduces any changes of input current without electrical connection to the input. The LED and the phototransistor are encapsulated as a single package. However, the proximity of input and output may lead to capacitive coupling and consequent transfer of noise. To avoid this, an RC network is frequently connected between the base and the emitter of the phototransistor. The capacitor provides a path for the transient current, and the resistor provides a discharge path.

There are other techniques of optical coupling, such as LED to a photodiode, or to a light-dependent resistor, or to a Darlington-connected phototransistor. The last refers to a phototransistor whose collector and emitter are connected to the collector and the base, respectively, of a second transistor. The emitter current of the input transistor is amplified by the second transistor. The result is an optical coupler of very high gain.

Incidentally, optical couplers may also be used for other purposes than noise protection, such as enabling a single source to trigger a number of SCRs without their mutual interference.

Other precautions that should be taken with all electronic equipment refer to dust, moisture, and heat. Dust can cause short circuits and prevent the necessary heat dissipation from components, thus causing overheating. Moisture can also reduce the resistance of insulating materials or produce corrosion that in turn causes breakdown. Heat can severely reduce reliability and longevity of electronic components.

All electronic and electric equipment should be mounted in an upright position and so that air can freely circulate around it. It should not be mounted over any heat-producing components, such as transformers, mo-

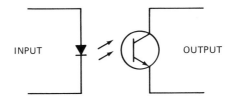

Fig. 9-19. Optical coupler.

tors, and so on. To protect them from dust, any doors to enclosures of electronic and electric equipment should always be kept closed unless there is a specific need to open them.

SCR SPEED CONTROL

Figure 9–20 shows an example of an SCR circuit that drives and controls a dc servomotor from a single-phase ac supply. The gate control circuit is ignored in Fig. 9–20, but shown separately in Fig. 9–21. It would produce a pulse signal every 180° of the sinusoidal wave-shape pattern of the ac source.

Usually, it suffices that the gate signal lasts only a few microseconds to trigger an SCR into the conducting state. Once the SCR is conducting from the anode to the cathode, it continues doing so until the voltage across these two terminals drops to zero or reverses. In Fig. 9–21, an RC network is provided to determine the point in the sinusoidal waveshape of the ac source at which the SCR is made to conduct. The gate control circuit is triggered when voltage V_g becomes positive, thus producing the breakover voltage of the SCR. This occurs as soon as the capacitor is charged and raises V_g to the necessary level. By adjusting the resistor, which would usually be done automatically by the servosystem, the breakover voltage is determined at which the SCR is triggered during the ascending positive slope of the sinusoidal wave.

A single-phase ac supply is also connected to terminals A and B in Fig. 9–20. Since this is alternating current, A will be alternately positive and negative, and B alternately negative and positive.

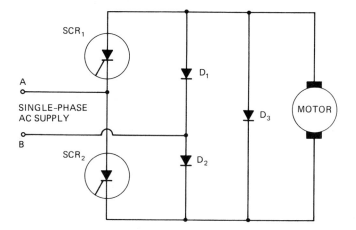

Fig. 9–20. Typical SCR speed control.

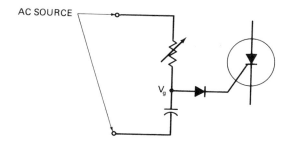

Fig. 9-21. Gate control circuit for SCR.

During the conductive condition of the SCRs, the situation is as follows: When A is positive, current flows through SCR_2, the motor, and diode D_1 and out through B. On the other hand, when B is positive, current flow is $B-D_2-motor-SCR_1-A$. Thus, the circuit utilizes each half-cycle without changing the direction of the current. It is *pulsating* dc, and since there are two pulses every 360°, the pulse frequency is 120 Hz for a 60 Hz supply. It can be rather easily smoothed out, although some ripple will usually persist.

Diode D_3 is a so-called free-wheeling diode. It is provided to dissipate between pulses the power stored in the inductance of the motor, which otherwise could keep the SCR on.

Using different current levels for the gate signals triggers the SCRs at various breakover voltages, as described above. This permits control of the part of the waveshape during which current flows into the load, which is indicated by the shaded areas in Fig. 9-22. The so-called conduction angle that corresponds to the breakover voltage is indicated by α. During the first 180° the current when triggered flows through SCR_2; during the second 180°, it flows through SCR_1. All of the current passes through the servomotor.

There are three different breakover voltages in Fig. 9-22. All are controlled by a gate control signal with three different conduction angles. In case (a), the conduction angle is $\alpha = 0$. Consequently, the average current, as indicated by the dashed line, is at its highest. In case (b), the conduction angle is $\alpha = 60°$, and in case (c), it is $\alpha = 80°$. The average current output decreases as the conduction angle advances. In each case, the gate control signal, once set, will repeat after every 180°. Thus, for case (b), it will repeat at 240° and then again at 60°.

The result is that such an SCR rectifies the ac supply, and by controlling the conduction angle regulates the motor speed.

The most serious disadvantage of SCRs is probably the relatively high form factor (p. 241). Another disadvantage is the rather high time constant

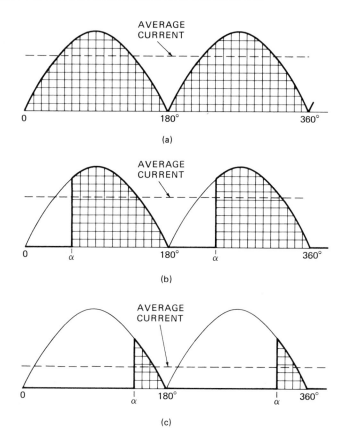

Fig. 9-22. Conductive modes of SCR circuit in Fig. 9-20.

due to the ripple of the rectified power. Often this is considered too high for the high dynamic requirements of robotic servos.

The SCR type speed control costs considerably less than the often preferred PWM speed controls. This advantage, however, may be offset by higher system costs, at least in applications above 2 hp. An even more important factor favoring SCRs is that they have greater peak current capabilities than the PWM speed controls, which are described below.

PWM SPEED CONTROL

A rectified ac current can be applied to a pulse generator that is able to produce a pulsating current of fixed frequency. The longer the pulses are, the greater is the average value of the pulses, as seen in Fig. 9-23. The

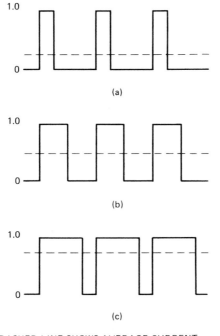

(a)

(b)

(c)

DASHED LINE SHOWS AVERAGE CURRENT

Fig. 9–23. Pulse width modulation.

amplitude remains unchanged, but the ratio of on-time to off-time varies. The method is called pulse width modulation, or PWM for short, and is widely used in robotic servosystems. Pulse frequencies are often 20 kHz or higher to exceed the upper limit of audible frequencies; otherwise audible noise of PWM systems can become pronounced. The ASR Servotron, for example, operates at 25 kHz, and the BLM 70/6 servodrive from Performance Controls, at 20 kHz.

Figure 9–24 shows a pulse generator suitable for pulse width modulation.[3] Two npn transistors, Q_1 and Q_2, are used as amplifiers. When Q_1 conducts, it produces a voltage drop across R_1. This makes the collector voltage V_c low. V_c is coupled to the base of Q_2 through capacitor C_1, and since it is low, Q_2 is not conducting. Because of the R_2C_1 network, C_1 will charge gradually. As it does, voltage V_b increases to a point where Q_2 starts conducting. At the same time, Q_1 stops conducting because of the low voltage coupled through C_2. This sequence would be repeated 20,000 times a second for a 20 kHz pulse frequency. The on- and off-times are determined by the two RC networks and can be adjusted by variable resistors R_2 and

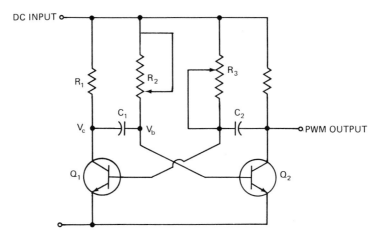

Fig. 9–24. Pulse generator. (*From Wilson, Ref. 3*)

R_3. The two resistors are mechanically coupled so that increasing one resistance automatically decreases the other.

CONTROL OF ACCELERATION AND DECELERATION

There are several ways to control acceleration and deceleration of a servomotor. Perhaps the simplest way is the combination of an integrator and a Zener diode shown in Fig. 9–25. The integrator consists of a resistor R and a capacitor C. This arrangement produces a voltage across the output that changes as an exponential function of time, as described before. The rate of change of the output can be determined by choosing the corresponding values of the resistor and the capacitor.

The Zener diode has a breakdown point to prevent the voltage from exceeding the breakdown voltage for which the Zener diode is rated. In addition, potentiometer P serves to adjust the maximum voltage of the output of the ramp generator.

To start the servomotor, a solid-state switch is turned on, and as the voltage across the potentiometer increases, the motor gradually gathers speed. When the voltage reaches the breakdown point of the Zener diode, the output voltage stabilizes, and the motor has reached its running speed.

To stop the motor, the switch opens, the capacitor unloads its charge, and the motor decelerates in the same exponential fashion as it originally accelerated.

A more elaborate arrangement may be used.[4] Thus, in Fig. 9–26 the input

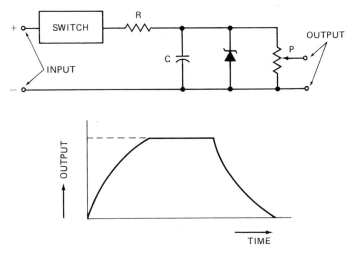

Fig. 9-25. Simple ramp generator.

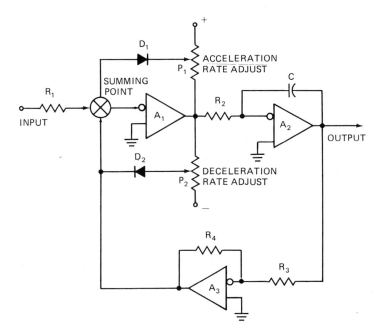

Fig. 9-26. Schematic of ramp generator.

is again dc controlled by an on–off switch, which is not shown in this case. The input signal is applied over the summing point to an inverting amplifier, A_1. The output from A_1 is applied to another inverting amplifier, A_2, and from there to the output of the ramp generator. A third inverting amplifier, A_3, feeds the output back to the summing point. To adjust acceleration and deceleration rates independently, two potentiometers, P_1 and P_2, together with diodes D_1 and D_2, are provided.

The amplifiers used for this circuit are usually 741 operational amplifiers. This type is one of the most widely used of the so-called operational amplifiers or "op amps." The term operational amplifier is somewhat misleading. It is derived from the device's original use for mathematical operations in analog computers. The versatile op amp has long surpassed its early applications.

Op amps use a negative feedback around the high-gain dc amplifier. The arrangement assures a highly accurate gain characteristic. Gains may have values in open loop of 100,000 or more. Of course, with the negative feedback, this gain becomes greatly reduced. In fact, it becomes proportional to the ratio of input resistance (in the case of amplifier A_1, this is R_1) to the resistance of the feedback circuit.

The 741 amplifier has two connections: one for inverting, the other for noninverting. In the case of Fig. 9–26, all three amplifiers are used in the inverting mode. Thus, if the input to one of these amplifiers is logic 1, its output is logic 0, and vice versa.

The diodes, together with their corresponding potentiometers, make A_1 a variable gain amplifier. By adjusting these potentiometers, the acceleration rate and deceleration rate can be independently adjusted.

Consider that the control switch is on, and, hence, the input is logic 1. This may appear to make the output of A_1 equal to logic 0 and the output of A_2 equal to logic 1. However, the latter condition will be delayed, since A_2 is connected as an integrator because of R_2 and C. The result is that initially a ramp output with a positive incline is produced.

The output signal is also applied to amplifier A_3. The result here is that, as long as the acceleration ramp is generated, A_3 reverses the signal and produces a ramp with a negative incline. This negatively tending ramp signal is a feedback signal applied to the summing point. It will gradually reduce the error signal to A_1. The feedback of the ramp signal produces a straight linear output ramp. Any variations of A_2 are immediately corrected by the negative feedback loop through A_3.

When the control switch is turned off, the capacitor discharges into the ramp generator and produces the reverse action, decelerating the servomotor at a rate determined by the adjustment of potentiometer P_2.

DIGITAL-TO-ANALOG AND
ANALOG-TO-DIGITAL CONVERSION

A digital-to-analog converter (DAC), as well as an analog-to-digital converter (ADC), is a frequently encountered component in robotic controls.
Feedback signals, for example, those from tachometers, are often analog,
while the electronic circuitry operates in the digital mode.

Inversely, the output to motors is probably in the analog mode, requiring
a reconversion from digital to analog.

DAC

The simplest form of a DAC is shown in Fig. 9–27. A voltage V_{in} applied
to resistor R_a produces an output voltage $- V_{out} = V_{in} (R_F/R_a)$, where R_F
is the feedback resistor of the inverting amplifier. If the same voltage input,
V_{in}, is applied to all resistors, then the output voltage is given by:

$$V_{in} R_F \left(\frac{1}{R_a} + \frac{1}{R_b} + \frac{1}{R_c} + \frac{1}{R_d} \right) = - V_{out} \qquad (9.2)$$

The four input resistors are chosen such that each has a resistance twice
the value of the preceding one. The digital signal is applied in binary code,
with each resistor corresponding to one bit. Thus, the decimal value 11
becomes binary 1011, which means that resistors R_a, R_c, and R_d are "on,"
while R_b is "off."

Let $R_a = 0.5$, $R_b = 1$, $R_c = 2$, and $R_d = 4$ with all values being in megohms, while the feedback resistor R_F has a resistance of 0.2 megohm. As-

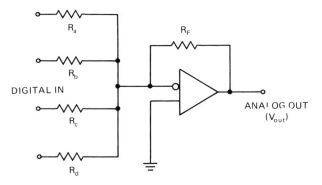

Fig. 9–27. Simple digital-to-analog converter.

suming that V_{in} equals 12 Vdc, then the resulting analog voltage obtained from the DAC is $(12 \times 0.2)(1/0.5 + 1/2 + 1/4) = -6.6$ Vdc. In fact, it can be readily calculated from Eq. 9.2 that by using a straight binary code for the above values, the analog voltages listed in Table 9-2 result.

A four-bit DAC, as used in this example, would be able to convert no more than 16 discrete input values. However, if it were expanded to, say, a 12-bit DAC, then the resolution would rise to $2^{12} = 4096$ discrete input values.

While the above-described principle is used in all DACs, the actual circuits are considerably more complex. They include voltage references, switching transistors, and so forth. Several companies produce monolithic DACs that combine all the circuitry on a single chip. Figure 9-28 shows the diagram of the DAC 710, a monolithic 16-bit DAC from Burr-Brown Corporation, designed specifically for closed-loop servo control in robots, numerical controllers, and other applications.

The chip is enclosed in a 24-pin hermetic DIP that measures overall about 1.25×0.52 in. A combination of current switch design techniques accomplishes monotonicity of 15 bits for ± 2048 around logic 0. This assures that the output remains a continuously increasing function of the input over the entire specification temperature range of 0°C to 70°C.

The D/A output is ± 2 mA. As digital inputs change from one adjacent code to the next, the change in output will not exceed $\pm 0.006\%$ of full range. The output gain is adjustable with an external potentiometer. The pin connection for the potentiometer is provided on the DAC as shown.

Table 9-2. Digital-to-analog conversion.

DIGITAL	BINARY	V_{out}
1	0001	−0.6
2	0010	−1.2
3	0011	−1.8
4	0100	−2.4
5	0101	−3.0
6	0110	−3.6
7	0111	−4.2
8	1000	−4.8
9	1001	−5.4
10	1010	−6.0
11	1011	−6.6
12	1100	−7.2
13	1101	−7.8
14	1110	−8.4
15	1111	−9.0

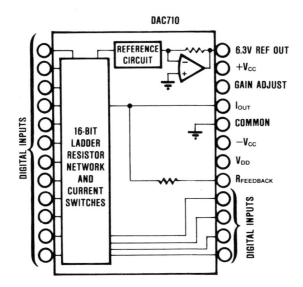

Fig. 9–28. Monolithic digital-to-analog converter. (*Courtesy of Burr-Brown Corp.*)

While the DAC 710 produces a current output, there is also a unit, the DAC 711, that converts the digital input into an analog voltage output of ±10 V. In this case, the chip contains an operational amplifier for providing the voltage output.

ADC

There are many types of ADCs. The one shown in Fig. 9–29 illustrates the principle[5] that is found in most of them. For the sake of simplicity, the diagram is limited to four bits.

The essential elements are:

- Four clocked flip-flops, *A*, *B*, *C*, and *D*, which require that a clock pulse be applied to change the state of the flip-flop.
- One flip-flop, *E*, that has to be on logic 1 to permit the clock signal to pass through AND gate *d*.
- Three AND gates, *a*, *b*, and *c*, each of which is made to pass the clock signal, provided the preceding flip-flop, *D*, *C*, or *B*, is on an output of logic 1.
- Three level converters that produce an output of -16 Vdc when the flip-flop is on logic 1, and of 0 V when the flip-flop is on logic 0.
- A high-gain inverting dc amplifier that is logic 0 when the input is logic

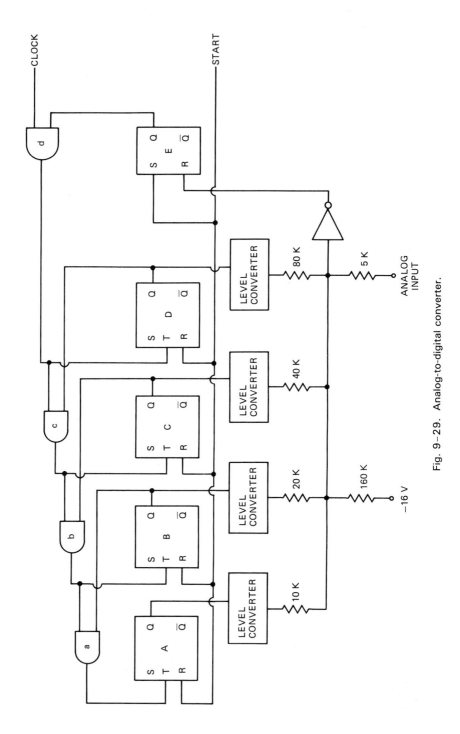

Fig. 9-29. Analog-to-digital converter.

304

1, and vice versa. It also becomes negative when input becomes positive, and vice versa.
- Resistors with various values.

The four clocked flip-flops represent a binary counter, providing the digital equivalent of the analog input. When counting begins, flip-flop D goes to logic 1, which in turn triggers flip-flop C with the next clock pulse, and so on. Initially, all four flip-flops are on logic 0. If the input signal is less than 0.5 V, the output of the amplifier will be logic 1. This will set flip-flop E on logic 0, and, consequently, gate d will not pass the clock signal.

If the input signal is greater than 0.5 V, the amplifier output goes on logic 0, and as the start signal comes in, gate d will let the clock signal pass. Now the counter will count, and as it goes through D–C–B–A, the level converters change from 0 to -16 V. The resulting negative current that passes into the resistors will increase until the input to the amplifier becomes negative, thus stopping the counter.

If, for example, the analog input is 12 Vdc, the input to the amplifier will be positive until the counter goes to 1110. At this point, the clock pulses will stop. The binary number 1110 corresponds to the analog 12 of the input.

BRAKING CIRCUITS

If a motor is to be brought to a rapid stop and kept there, brakes are needed. The simple disc brake that pulls two braking pads against a rotating disc by means of a solenoid is effective in stopping the motor but may also jolt it. Since this is undesirable, such brakes are used only as a supplement when the motor is completely or almost at a standstill. Before this happens, dynamic braking is used. This means turning the motor into a gernerator and discharging its energy into a fixed resistor, thus dissipating its energy. A number of circuits are rather effective. Thus, Electro-Craft Corp. states that it takes about 400 msec to stop its E-650 control system by dynamic braking.

Figure 9-30 shows a simple braking circuit for a permanent magnet motor. A pnp transistor is used. As long as the switch is closed and the motor is running, the base of the transistor is at the same voltage as emitter E. Consequently, no control current flows from the emitter to the base, and no working current flows from the emitter to the collector. Once the switch opens, the motor discharges its energy into the circuit. The rectifier diode blocks the reverse flowing current, the voltage on the base drops, and control current now flows from the emitter to the base. Consequently, the working current flows from the emitter to the collector and through the

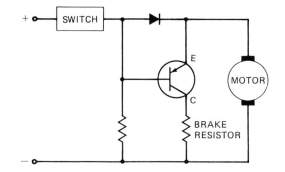

Fig. 9-30. Braking circuit for permanent magnet dc motor.

brake resistor, thus bringing the motor to a standstill. The time required to stop the motor depends on the motor inertia and the magnitude of the brake resistor.

POWER SUPPLIES

The main purpose of the power supply for servo control systems is to convert ac line voltage from the common power line into dc. It may also invert the dc back to ac when ac motors are used.

There are regulated and unregulated power supplies. Within limits, a regulated power supply will maintain a constant output voltage or current in spite of changes in line voltage, output load, ambient temperature, or time. An unregulated power supply will not compensate for such changes.

Figure 9-31 is a block diagram of the SAFI power supply of the McDonnell Douglas Electronics Co., which is used with its rotary motion systems. Like any basic power supply, it has a transformer that converts the incoming voltage into a convenient secondary voltage, or as in this case into two voltages. In fact, any number of voltages could be tapped off from the same transformer.

The 90-Vdc output of the power supply unit is unregulated and may change by ±20 volts. It is applied to the servomotor. The 15-Vdc and 5-Vdc outputs are regulated. They are used for the electronic amplifier and control system. The 15-V output will not change more than ±0.2 V, and the 5-V, not more than ±0.1 V.

Rectifiers

For full-wave rectification of ac current, the bridge rectifier is quite common. Four diode rectifiers may be connected as in Fig. 9-32. During one

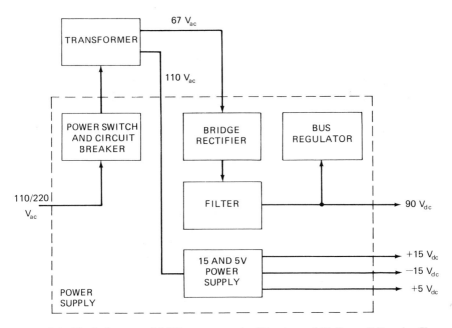

Fig. 9–31.   Block diagram of SAFI power supply. (*Courtesy of McDonnell Douglas Electronics Co.*)

half of the cycle of the ac input, point *a* is positive. Current will then flow via *a*, *b*, and *c*, then through the external load and back through *d* and *e* to *f*. The diodes prevent the current from flowing any other way.

During the next half of the cycle, *f* is positive, and the only path for the current is now via *f*, *e*, and *c*, then through the external load, and back through *d* and *b* to *a*.

In each half cycle, the current will flow through the external load in the same *c–d* direction. In this sense, it is direct current, albeit with a heavy ripple. With a 60-Vac supply, there are 120 such ripples each second. As

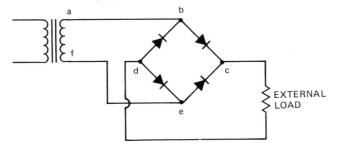

Fig. 9–32.   Bridge rectifier circuit.

pointed out earlier, a filter may considerably reduce the ripple. The filter may take various forms: from a simple RC network, as described above, to the more elaborate inductance–capacitance combinations.

Regulated power supplies usually require no more than a single-capacitor filter, since the regulator follows the filter and effectively rejects most of the ripple. However, an additional choke (see below) may be used just ahead of the motor. This is not only for additional filtering of the remaining ripple but also to limit the current that will flow during motor reversal.

Chokes

A choke consists basically of an inductance coil that impedes the flow of current impulses by means of its self-inductance.

Self-induction is the property of a conductor to induce an emf in itself whenever the magnitude of current flow starts to change. This induced back emf increases with frequency and is always in such a direction that it opposes any change in current flow. While any wire has self-induction, though usually of only negligible magnitude, the property is very much enhanced when the wire is formed into a coil.

Self-inductance is the amount of emf produced by such a coil. It is proportional to the square of the number of turns of the coil and inversely proportional to its length. The back emf induced in the coil, which opposes the emf that produces it, is proportional to the rate of change of the current. Thus, it can block rapidly changing currents quite effectively. The changes may be caused by ripples or by sudden reversals in motor rotation. The choke coil is installed in series with the line carrying the current. Frequently, the effect of the choke coil is further enhanced by added capacitors across the lines, usually one before and one behind the choke coil. The arrangement is usually referred to as a π-filter.

Regulation and Regulators

There are line regulation and load regulation. Line regulation is the extent to which output voltage can be kept constant in spite of variations in line voltage alone. Load regulation is related to variations in the connected load at constant line voltage. Both load and line regulations are generally expressed in percent of output voltages.

Often, specifications refer only to regulation and do not distinguish the types in the above sense. In such cases, one must assume that tests have been made with the rated maximum load and the rated ac voltage between the transformer and the rectifier. Thus, if the unregulated section of the above-mentioned SAFI power supply (Fig. 9–31) has, as mentioned, an input of 110 Vac and an output of 70 to 110 Vdc, the load and line regulation is obviously 44.4%.

On the other hand, for the regulated 15 Vdc, the tolerance is rated at ±0.2 V. This is the lumped voltage error for the supply, which includes adjustment error, line and load regulation, and so on. The line regulation for an input voltage from 104.4 to 132 Vac is equal to or better than 0.005 V. Hence, the line regulation is 0.005/15 = 0.0003 or 0.03%.

Load regulation from no load to full load is equal to or better than 0.010 V. Hence, the load regulation is 0.010/15 = 0.00067 or 0.07%.

Unregulated Power Supplies. The basic concept, as shown in Fig. 9–33, consists of adding a filter capacitor and a high-resistance bleeder resistor to the bridge rectifier. The purpose of the resistor is to discharge the capacitor after the circuit is de-energized. The ripple, in spite of the capacitor, amounts to 5 to 10%.

The advantage of the unregulated power supply is its low cost and the fact that efficiency is high, since power losses in the components of the unit are minimal.

Series Regulator. There are many forms of regulators. The one most widely used in control systems is the series regulator. Its purpose, as the name implies, is to improve the regulation of the power supply by means of an element in series with the outgoing line. Figure 9–34 illustrates a typical series regulator. The transistor is controlled by the difference between a voltage reference and the output voltage. Any deviation of this difference from zero will produce an adjustment of the voltage drop across the transistor. The sensitivity and speed of this response are such that even the remaining ripple is smoothed out to a large extent.

The voltage reference may simply consist of a Zener diode. Both the reference and the output voltage are applied to a differential amplifier. Its

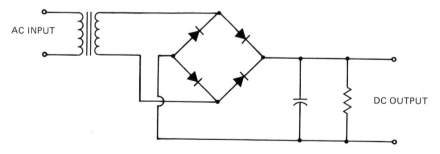

Fig. 9–33. Unregulated power supply.

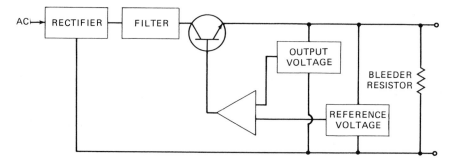

Fig. 9–34. Series regulator.

output is connected to the base of the transistor and regulates its resistance between the collector and the emitter, thus maintaining the output voltage at a closely controlled constant level.

While the efficiency of an unregulated power supply is 80 to 95%, the series regulator drops it to 30 to 50%. The regulation, however, is excellent, as shown by the example of the SAFI power supply.

Regulators must be protected against overvoltages. A suitable additional circuit, as in Fig. 9–35, which is often called a crowbar, is used for this purpose. It instantaneously throws a short circuit (crowbar) across the output terminals when a preset voltage is reached. This short circuit blows the power supply input fuse, shutting down the supply. The SCR is made conductive by the trigger circuit that responds to the excessive voltage.

Motorola produces a crowbar that uses a voltage–actuated MOSFET to trigger an SCR in case the bus voltage exceeds 6.2 V or the heat-sink temperature exceeds 125°C. The circuit occupies 60% of an IC chip that measures 0.18 × 0.18 in. The TO-220 package in which the IC is housed attaches directly to the heat sink of a printed circuit board. Trip time is 5 μsec. A lead is provided for remote adjustment of the voltage trip level or delay time; and the SCR can also be tripped by a remote signal.

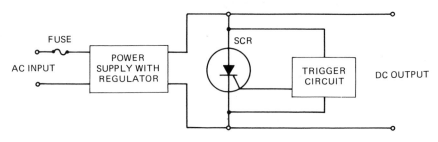

Fig. 9–35. Overvoltage protection.

SERVO CONTROLLERS

The advent of microprocessors brought about the separation of microprocessor-based servo controllers from the power amplifiers that are designed to supply the necessary power to drive the servomotor. An example is the C211 positioning controller from the Motion Control Division of Gould Inc. It consists of several printed circuit boards that are inserted in a motherboard. The C211 can be used with power amplifiers of many types and sizes.

One board contains a microprocessor with 32K bytes of EPROM* storage, 2K bytes of RAM storage, and 2K bytes of EEPROM storage. The position control program to be implemented is stored by the user in the EEPROM.

A second board contains the position feedback signal conditioning circuits. A resolver on the motor shaft is standard feedback, and the necessary signal conditioning for it is provided. A slightly different version for use with an encoder instead of resolver is also available. In either case, the feedback data are used to calculate velocity and provide the necessary rate control for a high-gain stable control system. The fact that velocity is computed from position feedback eliminates the need for a separate velocity feedback device, which is another example of the versatility of microprocessors.

Three additional boards provide various optional customized control functions. The advantage of this building-block approach is its great flexibility in tailoring speed and position control of a servomotor to a specific task.

The DMC 100, a product of Galil Motion Control, is another servosystem that separates the servo control and amplifier,[6] and thus permits using the same servo control for any size dc motor with a speed range of 30,000:1. The DMC 100 can also control brushless dc motors. However, in this case commutation of the motor must be done by the power amplifier.

Functional elements of the DMC 100 are shown in Fig. 9–36. They are divided into three groups: communication, control, and interface. All three groups are mounted on a printed-circuit board or supplied as separate chips.

The communication elements enable the user to command the system and to receive status information. Full communication with other equipment can be established via a standard RS232 serial interface and STD bus (see below under "Standards"). In addition, some functions can be performed by local switches.

---

*The various forms of memory, such as ROM, RAM , EPROM, and EEPROM, are discussed under separate headings.

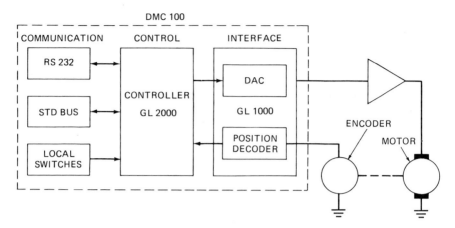

Fig. 9–36. The DMC controller. (*Courtesy of Galil Motion Control*)

Control is performed by an integrated circuit, the GL2000. It includes closed-loop position control, a digital filter for stability, and motor command generation. As already described, such microprocessor servo systems are so versatile that they usually do not need velocity feedback. In the case of the DMC 100, the position control circuit includes a digital filter as lead compensation. This eliminates separate velocity feedback and provides the same dynamically stabilizing effect internally. Gain and lead compensation can be adjusted for optimum system stiffness and damping.

Interface functions are performed by a second integrated circuit, the GL1000. It includes two parts: a digital-to-analog converter and a position decoder. The DAC is the interface between the digital output of the controller and the analog input to the power amplifier that drives the motor. The motor command is an analog signal of ± 10 V. Pulse-width modulation can also be accomodated.

The position decoder is the interface between the encoder and the controller. It converts the encoder signal into an input for the controller. A resolver can also be used for feedback, provided it has eight bits of feedback in parallel.

The controller permits programming of position, velocity, or torque. In the position mode, the motor will advance a specified distance and then stop. This distance can be represented as absolute position or as relative distance from the current position. The motion will consist of a preset acceleration and deceleration and also a slew velocity, that is, a maximum constant speed, between acceleration and deceleration, provided the distance traveled is long enough to accommodate such slewing.

In the velocity mode, the motor will accelerate to a specified slew speed. It will hold this speed until a stop signal is received.

In the torque mode, the motor applies a constant torque unless its velocity exceeds a limit specified by the user. Whenever the velocity limit is exceeded, the DMC shifts to velocity control. As soon as the torque can again be applied without violating this constraint, the DMC switches back to a torque command. This is very useful in robotic applications where a motor has to slew at a constant speed and then apply a known torque.

In all three control modes, it is possible to change the velocity level while the motor is moving. When this is done, the motor will change speeds at the programmed acceleration rate.

After the motor reaches the position for which it is programmed, there are three holding modes, in addition to the possible application of brakes.

In the servo mode, the system maintains the stopping position by correcting for any position errors on a continuous basis.

In the deadband mode, the motor is shut off as long as it is within a certain distance of the desired position. If the position error exceeds this distance, then the servo mode takes over to move the motor back to its programmed position.

In the third mode, the motor is shut off. No further correction takes place until a new position signal is received.

Another possibility is the use of repeat modes. In this case, the user specifies the number of times the move is to be repeated. Unlimited repetitions may also be programmed. A pause lasting from 1 msec to 30 sec may be inserted between moves.

Motion can be terminated by a command from the robot controller or from local inputs. In all but emergency termination modes, the motor will be decelerated gradually to a stop and then will assume one of the three holding modes mentioned above.

Three digital outputs provide status on motion and fault detection. These outputs appear either as LED lights or as signals to be transmitted to other terminals. Certain information can be requested at any time, including state of the system, state of the local inputs, current motor position, and current motor velocity. This also means that the robot controller can command the DMC 100 to send confirmation when a move is complete.

The home position (or origin), at which the absolute position of the motor is considered to be 0, can be defined by two methods. In the first, the user puts a command into the controller that defines the current motor position to be the origin. The other method consists of requesting the DMC 100 to search for a transition in a sensor output. If the output of the sensor is initially high, the controller will cause the motor to slew in a positive direction until it becomes low. Likewise, if the initial condition is low, the search will be in a negative direction until a high state is attained. The transition point is then defined to be the origin.

One other servo controller[7] is of interest in this connection. It is from Finell Systems Inc., and a block diagram is shown in Fig. 9–37. The FPC-1800 contains a microcomputer, digital-to-analog converter, decoder, and reference unit. It can interface to any host computer or master controller through an eight-bit bidirectional bus. Any commercially available or custom servo amplifier can be used with it. Feedback is by means of an encoder mounted on the motor shaft. The controller package is not greater than $3.40 \times 3.96 \times 0.48$ in.

The FPC-1800 is based on an Intel 8051 microcomputer and includes two programmable logic circuits and 4K bytes of ROM, (see note on p. 311), which stores the software for the control algorithm. Parameters are entered for the description of the motor, encoder, amplifier, and load. Also programmed by the user are position and velocity of the servomotor. The host has access to real-time data at any moment, and is thus able to verify control status whenever desired. It can also effect changes in the servomotor's velocity while the system is in motion.

Motion instructions include both relative and absolute position commands. Absolute move instructions are referenced to a HOME position.

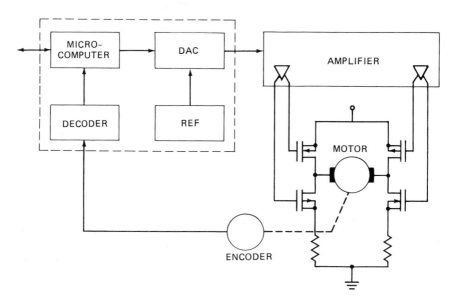

Fig. 9–37. FPC-1800 servocontroller. (*Courtesy of Finell Systems Inc.*)

The servo system can be repositioned directly to any positive or negative position with respect to this reference. A STEP instruction allows the servo loop to be repositioned in increments of programmable length. Relative position commands allow the system to move at a controlled velocity until an external signal is received or a stop instruction is issued.

It is worthwhile to quote here what James Harnden[7] has said about the future of servo controllers:

Future controllers will not have to be programmed with servo loop parameters to achieve optimal compensation. They will be sophisticated enough to "sense" the systems response to their control signals and adjust the compensation accordingly. This opens applications such as robotics to an era of true adaptivity. No matter what the load or arm extension, the servo loop can maintain optimal damping. This is the future of distributed intelligence in servo control, moving the "responsibility" out of the central controller. Instead of involved, costly host software tracking the relationship of each axis to all others and being responsible for overall system compensation, each axis will be allowed to react to the environment and adjust compensation locally. Host software development can concentrate on achieving sophisticated tasks of industrial automation as distributed intelligence controllers evolve precise, self-reliant closed loop motion controls.

MICROCOMPUTERS

The typical computer consists of five basic sections. As shown in Fig. 9–38, they are:

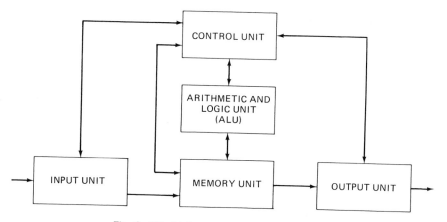

Fig. 9–38. Main components of a computer.

1. The control unit or control section, which directs the sequence of operation, interprets coded instructions, and sends the proper signals to the other computer circuits to carry out the instruction.
2. The arithmetic and logic unit, or ALU, which performs arithmetical and logical operations on the data that are transmitted to it from either the control unit or the memory.
3. The memory unit or main memory, in which information is stored in machine language, as will be discussed below. This information is to be called upon by either the control unit or the arithmetic unit. Additional external memory capacity is often provided.
4. The input unit, which receives outside information and distributes it to either the control unit or the memory unit. The input signal may be in the analog mode, in which case it will be converted into a digital signal by an analog-to-digital converter which is part of the input unit.
5. The output unit, which transfers data to an external device or from internal storage to external storage. To transmit signals in the analog mode, a digital-to-analog converter in the output unit will take care of the conversion. In discussions or schematics, input and output units are often grouped together as I/O units.

The control unit, ALU, and main memory, together, are frequently referred to as the central processing unit, or CPU. A microprocessor is a CPU in miniature that is usually contained on a single chip.

A microcomputer is a device that is built around a microprocessor. It includes a clock, additional memories for instruction and data storage, I/O units, and so on. A microcomputer consists of one to four chips. Generally, it is designed for specific applications.

Microcomputers, like microprocessors, use integrated circuits (ICs) that are usually based on NMOS technology. Another approach, however, is the use of high-performance MOSFETs, which were mentioned above. They operate on a 5-V power supply, like the NMOS devices, and are capable of passing a signal from input through the microcomputer to the output in 1 nsec. An example is the single-chip 16-bit microcomputer MK68200, which was designed by Mostek Corp. It is capable of data rates of up to 1.5 megabits/sec. Included are three 16-bit timers that provide interval timing, pulse-width measurement and generation, and other timing and counting functions.

Bit Sizes

Computer word length, that is, number of bits per word, is an important factor in determining whether a given computer is fast enough for a specific application. In general, longer word lengths produce greater information processing speed. It is the design of the CPU or microprocessor that determines the word length or bit size.

Four-bit microprocessors have been used widely for industrial controls. They contain all the circuits within one or two chips. Four-bit computers generally have a large number of input and output lines to accommodate external switching and control devices. They are limited in the amount of memory and I/O devices the computer can address.

Four-bit processors used to be the lowest in cost. However, as eight-bit microprocessors are increasingly designed on CMOS technology, the cost difference vanishes. Also, the energy consumption of eight-bit microcomputers is considerably less than that of four-bit processors.

Indicative of this development is the MC68HCO4P2 CMOS-microprocessor made by Motorola Inc. It is a single-chip, eight-bit computer that contains a CPU, a clock, memory, I/O circuits, and timer functions. Its internal power dissipation is less than 15 mW. In addition its "stop-and-wait modes" shut down inactive sections of its circuits and thereby reduce standby power consumption by about 75%. The chip has 32 bytes RAM and 1024 bytes ROM. A companion version, the MC68HCO4P3, has 124 bytes RAM and 2048 bytes ROM.

Sixteen-bit microprocessors perform relatively rapid calculations as well as logical operations. Clock rates range from about 1 MHz to 16 MHz. Processor manufacturers such as Mostek, Intel, Motorola, and National Semiconductor have developed operating system software that is among the fastest available. An example is the above-mentioned MK68200 from Mostek Corp., with its data rates of up to 1.5 megabits per second (Mbits/sec).

Full 32-bit microprocessor chip sets have also become available, such as the 68020 microprocessor from Motorola. On a surface of about 0.27 × 0.29 in., it combines some 180,000 NMOS transistors and 20,000 CMOS transistors. Power dissipation is 2 W maximum. Its maximum clock rate is 16.67 MHz. It can execute 2 to 3 million instructions/sec at a CPU cycle time of 125 nsec.

Memories

Memories store information. Actually, the ALU of a computer can operate only on the basis of stored information. This may be information that is temporarily on hold until it is replaced by some other information, or permanent information in the memory required for the operation of the ALU.

Before the computer can read a word, it must locate it. The time interval between the instant of calling for data from a storage device and the instant of obtaining it is called access time.

ROMs. ROM (read-only memory) storage is primarily for storing program instructions. The information is stored permanently or semipermanently, and can be read but not altered in operation.

The ROM is essentially a very large gridwork or array of gates. Depending on the input, it responds with a definite set of outputs. Diodes, which are arranged as in Fig. 9–39, can be used as gates. Here, the memory is used to provide binary numbers as outputs for the decimal numbers that are applied as inputs. The range of decimal numbers for which the corresponding binary number can be read, goes, in this case, from 0 to 9.

It can be seen from the connections in Fig. 9–39 that if the input is decimal 7, the stored output is binary 0111; if it is decimal 3, the stored output is 0011.

Along with the diagram, Fig. 9–39 shows a truth table. Any ROM can be represented by a truth table. It gives a clear representation of the possible combinations of the storage device.

As is obvious from the above example, the ROM can well be considered

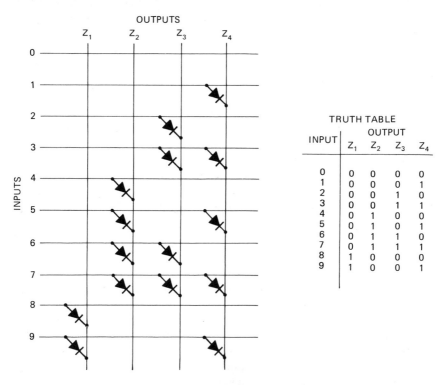

INPUT	OUTPUT			
	Z_1	Z_2	Z_3	Z_4
0	0	0	0	0
1	0	0	0	1
2	0	0	1	0
3	0	0	1	1
4	0	1	0	0
5	0	1	0	1
6	0	1	1	0
7	0	1	1	1
8	1	0	0	0
9	1	0	0	1

Fig. 9–39. Principle of ROM storage device.

a reference table, be it for trigonometric functions, entropies, binary numbers, or any other functions. The reference table characteristic even applies to the largest ROMs, although the combination of inputs and outputs may be expanded far beyond the simple example of Fig. 9-39. The diodes may be replaced by MOS transistors, and the entire ROM may be placed on a chip, as it usually is, but the principle remains the same.

The basic ROM is not erasable and cannot be reprogrammed. The programming is done in its production, where photographic masks are used to reproduce the program from wafer to wafer. Such ROMs are called mask-programmed.

The MCM68370 from Motorola is such a mask-programmed ROM. The block diagram in Fig. 9-40 shows its memory matrix of 8192 × 8 bits. The ROM is based on NMOS technology. There are four inputs applied to a NOR gate. The output of the NOR gate is logic 1 only when none of these inputs is active. Three of them are chip-select inputs that prevent the input or output of data unless one of them is active. The fourth is a chip-enable input that, when the ROM is inactive, puts it in a reduced-power standby mode, reducing the power consumption from the active 80 mA to a standby 15 mA. Hence this is a three-state device, since there are not only "zeros" and "ones" but also a high-impedance standby condition.

The address connections pass through the address decoder to the memory

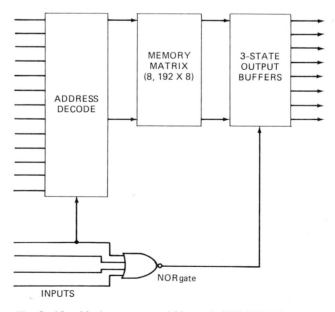

Fig. 9-40. Mask-programmed Motorola MCM68370 ROM.

matrix, and from there to an output buffer that can hold the output and transmit it in organized fashion to the rest of the circuit. The maximum access time from address and chip-enable is 200 to 300 nsec, depending on which of three models is chosen. The entire ROM is contained in a DIP package of about 1.44 × 0.55 in. The active levels of the chip-enable and the chip-select inputs are defined by the user.

It is a limitation of all mask-programmed ROMs, that in case the program to be stored in the memory is not obtainable off the shelf, and only a limited number of such devices are required, it would not be economically feasible to develop the production details for the particular application needed. In such cases, a PROM (programmable ROM) may be the answer. This would mean that the unit in Fig. 9–39 would be provided with diodes at all 40 crossing points of the matrix, and the connections would be made with fusible links. It would then be up to the user to burn off the connections to all diodes not used for the program at hand. Special equipment for doing this is available.

There is also the situation in which a ROM is desired, but it will be necessary at some future time to change the program of the ROM. Erasable programmable read-only memories are required here. They are available under such names as EPROM (erasable PROM), EEPROM OR E²PROM (electrically erasable PROM), and EAROM (electrically alterable ROM).

The EPROM is erased by exposing it to concentrated ultraviolet light for 10 to 20 minutes. A transparent quartz window over the IC is provided to apply the UV light from the outside. After exposure, the EPROM consists entirely of logic 0s or logic 1s. It can then be reprogrammed by electrical means. Here, too, a variety of equipment for erasing and writing is offered.

The electrically erased EEPROM has advantage over the EPROM that it can be erased in one second instead of 10 to 20 minutes. EEPROMs can be erased and reprogrammed up to a million times.

The EAROM, another electrically erasable device, is built using metal-nitride-oxide semiconductor technology. While the access speed of EPROMs and EEPROMs is slower than that of the ROMs based on MOS technology, the EAROM can compete with the latter in this respect.

One important characteristic of any ROM storage device is that it does not require a battery as standby in case of power failure. It is nonvolatile; its contents are not erased or changed when power is turned off. This is not true of RAMs. Thus, the control program for ASEA robots is usually stored in an EPROM, while the user program is stored in a RAM. Because of the volatility of the RAM, a battery backup is provided to protect the contents in case of power failure. The memory capacity of the ASEA robot controller can be readily increased with additional memory, including floppy discs.

RAMs.   The random access memory (RAM) serves not only for reading information, as does the ROM, but also for writing it in. Therefore, it is also referred to as read/write (R/W) memory.

The smallest RAM is a single flip-flop. Inputs into the flip-flop change its outputs, as described on page 282. Whatever the condition of the flip-flop is, it can be read at any time. By applying another signal, its condition can be changed, and the information that is read from it will be different. In other words, it is possible to erase the information whenever that is desired, and to write new information into this RAM.

RAMs constructed from flip-flops are referred to as static RAMs or SRAMs, in contrast to dynamic RAMs or DRAMs, described below.

Dynamic RAMs are designed on the basis of MOS transistors. The reason for this is primarily the high gate capacitance of MOS transistors due to the oxide that separates the metal gate from the semiconductor material (see Fig. 9–11). This capacitance can be used to store information for a short time. It can also be stored permanently, provided that power is not interrupted and the capacitor charge is renewed every 1 or 2 msec.

Figure 9–41 shows a single cell of a matrix arrangement similar to the individual matrix cells in Fig. 9–39. Three MOS transistors, M_1, M_2, and

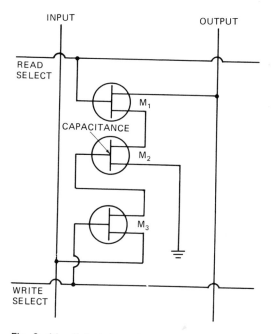

Fig. 9–41. Cell of dynamic RAM memory matrix.

M_3, are used. The "read select" line runs on the top of the illustration, the "write select" line on the bottom, the "input" line on the left, and the "output" line on the right.

To *write* into the storage cell, a binary signal "1" is sent over the write select line. This turns M_3 on, enabling it to conduct from the input line to the gate of M_2 and to charge the M_2 capacitance.

To *read* the content, the binary "1" is applied to the read select line. If current flows to the output line, this means a "1" is stored in M_2; and if not, the signal is "0."

The capacitance of M_2 is the real storage. Restoring its charge periodically requires additional circuitry, which is not shown.

Figure 9–42 illustrates how a microprocessor addresses a typical RAM memory. In this case, the microprocessor needs information stored in location 603 of the memory. The number 603 in straight binary code (machine language) of 16 bits reads 0000 0010 0101 1011. This number is transmitted over the 16-bit address bus to the decoder of the memory. The decoder "recognizes" the combination of input bits and passes them on to the requested address location.

At the time when the microprocessor addresses the memory, it simultaneously transmits a signal "1" over the control bus. This signal activates the memory and enables it to read and transfer to the processor the information contained in location 603. If the situation were reversed, and the

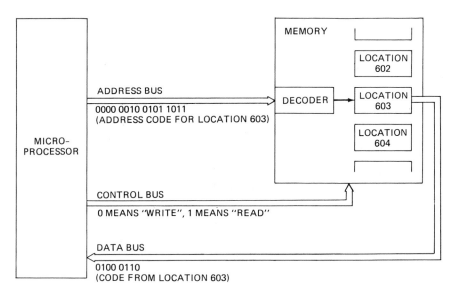

Fig. 9–42. Fetching information from memory.

microprocessor needed to store (i.e., to write) information in the memory rather than read it, then the initial signal over the control bus would be a "0" instead of a "1."

The dynamic RAM cell occupies considerably less space on a chip than the flip-flop of a static RAM. It also needs less power. Thus, as far as low-power requirements and high-density packaging at a very low cost per bit stored are concerned, the dynamic RAM is superior to the static RAM, which usually requires more power and is less densely packed. Both are accommodated on a chip, but the latter with less memory capacity for the same size chip. Thus, the decision between static RAM and dynamic RAM may often come down to this question: How much storage capacity is needed?

One fact, however, applies to both static and dynamic RAMs: If the power fails, the contents are lost. This necessitates emergency power supplies, which usually consist of a lithium battery small enough to be part of a printed circuit board. Since the dynamic RAM requires very little energy, it requires very little standby power.

Performance differences between static and dynamic RAMs are constantly changing. The above-mentioned static RAM from Cypress Semiconductor Corp. is a case in point. The 7C122 is a static RAM with a memory matrix of 256 × 4 bits and an access time of 15 nsec, which is competitive with the speed of any available RAM. It also cuts power dissipation into about half of the best previously obtainable by a static RAM.

Mass Storage

Peripheral devices for deposit and retrieval of large amounts of data are referred to as mass storage devices. Cassettes, floppy discs, and magnetic bubble memories are the most frequently used mass storage devices in robotic control.

Cassettes and floppy discs permit storage of a number of different programs that can be inserted in a robot controller to execute a previously developed program. They are well known because of their extensive use in personal computers, radios, and so on.

Magnetic bubble memories (MBMs) have the advantage of eliminating all moving parts. There is practically no wear. They are still the most expensive mass storage devices, but have the advantage of being particularly rugged and flexible. Intel's MBM stores one megabit (1 Mbit) on a chip that measures 0.8 × 0.8 in. It can transmit 300,000 to 400,000 bits/sec.

Allen-Bradley's Series 8200 Robot Control offers an MBM option. It consists of a bubble memory panel and plug-in bubble memory cartridges. Each cartridge provides 64K words of nonvolatile storage.

Magnetic bubbles are small cylindrical domains. These are regions of microscopic size in which all the spinning electrons are parallel to one another, resulting in maximum magnetization. The cylindrical domains are formed by a stationary external magnetic field in single-crystal thin films of synthetic ferrites or garnets that are grown on a substrate of gadolinium gallium garnet. Once the bubbles are formed, they can be moved along a path defined by a deposited layer of metal on the surface of the single-crystal film. The presence of a bubble corresponds to a logic 1 and the absence to a logic 0. When a bit is to be read, it travels under a conductor that creates two bubbles from the original one. The bubble pair then travels under a magnetoresistive detector to produce an output voltage in the range of millivolts.

STANDARDS

Connections within electronic devices as well as between computers, controllers, and the various peripheral devices are by bus. The bus consists of one or more conductors used as a path over which information is transmitted. Within a microcomputer three single or multi-wire bus lines will usually be found: the data bus, the address bus, and the control bus.

The data bus, which was shown in Fig. 9–42, transmits data to or from the I/0 unit to the microprocessor and the memories. The address bus serves for entering or retrieving information, between the microprocessor and memories. The control bus may convey a mixture of signals designed to regulate system operation. In a typical case, the data bus would transmit 8-bit information and the address bus 16-bit information, whereas the control bus may only convey 1-bit signals. A bus diagram, such as the one in Fig. 9–42, usually , but not always, shows bus connections as double lines.

Printed-Circuit Boards

The pc (printed-circuit) boards are inserted in slots of a cage that is provided with connectors in its backplane to match the connectors of the boards. To make the boards interchangeable, dimensions of slots and methods of connecting and wiring in the backplane correspond to specific standards.

STD Bus.   The STD bus is used as a standard by over one hundred manufacturers of pc boards to unify physical and electrical aspects of 8-bit microprocessor systems. An industry association called the STD Manufacturer's Group has the responsibility for the protocol of the STD bus.

Typical features of the STD bus include several memory sockets for ROM, EPROM, or RAM, and an 8-bit output port for a printer connection. Thus, the RSTD-PROPOS, a servomotor control system from Renco Corporation, contains on dual STD bus cards an 8-bit processor and four 8K-byte memory sockets: 2K RAM, 16K ROM, and 2K EEPROM.

Though the STD bus was created as a bus for 8-bit computers, it is also used with some 16-bit computers, particularly the 8088 Intel processor found on the IBM Personal Computer.

Eurocard.   The STD bus system and others use flat edgeboard finger connectors. Some industrial applications, however, use the so-called Eurocard, a European standard that has pin connectors and comes in several different board sizes. One commonly used size measures 160 × 100 mm (about $6\frac{5}{16}$ × $3\frac{15}{16}$ in.) and provides 32-pin connectors.

Users of the Eurocard claim that the pins are more reliable than the finger connectors. Companies such as Rockwell International, Intel, and others have accepted the Eurocard standard for many of their products.

16- and 32-Bit Bus Systems

There are three main bus standards available that handle 16-bit microcomputers: the Versabus used by Motorola, the Multibus used by Intel and National Semiconductor Corp., and the Q-bus used by Digital Equipment Corp. The advent of 32-bit computers led to bus standards that make 8-bit and 16-bit components compatible with those that are 32-bit. Quoting L. Teschler:[8]

These buses provide a large number of address lines for multi-megabyte memory addressing, and specify protocols designed to insure orderly multiprocessing, where several computers tied together want access to data paths simultaneously.

One such bus is the VME bus by Motorola, another the Multibus II developed by Intel. Both designs are simply different sizes of the above-mentioned Eurocard standard.

Multibus II

The Multibus II* has an open-system architecture[9] that, like the earlier Multibus I, accepts other specialized bus structures. It consists of five bus structures, which are shown in Fig. 9–43. Their functions are as follows:

---

*Trademark of Intel Corp.

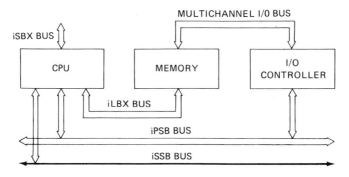

Fig. 9-43. Multibus II architecture. (*Courtesy of Intel Corp.*)

- The iPSB* (parallel system bus) is intended for board-to-board inter-communications. It includes a multiplexed 32-bit address and data bus capable of transferring 32 bits of data simultaneously by transmitting each bit on a separate wire. Data can be transferred at rates of up to 40 Mbytes/sec.
- The ISSB* (serial system bus) is capable of transmitting board-to-board messages identical to those on the PSB, but in a serial mode with sequential bit-by-bit data transmission at 2 Mbits/sec. It permits using lower-cost, smaller serial-line-connected boards.
- The iLBX II* (local bus extension) transfers data at up to 48 Mbytes/sec. It allows a single CPU board to access RAM and EPROM storage devices over a separate direct-line bus. By providing local memory expansion to 64 Mbytes and a clock rate to 12 MHz, the LBX II eliminates priority problems when several microprocessors use the system bus for execution memory.
- The SBX is an expansion bus that serves for connecting inexpensive I/O modules to the CPU board.
- The Multichannel* I/O bus transfers data at up to 8 Mbytes/sec to and from intelligent I/O devices remotely located. It supports up to 16 Mbytes of memory per device.

Interfaces

Interfaces allow two incompatible units to interact with each other. They are the hardware for linking two units of electronic equipment. Thus, the connection between a controller and a feedback input is a typical interface.

---

*Trademark of Intel Corp.

Interfaces may contain considerable electronic circuitry of their own, as will be described below.

Interfaces are of two basic types: serial interfaces and parallel interfaces. Serial interfaces transmit data serially bit by bit along a single path. Parallel interfaces simultaneously transmit 8 or 16 bits over that many wires. They are used only for relatively short transmission lengths that do not exceed more than a few feet.

The possibility of putting a microprocessor on a single chip and combining it on a single pc board with ROMs, RAMs, I/O ports, clock, and so on, has lead to the single-board computer (SBC). Both the SBC and the single-board system interface are connected to the system bus. In this case, the interface board can contain analog-to-digital converters, triacs that function as electrically controlled switches for ac loads, relays, optoisolators, or any other circuit components required to interface the microcomputer to the actuators and sensors of the robot. Operating in this way, the SBC becomes a single-board computer controller.

Transmission of data may be by twisted wire, coaxial cable, or fiber optics. A twisted pair can handle frequencies of 10 kHz or more. Coaxial cables not only offer excellent shielding against noise, but they can carry frequencies of 300 to 400 MHz. Fiber optics are inherently immune to noise and their frequencies extend into the gigahertz range.

In providing such connections, some bus standard is usually adhered to. The standard is the guarantee that the various components can be readily interconnected even if they originate from a variety of manufacturers, provided they all use the same standard.

The Electronic Industries Association (EIA) has issued standards that govern the interface between data processing terminal equipment and data communication equipment. The most important standards for robotics are standards RS-232, RS-422, R-423, and RS-449.

RS-232, also referred to as RS-232c according to the latest version uses serial binary data signals and defines the various signal characteristics and a 25-pin connector interface, as shown in Table 9-3. A number of pins, as can be seen, are not standardized, and thus may vary from manufacturer to manufacturer.

RS-232 is intended for use over a maximum range of 40 ft and transmission speeds in the range from zero to an upper limit of approximately 10,000 to 30,000 bauds, that is, 10,000 to 30,000 bits/sec.

EIA Standard RS-449, as well as RS-422 and RS-423, are intended gradually to replace RS-232. They are designed to be compatible with equipment using RS-232, but take advantage of more recent developments in IC design. The purpose is to reduce crosstalk between interchange circuits, permit a greater distance between system units, and permit transmission speed of up to 2 million bits/sec.

Table 9-3. Pin assignments according to RS-232.

PIN NUMBER	DESCRIPTION
1	Protective Ground
2	Transmitted Data
3	Received Data
4	Request to Send
5	Clear to Send
6	Data Set Ready
7	Signal Ground
8	Data Carrier Detector
9 & 10	Reserved for data set testing. These two pins shall not be wired in the data processing terminal equipment.
11 to 14	Unassigned
15	Transmitter Signal Element Timing (Data Communication Equipment Source)
16	Unassigned
17	Receiver Signal Element Timing (Data Communication Equipment Source)
18 to 19	Unassigned
20	Data Terminal Ready
21	Unassigned
22	Ring Indicator
23	Unassigned
24	Transmitter Signal Element Timing (Data Communication Equipment Source)
25	Unassigned

RS-449 specifies functional and mechanical aspects of the interface, such as the use of two connectors having 37 pins and 9 pins instead of a single 25-pin connector. RS-422 specifies the electrical aspects for communication over balanced lines where the currents on the lines are equal in magnitude and opposite in direction at any given point in the connecting cable. It covers data rates of up to 10 million bits/sec. RS-423 does the same for unbalanced lines at data rates up to 100,000 bits/sec.

Bitbus

The concept of the Bitbus* was from the outset to interconnect components of a control system, to improve system performance and reliability, and to reduce total system and maintenance cost, particularly in those applications that use a number of different control components at physically separated locations. Since the robot fits into this category, Intel worked closely with several robotic firms in the development of the Bitbus. As P. I. Wolochow[10]

---

*Trademark of Intel Corp.

has said, "Combining the strengths of existing hardware and protocol standards with complete firmware and software support, the Bitbus interconnect provides a simple, standard technology for connecting distributed controllers."

The Bitbus interconnect is based on the individual single chip 8044 microcontroller from Intel, which contains two functional elements: an 8051 microprocessor and an SDLC Serial Interface Unit.

SDLC stands for another standard: IBM's Synchronous Data Link Control. It assures timing at the beginning of each message by synchronizing characters and thus maintains a constant time interval between successive bits, keeping all equipment in the system in step.

Figure 9–44 shows a Bitbus interconnect with a typical robot workstation, two or more robots, a conveyer belt, and a central work-cell controller. The internal connections of the individual microcontrollers are by Multibus; the external interconnections, by Bitbus. A standard wrist/end effector interface at each robot (*A*) lets users select end-effectors from a variety of sources. The other standard interface (*B*) ensures a coordinated work-cell

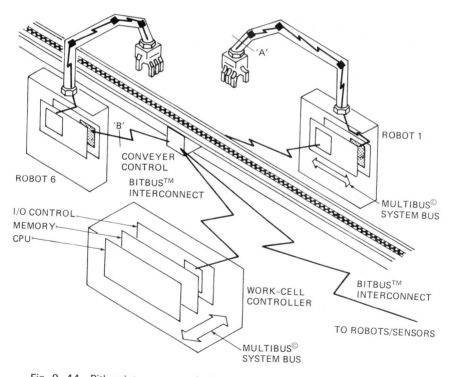

Fig. 9–44. Bitbus interconnect of robot work station. (*Courtesy of Intel Corp.*)

by providing a way to connect robot, conveyers, and so on, to the same work-cell controller.

A Bitbus message extends the standard SDLC format. It consists of 5 bytes for polling, status, and control, followed by up to 13 bytes of data. Polling consists of sending signals to the various control nodes, asking one after the other "Do you have anything to transmit?" This prevents simultaneous transmission of data from two different nodes. The status bytes inform about existing conditions, and the control bytes allow I/O connections on remote Bitbus boards to be activated for such functions as energizing solid-state relays to start a servomotor.

The 8044 microprocessor contains 4K bytes of (P)ROM, 192 bytes of RAM, clock, timers, interrupt controllers, and memory expansion bus. In Bitbus applications, the ROM is filled with permanent instruction routines that support the message-passing protocol, interact with user application tasks, and perform a series of self-diagnostic functions.

Each Bitbus controller can act as either a master or a slave. This provides a simple method for allowing backup master controllers. For example, a backup master can be programmed to wait for a poll from the primary master every second. If the master missed several polls, the backup would take control and switch itself from a slave to a master.

The Bitbus microcontroller interconnect is supported by a number of additional components. In Fig. 9–44, the Multibus-based robot controller contains the iSBC* 286/10 single-board computer, an iSBC 012CX memory expansion board, an iSBC 186/03 single-board computer, and two iSBX 344 Bitbus expansion modules. Each board performs a particular system function. The robot drive electronics are housed in a separate Eurocard housing mounted within the robot base. Bitbus connections link the robot controller to the robot, teaching pendant, and work-cell controller. An RS-232 interface is provided to support communication with existing display and control equipment.

Ethernet

Ethernet is a bus system for transferring large data blocks at high speed but also at relatively high cost. It can support up to 1024 processing stations over distances of up to 1.6 miles. Transmission speeds of 10 million bits/sec can be obtained.

Ethernets can be significant in robot applications. For example, connecting assembly and processing robots over an Ethernet interconnect can

---

*Trademark of Intel Corp.

provide real-time coordination of robot tasks.[11] Figure 9-45 shows an assembly operation for electronic chips. Here, robots test wafers, package the good devices into their plastic housing, and retest the final product prior to shipping.

Because of the Ethernet interconnect, the test station takes an active part in the production. Take, for example, the bond wires that are used to connect the bonding pads of the chips to the package pins. When the test station detects drifts of errors in bonding location, it automatically issues a recalibrate command to the robot performing the actual bonding operation.

Furthermore, other areas of the operation can be integrated with the chip assembly over the Ethernet bus. Accounting, quality assurance, inventory control, and so on, could all benefit from immediate data concerning manufacturing status. A continuous, real-time analysis of the entire manufacturing process thus becomes possible.

MAP

General Motors Corp. developed its Manufacturing Automation Protocol (MAP) to standardize communications among about 50,000 programmable devices in its operations. It will not tolerate installation of programmable devices that cannot communicate through this protocol. To create computer-integrated machining (CIM), General Motors has determined that MAP will cover all of the computers and terminals, machine tools, robots, and other automated equipment, regardless of producer, and permit their connection into one total manufacturing network.

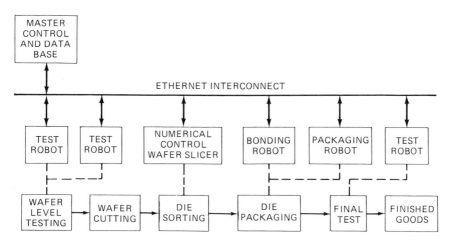

Fig. 9-45. Chip assembly operation.

While Ethernet is a probabilistic network that does not regulate when the terminal may gain access to the computer network, MAP is a deterministic method that determines the time slot for authorization of a terminal's usage of the computer network.

MAP is based on a seven-layer specification developed by the International Organization for Standardization (ISO) for open systems interconnection, together with standards from the National Bureau of Standards (NBS), the American National Standards Institute (ANSI), and the Institute of Electrical & Electronic Engineers (IEEE).

The lowest of the seven layers provides access to the medium of data transport, whether it be cable, fiber optics, or other means of carrying the data. This layer is based on standard IEEE 802.4. The next layer, based on IEEE 802.2 Class 1 link level control, refers to the data link that allows connection of adjacent devices. The third layer uses Internet, which has been accepted by ISO and NBS. It involves the connection, by network, of nonadjacent devices.

For the fourth layer, General Motors adopted the ISO and NBS class 4 protocol. It covers the transport layer over which data are transferred between two devices anywhere on the network. The fifth layer identifies each device, provides security of the network, and synchronizes data in the network. General Motors has not used other than ISO standards for this layer, but is considering NBS protocols. The sixth layer, so far, is also based on ISO standards. It restructures the data to and from devices with different languages. For the seventh and final layer, different protocols are being considered. It covers the application that services all applications programs, including direct numerical programs of machines on the factory floor.

Although this concludes the description of some of the more important standards used in electronic systems, it is by no means a complete review. This is particularly so because these standards are far from being settled, in view of the rapid progress of electronic technology. The purpose here has been merely to introduce the concept of standards as used in electronics in general and in robotics in particular.

REFERENCES

1. McNulty, T. C. Power MOSFETs—What the designer needs to know. *Electronic Products,* February 7, 1984, pp. 133–136.
2. Carlisle, B. H. A new breed of high-current MOSFETs. *Machine Design,* March 8, 1984, pp. 189–192.
3. Wilson, J. A. *Industrial Electronics and Control.* Chicago: Science Research Associates, Inc., 1978.
4. Pearman, R. A. *Power Electronics.* Reston, VA: Reston Publishing Co., 1980.

5. Bartee, T. C. *Digital Computer Fundamentals.* New York: McGraw-Hill Book Co., 1972.
6. Tal, J. Motion control by microprocessors. Published by Galil Motion Control, Mountain View, CA, 1984.
7. Harnden, J. A. Programmable dc motor controller facilitates flexible servo system design. *Powerconversion International,* November/December, 1983, pp. 46–51.
8. Teschler, L. Single-board computers pack much more punch. Reprinted from *Machine Design,* April 26, 1984. Copyright, 1984, by Penton/IPC., Inc., Cleveland, OH.
9. Cooper, S. The new 32-bit buses: Multibus II. *Electronic Products,* July 16, 1984, pp. 81–84.
10. Wolochow, P. I. Intel's Bitbus microcontroller interconnect. *Robotics Age,* June, 1984, pp. 30–36.
11. Webb, M. K. Local area networks and factory automation. *Design News,* February 6, 1984, pp. 179–183.

Chapter 10
PROGRAMMING AND CONTROLLERS

PROGRAMMING METHODS

Robots may be divided into the following four categories:

1. Robots with point-to-point open-loop control.
2. Robots with point-to-point closed-loop control, programmed by teaching.
3. Robots with continuous-path closed-loop control, programmed by teaching.
4. Robots with off-line programming.

Robots with point-to-point open-loop control are usually pick-and-place robots, often with only two or three degrees of freedom, which transfer items from one place to another. Controlling of intermediate points of a path is hardly available. Typically, actuators are pneumatic or hydraulic cylinders, and programming consists in adjusting stroke length by mechanical stops or limit switches that control stroke length. Only two positions for each axis are programmed. Response of a limit switch at the end of one motion triggers the next motion, so that not only reverse motion but also motion in different axes can be obtained. More complex motions can be programmed by rotating drums actuating switches at different drum positions or by pneumatic or any other locic circuits.[1] Also, programmable controllers (page 355) may be used. The number of programmable motions is limited to 15 to 100 steps, depending on the complexity of the robot.

Robots with point-to-point closed-loop control, programmed by teaching, are equipped with feedback sensors that signal the position of the robot end-effector back to the robot controller. Contrary to robots with open-loop control, each axis of these robots can be commanded to move to and stop at any point within its limits of travel. While being programmed, the robot is restricted to a slow speed, so that it can be easily and safely manipulated.

A hand-held control unit, the teach pendant or teach box, is provided with a keyboard and connected by cable to the controller. The robot is manipulated by using these keys. While it is moved by manual inputs from point to point, the feedback devices signal the corresponding point positions back to the controller. When a certain key is actuated on the teach pendant, the point is stored in the memory of the controller. Instead of the teach pendant, an operator's panel with keyboard and CRT (cathode-ray-tube) display may be used.

Robots with continuous-path control, programmed by teaching, are usually led manually through the cycle they are to perform. The speed of motion is again under "teach restrict." Such lead-through teaching is commonly used for spray painting, polishing, and arc welding where the exact path is executed manually and stored in the memory, so that it may subsequently be repeated over and over under automatic control. The programmer must be skilled in the task at hand and goes with the robot through the same motion he would apply for manual work. Sometimes, as in arc welding, automatic seam followers may teach the path rather than the human being. In either case, the path is entered into the controller's memory via the feedback devices and can be played back thereafter in unlimited repetitions leading the robot through the same path automatically. The speed with which the robot repeats is adjusted to the desired level.

Actually what is stored is not a "continuous" path. Rather, it is a sequence of points that, however, are so close together that they have the appearance of a continuous line.

Robots with off-line programming have the advantage that in order to prepare a program, the robot need not interrupt the work it is doing. Besides, where complex motions or hundreds of points are to be programmed, teach-in programming can be time-consuming and error-prone. With off-line programming, it is possible to utilize the program already stored in a CAD/CAM data bank and integrate the robot more closely into the total manufacturing system. In any case, off-line programming is possible without the robot. Once completed, the program is entered into the memory of the robot controller by various means, including discs and bubble memories.

TEACH COORDINATES

To obtain the smooth motion required of a robot with continuous-path control, it will usually have to move in all axes simultaneously. This is not necessarily so with point-to-point closed-loop control. Here, the different axes may move either sequentially or simultaneously, depending on the design of the robot. There are also methods to interpolate straight segments of the path with smooth transitions between segments.

The coordinates of the robot program determine location and movement of the end effector to a desired point. Associated with a programmed motion and its endpoint are two kinds of data:

1. The tool center point, or TCP, which is a point that lies on a vertical line extending from the wrist mounting surface to which the end effector is attached. The point on this line where the end effector contacts the workpiece is the TCP. The coordinates of unique locations of the TCP are programmed as the robot is taught its assigned task. The TCP follows the straight-line path as the arm moves from one programmed point to another. Some manufacturers refer to the tool tip instead of the TCP when defining the reference point.
2. The function, which specifies the operation to be performed when the robot reaches the programmed data point, such as closing or opening of the gripper.

The TCP of a Cincinnati Milacron robot, for example, is predefined by the programmer, who enters into the memory of the robot control the corresponding distance between the wrist mounting face and the TCP.

A robot may be programmed in any one of four coordinate systems:[2] rectangular, cylindrical, hand (or wrist), and world. Generally, the specific coordinate system can be chosen before programming begins, and can be changed at any time during programming. This permits programming one segment of the robot in one coordinate system, and switching to another for the next segment.

When the rectangular system is selected, the robot will respond to a command given by a key or button of a pendant, by moving the TCP along the X, Y, and Z axes. All axes are referenced to a point of origin, which is the home position for each robot axis. The rectangular coordinate system is used in most applications.

With cylindrical coordinates, the robot moves the TCP through a rotational motion about the Z-axis, along a radius in and out from the Z-axis, and along the Z-axis. This coordinate system is used frequently for machine loading and unloading and makes it easier to move the robot between two distant points in the work envelope.

The hand coordinate system (also called the tool coordinate system) is illustrated in Fig. 10-1 on the basis of Cincinnati Milacron's T^3 robot. In this case, the coordinates are those of the end of the robot arm, to which the robot control automatically adds the preprogrammed distance to the TCP.

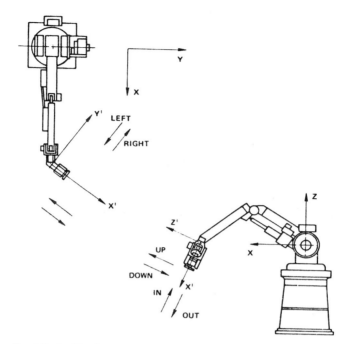

Fig. 10–1. Hand coordinate system. (*From Cunningham, Ref. 2*)

Consider a coordinate system that applies to the TCP with its axes X', Y', and Z'. Initially, these axes are parallel to the X-Y-Z coordinates of the robot. However, each time the TCP has been moved, a new X'-Y'-Z' coordinate is defined, and programming proceeds according to the new X'-Y'-Z' directions. When loading or unloading a machine, such motion, which is with respect to the orientation of the hand-held part, is helpful.

World coordinates are referenced to the earth, which means, in practical terms, to the shop floor or simply to the base of the robot. The X- and Y-axes are on the floor with Z pointing up.

ROBOT CONTROLLERS

To quote W. C. Carter:[3]

A robot controller is a device whose main task is to figure out the position each joint should assume in order that the (end effector) that the robot is holding will have its action point at the desired place with the desired orientation.

A robot controller thus has essentially three tasks: (1) to control one or more robot arms, where robot arm means the mechanical part of the robot (i.e., the robot itself without its supporting peripherals); (2) to provide programmable logic control for interfacing to other devices, such as teach pendants, mass memories, feedback devices, and so on; and (3) to communicate with higher-level devices that are intended to interfere with the running program or to call for a change in program of the robot controller.

The following discussion is based on Carter's article (see Reference 3).

Given a robot with a world reference frame, and with its joints and end effector in a given position, the tool tip can be determined. Assuming an articulated robot with waist, shoulder, elbow, wrist, hand, and fingers, the waist location is defined with respect to the world or base, next the shoulder is defined with respect to the waist, then the elbow is defined with respect to the shoulder, and so on. This succession from joint to joint in a forward direction is called the forward transform, and in a typical robot control language it is represented by:

FINGERTIP = WAIST:SHOULDER:ELBOW: WRIST: HAND:FINGER

The inverse transform, which is more frequently used, departs from the position the end effector is to assume and then analyzes the necessary joint positions. The fingertip can usually be assumed to hold a tool. It is the tool tip (or, as mentioned before, the TCP) that is really of interest. Therefore, a command to the robot to move its fingertip, could be expressed by:

MOVE BASE:FINGERTIP:TOOL

This permits significant simplification in programming. Suppose the robot is to drill with a 3-in.-long drill, and the drill is then changed to a 4-in. drill. By changing only the TOOL transform, this can be accomplished without having to change the whole program. On the other hand, if 5-in.-thick pallets were changed to 3-in. pallets, a simple change in the BASE transform would suffice.

Once the desired joint coordinates have been chosen, the space–time path to get to the desired point must be determined. There are several choices:

1. Independent motion of each joint to its desired location.
2. Synchronized motion of all joints, which requires the controller to make sure that each axis handles its move during exactly the same time interval, maintaining the synchronism of the various movements. The method results in a somewhat wavy path of the end effector, however.

3. Motion with linear interpolation to obtain a smooth path of the end effector. This requires:

(a) A line in space laid out to define the desired path.

(b) Verification of the position of the end effector on that line at regular intervals. Each verification is followed by a correction in the path. The frequency of the verification depends on desired speed, allowable acceleration, action at end of the line, and so forth. A typical approach might be to make the verification every 10 msec to some degree of approximation, and every 100 msec exactly.

(c) Generation of signals to be transmitted to the servo drives of the various joints for implementation of the desired motion.

MCS-60

The MCS-60 robot controller from Automation Intelligence is designed to cope with tasks as outlined above. It is based on the Intel line of microprocessors and on the Intel Multibus. The system has the capability to control up to six axes.

The controller is contained in a single cabinet. Auxiliary devices include a teach pendant, a CRT, a floppy-disc loader, and a vision subsystem. The cabinet is approximately 30 × 26 in. by 72 in. high and weighs about 300 lb, and is designed for a normal industrial environment, which includes temperatures of up to 122°F. Maximum power requirements, including six drives of 300 W each, is about 3 kW at 120 V, single phase, 47 to 63 Hz.

The control cabinet houses an IEEE P796 card cage, I/O modules, power supplies, vision preprocessor, and drives. The pendant, disc loader, and terminals are connected to the control cabinet by cables.

The P796 bus card cage houses several multilayer printed circuit cards, as follows:

- The RBDI, or Robot Basic Data Interface card, which provides the central MCS-60 processing functions. It is a single-board computer based on the Intel 8086 and includes three serial interfaces for the pendant control station and CRT terminal, together with either a debugging aid or a stand-alone vision system.

- The RMSC, or Robot Motion and Sync Card, which provides robot motion control functions.

- The RUTL, or Robot Utility Card, which performs basic interface functions from the front panel on the control cabinet. It also provides power monitoring, nonvolatile memory, and other miscellaneous functions.

- The RMEM, or Robot Common RAM Memory, which is common for all processors. In addition, each processor is provided with its own ROM and RAM memories.
- The RVIS, or Robot Vision Controller, a single-board computer that provides calibration and control functions as well as recognition algorithms for the vision system.
- The SBC 208, which is a floppy-disc controller.
- The RACC2, or Robot Axis Controller Card, which sends drive control information to the controller or receives it from there. It also drives the RAIC to control the drive regulating loops at the single drive level.
- The RAIC, or Robot Axis Interface Card, which consists of a motherboard with three daughter boards and provides the analog regulating functions to supply a reference voltage to three motor drive assemblies. It also includes digital I/O points for overvoltage or overtemperature sensors, and for limit switches indicating physical travel limits and calibration positions.

Figure 10–2 shows the functional arrangement of the MCS-60. The Basic Data Interface processor (BDI) is the principal controlling process for the man/machine interface, data handling devices interface, and other specialized interfaces. The VIS Vision System, if used, communicates directly with the GVS-41 vision preprocessor.

The MSC Motion Control System and BDI communicate with the RACC2 and I/O interface system over shared memory communication channels. I/O and analog I/O modules from the Numa-Logic line of I/O modules are connected to the Numa-Logic I/O bus.

The ACS, or Axis Control Subsystem, is provided with axis control, including position and speed loop closures, and management of certain digital I/O systems associated with the drives. The ACS uses one processor for communication and the other for axis regulation. The second processor communicates over an axis control bus to the cards of the AIC, or Axis Drives Interface. These cards provide analog speed position regulating capabilities. They also convert analog signals into digital signals for communication backup to the remainder of the system.

VAL® II

VAL II was created by Unimation Inc. for the operation of its UNIMATE and PUMA robots. It is referred to by Unimation as a robot language and control system with expanded computer logic and advanced communication capabilities. The VAL II internal CPU functions at rates 10 to 12 times faster than Unimation's earlier VAL model, and can hold twice as much

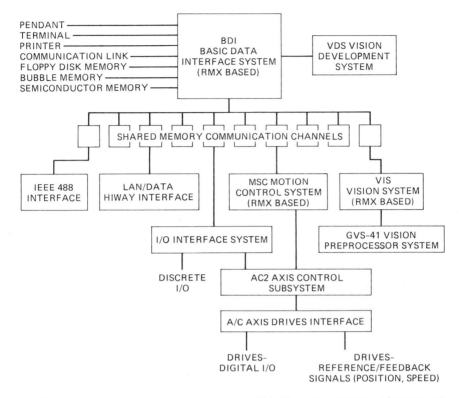

Fig. 10-2. Functional arrangement within the MCS-60 robot controller. (*Courtesy of Automation Intelligence, Inc.*)

memory. It can respond to critical events, such as an external stop command, within 28 msec.

Larson and Coppola,[4] in describing VAL II, differentiate between three levels of robot control systems:

- Level 1: These control systems offer robot users manual teaching capabilities and front-panel programming. Typical level-1 applications are spot welding and pick-and-place jobs.
- Level 2: These control systems use keyboards and CRT displays. Users can develop robot programs written in simple robot programming languages—programs that help them to enter motion, branching, coordinate transformation, and signal instructions. Palletizing and arc welding are typical applications.
- Level 3: These control systems use a robot programming language with

extended capabilities. They can handle the applications of levels 1 and 2 and also modify the robot arm's path from data transmitted through external sensing devices.

VAL II provides the functions required for level 3 systems. Its basic hardware comprises a printed circuit module, a servo control subsystem, a trajectory–human-interface subsystem, and a power distribution system.

The RAM unit has 64K bytes with battery backup. Another 64K CMOS memory contains the operating system. Also included is a double-sided disc unit that stores about 10,000 program steps per disc, representing an additional one megabyte of memory.

During operation, about half of each major clock cycle is devoted to arm-trajectory planning and computation. The remaining part is divided up into time slots for the CRT display, the supervisory computer, modification signals, main-program step execution, and process-control-program step execution. To ensure that the system can respond as quickly as possible, each task is scheduled to run for a certain amount of time during each major clock cycle. Any event will be received and acted upon within 28 msec of its occurrence.

The VAL II system provides a standard path-modification interface called Alter Mode. This means that the programmer specifies the coordinate system for path modification into either world or tool coordinates. All data received then move the tool tip relative to the selected coordinate system.

In Alter Mode, the programmer also selects whether or not the effects of the path-modification data are to be cumulative. In the noncumulative mode, only the most recent input data affect the tool tip location. In the cumulative mode, the effects of each data message are summed up and retained so that the tool-tip location reflects all past data.

The Alter Mode makes it easier to connect "smart" sensors to VAL II controlled robots. The sensors gather information, evaluate it, and convey it to the VAL II system. For example, a vision system can correct the position of a robot by using data received from its field of view. If the vision system notices that the robot's taught path does not correspond exactly to the profile of the workpiece, the VAL II system can be instructed to correct its motion so that the robot's path follows the path toward the piece correctly.

VAL II can be interfaced to a supervisory computer system through the Digital Data Communications Message Protocol (DDCMP) used by Digital Equipment Corp. in its network communication. The physical connection is a standard RS-232-C serial link running at 9600 bits/sec. This protocol provides error checking and automatic message transmission, as needed for factory communication. For each message sent, the initial portion consists of the current message number and the last received data-message number followed by a 2-byte redundancy check. If a message has not been received properly, it is retransmitted automatically without involving the VAL II layer of the system.

When a robot is to be coordinated with other machines or processes, it is necessary to connect various binary and analog status and control lines. The solution is usually to add a programmable logic controller to the system, and interface it to the various machines and to the operator control panel. To avoid the need for such a unit, and to facilitiate interfacing to the robot, VAL II allows a second application program to run concurrently with the main robot control program for the purpose of process control.

A process control (PC) program can execute all the standard VAL II program instructions, except for instructions that directly cause robot motion and action. It can read and write binary and analog I/O signals, read and modify program variables that are shared by the main robot control program, and use all computation facilities of VAL II. When active, the PC program is assigned a part of the clock cycle and runs concurrently with the main robot program.

TEACH PENDANT

Figure 10–3 shows the Allen-Bradley series 8200 robot control system.[5] It is a compact modular CNC that can be used with a 2- to 12-axis robot. Generally, the system consists of six main hardware components:

- Processor/power supply and I/O assembly (not shown in the illustration)
- CRT/keyboard
- Main control panel
- Secondary control panel
- Teach pendant (hand-held teach unit)
- Data cartridge recorder

The CRT/keyboard and data cartridge recorder are optional. A tape punch may also be used. Other optional items are a tape reader and a remote terminal.

The pendant, shown separately in Fig. 10–4, is equipped with an RS 422 serial interface that allows the use of a lightweight interconnection cable that connects to the main control panel. Standard cable lengths are 15, 25, and 50 ft. The weight of the pendant is 2.5 lb, and its size is 10 × 2.25 × 2.75 in.

To teach the robot its task programs, the robot arm is moved to various

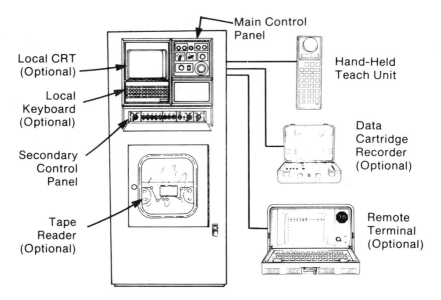

Fig. 10-3. Components of robot control system. (*Courtesy of Allen-Bradley*)

locations by using the jog keys of the pendant. Each step is recorded or stored in the robot control's processor RAM memory by pressing the pendant STORE key. An entire part program is constructed from a sequence of these individual steps.

The teach pendant is controlled by an onboard microprocessor and the EPROM memory. The microprocessor directs the communication between the robot control and the teach pendant. The microprocessor also monitors the switches, operates the status lights and displays, and provides diagnostic testing.

An emergency STOP pushbutton is prominently located on top of the pendant. Once it has been actuated, it has to be pulled to be released, in order to prevent accidental energizing of the robot.

Another safety feature consists of limiting the axis speeds during programming to approximately 20% of maximum speed.

The keyboard of the pendant has 32 keys, 16 of which are provided with light emitting diodes (LEDs) as indicators to show active modes and status during teaching. An eight–digit alphanumeric display on the pendant indicates the current step number and gives error-code or status messages. This includes built-in self-diagnostics to identify hardware faults on the alphanumeric display.

Three coordinate system keys are provided: JOINT, WORLD, and HAND. When world or hand coordinates are selected, it is also possible to

Fig. 10-4. Teach pendant. (*Courtesy of Allen-Bradley*)

determine as reference point either the tool center point (TOOL) or the wrist center point (WRIST). However, the world and hand coordinate systems can be selected only when the Allen-Bradley 8200 control is equipped with coordinate transformation capabilities.

Two modes of storing axis positions in the memory are provided. One mode, the absolute mode, consists of recording positions with the data relative to a present zero point. The other is the incremental mode, which stores positions relative to the last recorded data point.

Six-axis jog capability is standard. Two keys are provided per axis to allow jogging in either direction. The path generated by resultant axis motion is dependent on the active coordinate system; that is, straight-line motion occurs in the world coordinate system, and so on.

When active, the HOME key indicates that an axis will move to its home position when initiated by a jog pushbutton. The home LED turns off when the axis reaches its home position. This function provides the ability to initialize the robot to an absolute coordinate system. Each point in the working area is then defined by absolute coordinates relative to a predefined robot zero point.

The INC key is used to select the incremental jog mode. In this mode, the axis jog pushbutton moves the selected axis a predetermined distance. Two different increment lengths can be selected through the HI and LO pushbuttons.

Four function keys are provided for tool functions specified by the robot manufacturer. They are designated as follows: (1) gripper 1 open, (2) gripper 1 close, (3) gripper 2 open, and (4) gripper 2 close.

Once a program is completed and entered in the memory, it may again be connected to the teach pendant for editing. A group of keys is provided for this purpose as well as for teaching. They are:

- The STORE key, to save a position as a program step in either the teach or edit mode.
- The DLT POSN key, to delete a step when in the edit mode. Any position may be modified through combination of the DLT POSN and STORE functions.
- The STEP FOR key, to increment the active program step number to the next higher step. Axis motion occurs to the corresponding position when the LED of the REPLAY key is on.
- The REPLAY key, to enable/disable axis motion to the position relative to the active step number shown on the alphanumeric display.

TEACHING BY PENDANT AND CONTROL CONSOLE

The commands that can be entered by the pendant or box are limited, in spite of their versatility. The more sophisticated robot needs more elaborate methods of programming. Thus, in the following, the T^3 566 robot and its heavy-duty version, the T^3 586, from Cincinnati Milacron are used as examples of teaching a robot by means of the pendant and control console.[6]

Both robots are hydraulically powered, servo-operated, and computer-controlled industrial robot systems designed for maximum application flexibility. The arm is of a jointed configuration (Fig. 2–1), with six rotary axes of movement. These axes consist of the arm sweep, shoulder swivel, and elbow extension, along with pitch, yaw, and roll orientations of the wrist.

Up to 1750 programmed points can be stored in memory. With the addition of a floppy disc, storage becomes available for up to seven programs of 1750 points each. Interfacing can be provided for up to 52 input signals. A maximum of 44 output signals also are available. These signals can be used to control the end effector or to interface with surrounding equipment.

The controlled-path operation of the robot is shown in Fig. 10-5. It provides simultaneous control of all axes during both teaching and automatic operation but not during the manual mode described below. During teaching, the system lets the programmer position the robot to the desired location by commanding a direction of motion for the tool, as opposed to directing an individual axis. Also during teaching, the programmer is not required to generate the desired path, but only to identify individual path endpoints. During replay and automatic operation, the computer control directs the arm along a straight-line path between the programmed endpoints, at a specified velocity, with acceleration and deceleration spans provided.

The T^3566 and T^3586 robots are not only programmed by the teach pendant but also by the control console, as shown in Fig. 10-6. There are three modes of operation: manual, teach, and automatic. The control console is used to select the mode desired.

In the manual mode, the controlled-path concept, as mentioned above, does not apply. The programmer can move each axis of the arm independently of all the other axes. This is done through use of the pendant. The primary purpose of the manual mode is to move to the home position; that

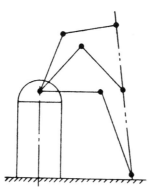

Fig. 10-5. Controlled-path operation.

Fig. 10-6. Control console. (*Courtesy of Cincinnati Milacron*)

is, the point of origin of the six coordinates. Before the programmer can enter either the teach or automatic modes, he must have moved to the home position.

Teach Mode

This mode is for teaching the robot its designated task. A system of coordinates is selected, and the axes move in simultaneous fashion to the desired point. In addition to locating the TCP and orienting the wrist, the programmer specifies the speed at which the arm is to move between the various programmed points and the function to be performed at each of these points. Standard functions available with the T^3 systems include but are not limited to the following:[7]

- Delay—Motion of the robot is stopped for a defined period of time before it is resumed.
- Wait—Motion of the robot is stopped until a signal is received from some external source. The function may also be made conditional upon some combination of various signals.
- Output—Motion is stopped, and a signal is issued to some auxiliary equipment. This signal may be pulsed or simply turned on and off.
- Continue—Motion is not stopped at the programmed point, but continues on to the next point. Passing through this point may or may not be signaled to the robot control.
- Tool—Motion is stopped until one or two defined tools (end effectors) have been attached to the robot wrist.
- Perform—The sequence of points in the trajectory of the robot is interrupted by this function. Upon encountering a sequence entry point for a relocatable or indexed sequence as described below, the robot checks for a signal or combination of signals to determine if it should enter the sequence path or continue along its present path.
- NOP—No operation is to be performed at the programmed data point.

Besides manual and teach modes, the automatic mode is provided. In this mode, the arm will continue to move through its taught cycle until either the END OF CYCLE STOP or the INTERRUPT button is depressed. The END OF CYCLE STOP stops the arm the next time it reaches the last point in its cycle, whereupon the operator can send the arm to its home position by pushing the RETURN button. The INTERRUPT button causes the robot arm to stop and to resume its motion when the CYCLE START button is depressed.

Whenever something goes wrong with the system, or the operator issues an unacceptable command, an error code appears in the upper left portion of the CRT. The code gives an indication of the nature of the error and serves as a guide in correcting problems.

There are a number of special commands that can be used to aid teaching, as described in the following paragraphs.

Data Display, Copy, and Modify

The DISPLAY feature lets the operator view the six coordinate values associated with the TCP. The command COPY permits duplicating any specific part of the program, and MODIFY lets the operator adjust the coordinates or other information associated with any point or sequence of points. All of these commands are carried out through the use of the CRT and keyboard.

Decision Making

Decision making allows a robot to deviate from its normal-path program and perform other tasks based on changes in external events.[8] A robot generally is part of a working environment. Its actions must be coordinated with functions of conveyers, machine tools, other robots, and so forth. It needs interconnections by limit switches to signal when parts are in position or by timing signals to synchronize it with other processes. Sometimes, it is desirable to have the robot controller communicate with other digital systems through interconnect buses. This, for example, permits decisions being made by another computer to be transmitted to the robot.

Development of more and more sophisticated sensors makes decision-making features an important part of advanced robots. Inherent programming possibilities provided in the robot control system are part of a decision-making process by interconnection with outside sensors. Some examples are described below.

Palletizing

Palletizing and similar operations frequently consist of positioning equal-size packages one above the other. Figure 10–7 shows the robot gripper in the process of doing this. Once the gripper has reached the top of the previously placed package, it has to release the package it is carrying. It then must proceed to fetch the next one for further stacking.

When items must be stacked or unstacked, only two points—one above the stack and one at the bottom of the stack—need to be programmed. A tactile sensor on the robot's gripper is used to send an interrupt signal whenever an object is obstructing the path. The interrupt signal causes arm motion to cease immediately, and the robot proceeds to the next step in its cycle. END OF CYCLE STOP stops the arm the next time it reaches the last point in its cycle, whereupon the operator can send the arm to its home position by pushing the RETURN button. The INTERRUPT button causes the robot arm to stop, and then to resume its motion when the CYCLE START button is depressed.

Relocatable and Indexed Sequences

Figure 10–8 shows the trajectory of a robot. The relocatable sequence consists of the program points NN.001, NN.002, and NN.003 and then a return to the main cycle. Sequence entry points can be provided and repeated at any point in the trajectory. Programming of a relocatable sequence elimi-

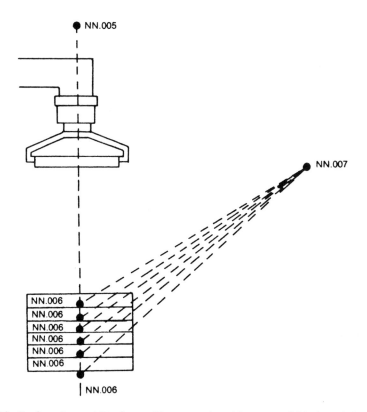

Fig. 10–7. Search capability in stacking operation. (*Courtesy of Cincinnati Milacron*)

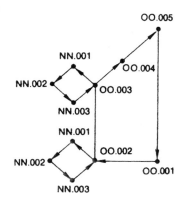

Fig. 10–8. Relocatable sequence. (*Courtesy of Cincinnati Milacron*)

nates redundant programming effort, and reduces the number of sequences that need to be stored.

An indexed sequence differs somewhat from the relocatable sequence. The sequence is cut into several segments. The first time an indexed sequence is requested, the robot moves through the first of these segments and then proceeds along the main cycle. The second time, it goes through the next segment. Thus, each time the indexed sequence is entered, the arm moves through a different segment of the sequence.

Relocatable and indexed sequences can be used together for palletizing operations. In this case, the indexed sequence identifies unique pallet locations while the relocatable sequence contains the redundant palletizing maneuvers.

Stationary-Base Tracking

A special feature of Milacron's T^3 robots is their ability to track objects on a moving conveyor while the arm remains stationary. This is called stationary-base tracking, and is illustrated in Fig. 10-9. Here, the robot is taught its assigned task while the conveyor is stationary. The conveyor is

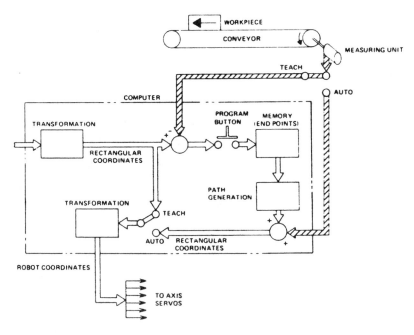

Fig. 10–9. Stationary-base tracking. (*Courtesy of Cincinnati Milacron*)

provided with a measuring unit—usually a resolver. If the conveyor runs parallel to the Y-axis of the arm, the reading obtained from the resolver is subtracted from the Y-coordinate value of the robot, and the difference is stored each time the PROGRAM button is pushed.

During the automatic mode, each time a set of rectangular coordinates is computed in the robot controller, the reading from the conveyor's resolver is added to the Y-coordinate value before the transformation to robot coordinates takes place. In this way, the robot is able to perform operations on a moving part without interruption while the conveyor is moving forward or backward at a constant or variable speed. Even a stopped conveyor does not interrupt the robot action on the part resting on the conveyor. In addition, the robot can work with four separate conveyors, if required, even if the conveyors run along three different axes.

Expanded Instruction Capabilities

A typical example of the variety of instructions that can be programmed into a robot is provided by the ASEA IRB robot system. Here, the instructions include the following:

- Movement between two points, at optional constant speed in mm/sec and three optional degrees of accuracy of positioning.
- TCP programmable for nine different positions relative to the wrist center. The exact positions can be determined by manual running of the robot.
- Programmed reference points. When the reference point instruction is executed, the part of the program that follows the reference point is carried out relative to the position and not relative to its orientation.
- Coordinate transformation by displacement in an optional direction and rotation around the vertical axis in the base coordinate system. Up to five different coordinate transformations can be programmed. Any of the transformations may be called at optional positions in the program.
- Circular movement defined by three points along a circular arc in space.
- Weaving movement. This is important for arc welding by the robot (see p. 446). Weaving is stored in a subprogram and superimposed on straight-line positioning.
- Program control of or by peripheral equipment, through inputs and outputs, or with the assistance of an internal register. Outputs can be set at "0" or "1" and can be inverted and pulsed. Positive and negative numbers can be entered in the register.

There are also logical instructions such as:

- Jump either conditionally or unconditionally to optional instruction addresses within a program.
- Wait in a program sequence that can be generated conditionally or unconditionally. The conditional wait time can be programmed for 0 to 320 sec. The unconditional wait time can be programmed for 0 to 100 sec.
- Interrupt before proceeding with the immediate execution of the next instruction or of one of five subprograms, depending on which of the digital inputs is activated.
- Pick and place parts of different shapes. Each shape pattern is programmed in the form of an individual subprogram and is treated independently of the others.
- Call-up of subprograms linked in three levels. The subprogram called can be repeated up to 99 times.
- Reading of optional program blocks from floppy discs during continuous operation.

The ASEA IRB robot system uses a joystick with three degrees of freedom. It is located on the hand-held programming unit (teach pendant) to control the movements of the TCP in rectangular or cylindrical base-oriented coordinates, or in rectangular wrist-oriented coordinates. When the program is executed, the movements are carried out in either straight-line or robot-axis coordinates.

An application program consists of a main program and an optional number of subprograms. It must contain instructions that specify speed, the coordinate system, the TCP, and the frame location. An individual program step consists of an instruction and an instruction number, either with or without additional information known as an "argument."

Thus, a program consisting of seven instructions, numbered 10 to 70, would be executed in the following manner:

```
10   SPEED 500 MM/S MAX SPEED 1000 MM/S
     (Basic speed for the following instructions.)
20   POS V = 100
     (Position at 100% of basic speed.)
30   WAIT TIME 3.5 S
     (Wait for 3.5 sec and then proceed with next instruction.)
40   POS V = 50 FIN
     (Position at fine point, small zero zone, at 50% of basic speed.)
50   CALL PROG 7
     (Call up subprogram 7.)
```

60 SET OUT 5
   (Set output 5.)
70 RETURN
   (End of program and return to 10.)

In addition to the standard program, additional programs are available, such as:

- Search with up to three sensors. The search stop can be delayed by 0.5 sec. Searching is carried out as either a linear search between two points, or by directional search commenced by linear scanning followed by free search in accordance with preprogrammed correction vectors.
- Control of speed by an external sensor. Time lag is automatically compensated for, if the required speed change is more than 25% of the programmed speed within 50 msec.
- Contour following in accordance with preprogrammed correction vectors. Up to three sensors can be used simultaneously.
- Addressing of up to 16 sensors. Sensors may be of the digital type, with up to eight bits plus one sign bit, or they may be of the analog type. Connections are made to the digital or analog inputs of the robot control system.

Another feature of the ASEA IRB robot system and other models is the ability for connection to an external computer either to serve as a program bank or to control the movements of the robot directly.

PROGRAMMABLE CONTROLLERS

The programmable controller, or PC for short (not to be confused with PC-personal computers), is based on solid-state digital logic. It is primarily intended to take the place of electromechanical relay panels in control applications. However, electromechanical relays have to be rewired when the program is changed. PCs, on the other hand, are simply reprogrammed by means of a keyboard and a CRT display. Because of their solid-state flexibility they are frequently designed to store complex procedures and sequences in their memory section and can be used for pick-and-place robots that are equipped with limit switches or other on-off position-feedback devices interconnected with the PC.

A PC may contain a sequencer capable of sequencing 999 steps with 64 contacts in each step. Each contact can be programmed to be either on or off for each step. In any one step, one group of contacts typically is programmed to indicate which robot axes are to move during the step, and in which direction.

Position sensors can feed back to the PC, and the status of each is compared with the corresponding sequencer contacts. When the status of 11 position indicators matches those of the sequencer step being executed, the sequencer is free to move to the next step.

A PC can be programmed by the teach box of the robot. To do this, the robot is moved to the desired position. Pressing the TEACH button calls up a preprogrammed instruction to copy the on–off status of all pertinent inputs to be copied into a sequencer step. By moving the robot through its steps sequentially, pressing the TEACH button at every endpoint, a PC is taught the sequence of movements.

PCs can also be equipped with arithmetic modules to adjust the acceleration rates and speed for each axis, so that all movements terminate simultaneously.

PROGRAMMING LANGUAGES

A number of languages are in use for programming robots, such as American Robot Corp.'s AR-BASIC and AR-SMART, Automatix's RAIL, GCA's CIMPLER, General Electric's HELP, Hitachi's HAL, IBM's AML, Unimation's VAL II, and so on. There are a number of serious endeavors afoot to create a standard language, particularly for off-line programming, but as Ottinger and Stauffer[9] point out:

> As an indication of the challenge presented in this area, a standard language used for off-line programming should have the following capabilities: (1) be general purpose, (2) allow computation from sensors, (3) work with other devices, (4) work in a real-time environment, (5) be able to obtain data when it is needed, (6) be able to check conditions to synchronize events, (7) be able to easily express manipulator positions in space, (8) support debugging and testing, (9) be independent of the physical configuration and kinematic features of a particular arm, and (10) be accessible to inexperienced operators.

It can readily be seen that this is a tremendous task.

In the following pages, programming languages and their use will be discussed in terms of increasing versatility—and complexity. However, it should also be realized that even the most complex-appearing language is nothing but a typing in of commands, the sequence and meaning of which are read from tables supplied by the manufacturer. What should not be underestimated, however, is that elaborate programs require considerable preparation time and editing, a challenging but often tedious endeavor, before they are ready to be put into practice.

Program Preparation

Programming languages as discussed here are high-level languages. They are problem-oriented programming languages as distinguished from the more complex machine-oriented programming language. The problem-oriented language is automatically converted into the machine language which implements the instructions.

A building-block approach is generally used to prepare a program. In doing so, a top–down design is frequently applied. It is also referred to by other names such as stepwise refinement, hierarchical design, structured programming, and so forth. A typical preliminary top–down design of a simple robot application program, as given by Richard W. Peck,[10] is shown in Fig. 10-10. Here, the task of the robot is to insert part A into part B. This stands at the top of the diagram. The task is then divided into three steps: (1) get part A, (2) move to near part B, and (3) place part A in part B. Steps 1 and 3 need additional breakdowns, and the grabbing and releasing of part A requires still further details.

The important principle is that in representing each new level of the program, the functional elements of the program become defined. The completed top–down design gives the steps the programmer can tackle one by one to produce a program.

Next the programmer will use certain logic steps. Peck recommends par-

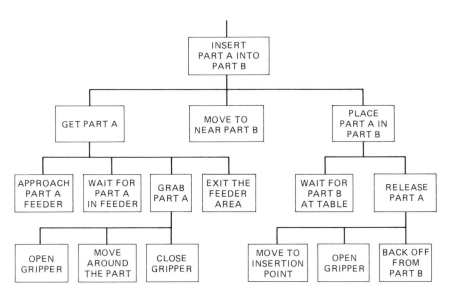

Fig. 10-10.   Preliminary top-down design of a simple robot. (*From Peck, Ref. 10*)

ticularly use of the WHAT-IF logic, such as WHAT-IF the part does not fit, or WHAT-IF the gripper drops the part, or WHAT-IF the gripper does not close. The last is shown in the flow diagram of Fig. 10–11.

Here, the first step is to pulse the digital output channel that closes the gripper, after which a waiting period of 100 msec is inserted. Then the question is asked: Gripper still open? If the answer is "no," then it is clear that the closing of the gripper has functioned properly, and the WHAT-IF logic can be stopped. If the answer is "yes," it means the gripper has not closed. The required step is then to pulse the digital output channel that opens the gripper. A delay of 50 msec follows. Now, the digital output channel that closes the gripper is again pulsed, a waiting period of 200 msec in inserted, and the logic returns to the question: Gripper still open? Another step in

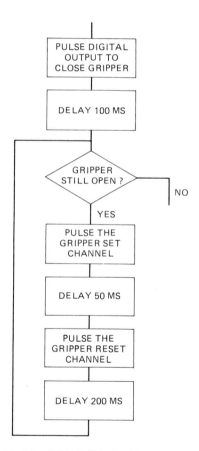

Fig. 10–11.   WHAT-IF logic. (*From Peck, Ref. 10*)

preparing the program is to determine common subroutines in the program. Subroutines are parts of a program that may be repeatedly used. They can be called up by the main routine of the program at any time. Thus, the previously described palletizing routine for stacking objects represents a typical subroutine.

On the basis of the above considerations, Peck has suggested a four-step preparation before a robot is programmed with high-level languages. These steps are:

1. Generate a chronological list of actions to be performed by the robotic system, assuming everything in the system functions perfectly.
2. Produce a preliminary top–down design based on the previously generated list of actions.
3. Generate a detailed design. Add to the design the WHAT-IF's, modifying the design to include these tests and their associated processing.
4. Identify common functions performed throughout the program or that already exist from previous applications, modifying the design to share the common functions and include the already existing modules.

A Teach-Pendant Language

AR-SMART is a proprietary command language that was developed by the American Robot Corporation. It is used with a teach pendant with a 20-key pad, as shown in Fig. 10–12. Each of these keys can have up to four functions. Shifting from one function to another is accomplished by three additional keys on the pendant. Two 16-character display lines are provided for alpha-numerical messages.

A joystick is used to position the robot during the teaching phase. The joystick permits control of the robot in three dimensions. Furthermore, it contains speed control, an axis select switch (arm or wrist), a run/stop switch, and a switch for cutting off the motor power.

The language contains a variety of commands. At the simplest level, programs can be generated by teaching points and walking the robot through a control sequence. More sophisticated programming techniques permit mathematical computations, operator interaction, conditional branching for subprograms, dynamic definitions of points, continuous path motion, control and interaction with external sensors including other robots, and multiple frame and tool definitions.

Most of the commands in AR-SMART consist of a single keystroke. An abbreviation of some sort representing the command is displayed as soon as it is entered, and any information required to execute the command is requested on the display. When additional information is entered, editing

Fig. 10-12. Teach pendant for AR-SMART. (*Courtesy of American Robot Corp.*)

keys are used to backspace, cancel the entered input, or cancel an entire command. Typical entries are:

BSP — backspace a digit when entering numbers
ENT — terminate a line of input or response to request
CAN — clear a line of input or cancel a command
YES — affirmative response to yes/no question
NO   — negative response to yes/no question

There are three modes in which the pendant can be used: immediate mode, learn mode, and program mode, which differ as follows:

- In immediate mode, commands are executed as entered. No record is kept of these commands.
- In program mode, commands are accepted from the teach pendant and stored for later use.
- In learn mode, the other two modes are combined; that is, commands are executed as entered, but also stored in the teach-pendant memory for later recall.

The keys on the pendant that are used for switching to any of these three modes are IMM for immediate mode, PGM for program mode, and LRN for learn mode.

There is also selection among three different motion operations. These are setup operations, position operations, and movement operations.

The mode of operation has to be determined before the robot can be moved. Examples are the command to start up the robot, or to calibrate it, or to determine whether the elbow should be kept in an up or down position. Typical commands for these purposes are:

CLB ⟨position number⟩—calibration procedure
STS—display robot status
⟨0:1⟩, that is, choice of 0 or 1
MVM ⟨0:1⟩—move made (0 = joint; 1 = straight line;
2 = circular)
ELM ⟨0:1⟩—elbow mode (0 = Up; 1 = Down)
FBM ⟨0:1⟩—front back mode (0 = Front; 1 = Back)
SSS ⟨interactive⟩—set software stops
SPD—set desired speed
SMV—set maximum speed

Another task is to define point, frame, or tool operations and to provide the necessary parameters. Points are associated with the trajectory of the robot. A point sequence will be the path followed by the end effector. Any point in the robot's workspace can be defined to be the frame of reference relative to which other points may be defined.

Tool operations use the TCP as reference. Point, frame, and tool operation may be defined analytically using the known coordinates and offsets, or they may be taught interactively. Among the symbols used are:

TDF ⟨tool vector⟩   —define a tool in terms of a point
STT ⟨tool number⟩   —set tool type
RCT ⟨tool number⟩   —recall a tool type to the registers
FDF ⟨point number⟩ —define a frame in terms of three points

SFT ⟨point number⟩ —set frame to a point
RCP ⟨point number⟩—recall a point to register

Movement operation are the actual commands to position the robot to a desired point. Such commands are:

MOV ⟨point number⟩—move to the specified point
STP   —stop executing commands until the arm reaches the desired point

There are also external control operations that permit either control or sensing of external devices. They provide access to external I/O lines, external A/D and D/A converters, and an optional gripper on the end of the robot arm. To incorporate them into the program, the following commands apply:

IN ⟨address⟩   —read data from the specified I/O line or A/D converter
OUT ⟨address⟩ —send data over specified I/O line or D/A converter
WAT ⟨number⟩—delay execution in milli-seconds
PAS ⟨number⟩ —wait for signal from the operator before proceeding

There are many more programming possibilities. To illustrate the method, however, a sample program is described below.

A "source" pallet arrives on a conveyor that stops at a designated location. As shown in the diagram of Fig. 10–13, the source pallet contains four horizontally aligned pins. The "destination" pallet can carry nine vertically aligned pins. The robot must take out the pins from the source pallet, turn them from the horizontal into a vertical position, and locate them in the destination pallet. After each pallet has been unloaded, a signal is given to the source conveyor to move on to the next pallet, while the robot awaits its arrival. A pallet is not sent on its way until it is completely empty or completely full.

The orientation of the pallets has been taught to the robot such that the Z-axis of the pallets is parallel to the center axis of the pins. This means that for the incoming pins the Z-axis is horizontal, while for the outgoing pins, it is vertical. This relieves the programmer from any further consideration of the differences in orientation among pallet, pins, and robot.

The following lines for external devices are provided:

- Output Line 10—A 100-msec pulse over this line causes the source conveyor to continue.

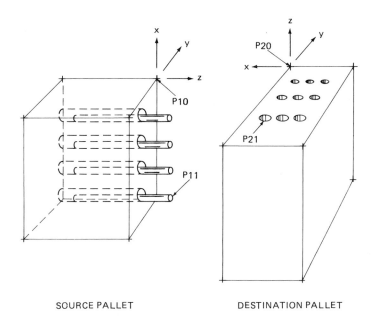

SOURCE PALLET   DESTINATION PALLET

Fig. 10-13. Diagram of sample program. (*Courtesy of American Robot Corp.*)

- Output Line 20—A 250-msec pulse over this line causes the destination conveyor to continue.
- Input Line 11—A new source pallet is in place when the signal over this line is on binary one.
- Input Line 21—A new destination pallet is in place when the signal over this line is on binary one.

There are additional points that had to be taught initially, such as:

P1 —a "safety" point at $Z = 2$ in. on both frames; that is, the frame of the source as well as of the destination pallet

P2 —point of origin on both frames

P10—frame of source pallet; that is, its location and orientation

P11—position of first pin on the source pallet

P12—distance from one pin to next on source pallet

P16—internally, absolute source pallet frame of first point

P17—internally, absolute source pallet frame of current point

P20—frame of destination pallet; that is, its location and orientation

P21—position of first receiving hole on destination pallet

P22—distance to next following receiving hole on destination pallet along the X axis

P23—distance from end of one row of receiving holes on destination pallet to beginning of the next

P26—internally, absolute destination pallet frame of first point

P27—internally, absolute destination pallet frame of current point

T10—TCP of the gripper

Registers in the memory that are used in the program are:

R10—number of pins remaining in the source pallet

R20—number of empty slots in the destination pallet

R22—number of pins placed in destination pallet

R30—number of destination pallets loaded

The actions to be taken in the program are as follows:

- Setup: (a) Clear appropriate registers; (b) request a source pallet; (c) request a destination pallet.
- Main loop: (a) Display number of destination pallets loaded; (b) move to safety point near source pallet; (c) compute next location on destination pallet during move; (d) remove pin from source pallet; (e) if source pallet is empty, request another one; (f) move to safety point near destination pallet; (g) compute next location on source pallet, and wait until it gets there; (h) load pin into destination pallet; (i) if destination pallet is full, increment number of pallets loaded and request a new destination pallet; (j) continue main loop.

To execute all this, requires the following commands (the ⟨⟨⟨ . . . ⟩⟩⟩ indicates titles that separate the different steps and is not part of the program):

⟨⟨⟨ START OF THE PROGRAM ⟩⟩⟩

```
=              clear any pending math operations
num 0          clear the accumulator register
sto r30        number of pallets loaded is zero
sto r10        no more source pins
sto r20        no room in destination pallet

num 1
out p11        request a source pallet
out p21        request a destination pallet
```

```
wat 100          delay for source pallet
num 0            clear accumulator to drop request line
out p11          clear source pallet request
wat 150          delay additional time for destination pallet
out p12          clear destination pallet request

stt t10          establish the current TCP

sft p10          source base frame
rcp p11          first pin
dpt p16          make this a new absolute base
abs p16
rcp p16          and establish current position at it
dpt p17
sft p20          destination base frame
rcp p21          first pin
dpt p26          make this a new absolute base
abs p26
```

⟨⟨⟨ THE MAIN LOOP ⟩⟩⟩

```
lab 1100         top of the main loop
dsp r30          show number of destination pallets loaded
```

⟨⟨⟨ MOVE TO SAFETY POINT OVER SOURCE PALLET ⟩⟩⟩

```
spd 40           move as fast as possible
mvm 0            joint mode
sft p 17         establish source frame
mov pl           move to safety point
```

⟨⟨⟨WHILE MOVING TO SAFETY POINT, COMPUTE NEXT LOCATION ON
DESTINATION PALLET⟩⟩⟩

```
rcl r20          still working on this destination pallet?
ifg O
gts 1105         yes, figure out location of next point
```

. . . new pallet . . .

```
rcp p26          reset current point to first one
dpt p27
num O
sto r22          nothing in this row
gts 1109
```

. . . see if current row or next one . . .

```
lab 1105
sft p27          establish current frame as current destination
rcp p22          assume next pin is in the current row
rcl r22          see if it really is
+
num 1
ifl 3            is this the third time through?
gts 1106         no continue ahead
rcp p23          yes, proceed to beginning of next row

lab 1106
sto r22          reset record of number of pins in row
dpt p27          relative point
abs p27          new frame base for next pin
lab 1109         . . . end of next pin computations
```

⟨⟨⟨ CHECK TO SEE IF NEW SOURCE PALLET WAS REQUESTED ⟩⟩⟩

```
rcl r10
ifn 0
gts 1120         if r10 ! = 0, continue program
```

⟨⟨⟨ WAIT FOR A NEW SOURCE PALLET ⟩⟩⟩

```
stp              stop motion until pallet is there
lab 1105
in p11           read source pallet sensor
ife 0            as long as it is zero, wait
gts 1105
num 4            reset number of items on source pallet
sto r10
```

⟨⟨⟨ REMOVE PIN FROM SOURCE PALLET ⟩⟩⟩

```
lab 1120
sft p17          reset frame just in case
spd 5            slow down to remove pin
mvm 1            straight line movement
stp              stop over the pin
num 1            activate gripper
mov p1           remove pin and back out to safety position
```

⟨⟨⟨ MOVE TO SAFETY POINT OVER DESTINATION PALLET ⟩⟩⟩

```
spd 40           move as fast as possible
mvm 0            joint mode
sft p27          establish destination frame
mov pl           move to safety point
```

⟨⟨⟨ WHILE MOVING TO DESTINATION PALLET, SEE IF WE NEED A NEW SOURCE PALLET ⟩⟩⟩

```
rcl r10          number of pins remaining in source pallet
—
num 1
=                number of pins now remaining in source pallet
sto r10
ifg 0            if there is anything left
gts 1130         continue ahead
num 1            otherwise request pallet
out p10
wat 150          wait for source line to begin to move
num 0
out p10
rcp p16          reset source frame to original
dpt p17
gts 1135         continue with program
```

⟨⟨⟨ COMPUTE NEXT X-LOCATION ⟩⟩⟩

```
lab 1130
sft p17          set base to current frame
rcp p12          increment to next X-location
dpt p17          copy it to a new point
abs p17          create new frame
lab 1135         done computing next source point
```

⟨⟨⟨ CHECK TO SEE IF NEW DESTINATION PALLET WAS REQUESTED ⟩⟩⟩

```
rcl r20          was a new destination pallet requested?
ifg 0
gts 1150         if r20 ! = 0, continue program
```

⟨⟨⟨ WAIT FOR NEW DESTINATION PALLET ⟩⟩⟩

```
stp              stop motion until pallet is there
lab 1145         read source pallet sensor
ife 0            as long as it is zero, wait
gts 1145
num 9
sto r20          number of slots remaining on destination pallet
```

⟨⟨⟨ INSERT PIN IN DESTINATION PALLET ⟩⟩⟩

```
lab 1150
sft p27          re-establish destination frame
spd 5            slow down to remove pin
```

```
mvm 1          straight line movement
mov P2         insert pin
stp            stop fully inserted
num 0          release gripper
mov pl         back out to safety position
```

⟨⟨⟨ SEE IF A NEW DESTINATION PALLET IS NEEDED ⟩⟩⟩

```
rcl r 20
—
num 1
=
sto r20
ifg 0          if ⟩ 0, system is ready to continue
gts 1160

num 1
out p20        activate i/o request
wat 250
num 0
out p20        clear i/o request

rcl r30        update number of pallets loaded
num 1
sto r30
```

⟨⟨⟨ END OF MAIN LOOP ⟩⟩⟩

```
lab 1160
gts 1100
```

In addition to AR-SMART, an enhanced version, AR-SMART XT, was developed by American Robot Corporation. Added capabilities, such as automatic error recovery and computer communications, were provided. Furthermore, AR-SMART XT enables American Robot's MERLIN Intelligent Robots to detect when the robot collides and to stop motion instantaneously, executing a sequence of commands selected by the user to shut down the application safely.

AR-SMART XT allows a host computer to make strategic decisions, to command the robot to a position in space, and to determine which subroutine is to be executed. Circular interpolation enables a robot to interpolate a circle after only three points have been entered into the system. This is particularly useful in complex motions in welding, sealing/gluing, routing, and cutting applications. All these functions are programmable, but it has been shown that careful preparation of the many steps to be programmed is necessary before the robot is capable of executing complex tasks automatically.

OFF-LINE PROGRAMMING

As programs grow more complex and their preparation is more time-consuming, it becomes necessary to develop a method that permits off-line programming. This enables programming to occur away from the distractions associated with the manufacturing atmosphere, and without interrupting the operation of a robot that is actively engaged in its routine but is earmarked to be reprogrammed. Also, the integration of the robot and other machines into an integrated production process combined with CAD/CAM facilities, calls for handling of the robot programs by specialists, who prepare a program systematically and then insert it into the robot when the need arises. In off-line programming it can be assumed that the robot forms part of a work cell, and that the programmer needs a graphic representation on his CRT screen, not only of the robot and its motions but also of the work environment.

A number of systems have been developed for such off-line programming. Among them are the RoboTeach from General Motors, the Robographix from Computervision Corp., the Robot-SIM from General Electric's Calma Co., the GMF SmartWare from GMF Corp., the programming system by Intergraph Co., the AutoMod and AutoGram from AutoSimulations, Inc., the I-GRIP from Deneb Co., the PLACE, COMMAND, and BUILD modules from McDonnell Douglas Corp. (McAuto), and many others. The following discussion describes two examples: RoboTeach and Robographix.

RoboTeach

This short description of the RoboTeach system is based on a paper by Robert Tilove, Vadim Shapiro, and Mary Pickett.[11]

The system seeks to combine the advantages of off-line robot programming languages with interactive work cell layout and program simulation. It includes selection of robots and endeffectors, specification of physical arrangement of robots, studies of the ability to reach an object, and so on.

RoboTeach uses a database that contains geometric descriptions of tools, parts, fixtures, and other design or manufacturing data, such as desired weld paths for automotive spot welding applications and so forth. The database is contained in General Motors Corporate Graphics System, CGS. CGS supports a large variety of interactive graphic systems for computer-aided design and analysis, including GMSolid, a modeling system that was developed at General Motors primarily for the purposes of CAD/CAM. Design data generated in other systems, including those that are commercially available, can be transferred to CGS for additional processing.

RoboTeach contains a work cell layout subsystem, a program generation and editing subsystem, and a simulation subsystem.

The work cell layout subsystem enables the user to select robots for inclusion in the cell model, specify the physical arrangement of objects, test points for reachability, and so on.

The program generation and editing subsystem supports the specification of robot programs. Graphical displays of the design data and interactive techniques are used wherever possible; for example, to specify values for geometric parameters in the program, to select program statements from menus, and so on. The program generation subsystem produces the specification for the task sequence of the robot. Task sequences are stored in the database along with the robot designation and other data describing the work cell.

The simulation subsystem makes use of the modeled robots, task sequences, and other stored data. It graphically simulates the effect of executing a number of robot programs concurrently. In addition, it performs a certain amount of automatic verification, produces cycle time estimates, and so forth.,

Work cell layout, program generation and editing, program simulation, and various computer-aided design systems are available in an integrated environment. An error detected at simulation time can be remedied through program modification, cell layout modification, or the redesign of tool or part fixtures. Programs developed and debugged in this manner then may be translated by a final subsystem from their internal form into the format accepted by the robot computer and transferred.

Robographix

The Robographix system was developed by Computervision Corp. and provides four major functions:[12]

1. Preparation of the work cell.
2. Creation of the robot work path.
3. Simulation of the robot in motion.
4. Production of a digital output to direct robot motion.

In creating the work cell, the programmer can select and locate memory-stored three-dimensional graphic models of robots, machines, conveyor belts, fixtures and tooling, and various other components. Currently included in the Robographix software package are kinematic models of Cincinnati Milacron T^3566 and T^3586, ASEA IRb-6 and IRb-60, Automatix 600 and 800, and Unimation's Puma 560 and 760.

The data on these robots include link lengths, arm dimensions, and velocities. The availability of all these data not only permits graphic manipulation but also facilitates determination of the most suitable robot for a specific application. Data on other robots can be added.

Similarly, data on end effectors are stored and can be enhanced by the user. Special tooling can be modeled quickly by modifying standard configurations.

To assemble a work cell, the programmer calls up 3-D models of manufacturing equipment or creates new models. A picture like the one in Fig. 10-14 appears on the CRT. Although the illustration here shows the diagram in black and white, the screen shows it in full color. All the various

Fig. 10-14. Graphic display of work cell and robot. (*Courtesy of Computervision Corp.*)

pieces of equipment can be positioned in the work cell, and several variations can be tried to determine the best arrangement.

Once the work cell is defined, the robot is selected. A special "reach" command reveals the robot's working envelope. This permits proper location of the robot before programming begins.

To determine the dimensions of an end effector, the programmer first calls up from the database the part that the end effector will grip, paint, weld, or manipulate. Then, a suitable end effector is chosen. The point at which the end effector contacts the part can be defined, so that the part is being taken up at a particular location, orientation, and twist. If the point where the end effector grips the part needs to be checked, the programmer can zoom in for a close look. Later, during creation of the work path, the attachment angle of the end effector can also be specified, and be tested and modified under simulation conditions.

Next, a generic work path or robot program is created. Robographix provides generic programming capability so that no matter which robot is used, it requires only one path-creation method and programming language. Using either the point-to-point or continuous-path mode, users create, test, and modify a robot path in a fraction of the time previously required.

In the point-to-point method, the path is created by digitizing each point using an electronic pen and a digitizing table, or by using known locations in the work cell as reference.

In the continuous-path mode, the end effector automatically maintains a user-defined relationship with surfaces or other geometry on the part and with other work cell elements. In either mode, any section of the robot path can be graphically edited if the geometric relationships in the cell change, or if a collision between the robot and another element in the work cell is detected. Once the path is programmed, it can be simulated on the CRT, and changes can be made wherever necessary.

A Robot Language Processor (RLP) of the Robographix package translates robot programs into the language appropriate for the robot in use, such as VAL for Unimation robots, RAIL for Automatix, and so on. The RLP generates coded information and stores it for display on the CAD/CAM terminal. The robot path as translated by the RLP is then downloaded to the robot control system. Communication links can be used to send the completed program directly from the CAD/CAM system to the robot.

ARTIFICIAL INTELLIGENCE

One simple view of artificial intelligence (AI) is that it is concerned with devising computer programs to make computers smarter. William B.

Gevarter[13] has made the following comparison of AI with conventional programming:

Artificial intelligence

- Primarily symbolic processes
- Heuristic search (solution steps implicit)
- Control structure usually separate from domain knowledge
- Usually easy to modify, update, and enlarge
- Some incorrect answers often tolerable
- Satisfactory answers usually acceptable

Conventional computer programming

- Often primarily numeric
- Algorithmic (solution steps explicit)
- Information and control integrated together
- Difficult to modify
- Correct answers required
- Best possible solution usually sought

By heuristic approach is meant the application of empirical rules, that is, rules of thumb, until a suitable answer is found. AI programs deal with words and concepts and often do not guarantee a correct solution. As in human problem solving, some wrong answers are tolerable.

Robotic AI is based on world modeling systems, in which moves are described in terms of positions and motions of the objects being manipulated. This implies "world models," which are already evident in the graphic manipulation in off-line programming, where robot, workspace, and objects involved are symbolically manipulated. However, the principle of world modeling within the framework of AI goes beyond this. It becomes a task-oriented system in which the robot is assigned a task and must then itself develop the plan for accomplishing it.[14] In essence, the robot programs itself. Research on world modeling is being conducted at IBM, MIT, Stanford, and other research institutions.

REFERENCES

1. Holzbock, W. G. Designing with moving parts logic. *Hydraulics & Pneumatics*, July, 1979, pp. 59–62; August, 1979, pp. 64–66; September, 1979, pp. 100–106; October, 1979, pp. 166–170.

2. Cunningham, C. S. Robot flexibility through software. *Proc. Ninth International Symposium on Industrial Robots,* Washington, D.C., 1979, pp. 297–307.
3. Carter, W. C. Modular multiprocessor design meets complex demands of robot control. *Control Engineering*, March, 1983, pp. 73–77.
4. Larson, T. M., and Coppola, A. Flexible language and control system eases robot programming. *Electronics*, June 14, 1984, pp. 156–159.
5. Series 8200 Robot Control Specifications, 8200-2.1.4. Allen Bradley, Highland Heights, OH, 1983.
6. Tarvin, R. L. Design of a computer-controlled industrial robot for maximum application flexibility. *Proc. Swiss Automatic Conference*, Zurich, Switzerland, March, 1980; updated April, 1982. Cincinnati Milacron, Cincinnati, OH.
7. Hohn, R. E. Application flexibility of a computer-controlled industrial robot. Technical Paper MR76-603. Dearborn, MI: Society of Manufacturing Engineers, 1976.
8. Holt, H. R. Robot decision making, Second National Conference on Robots in Manufacturing. SME Paper No. MS-77-742. Dearborn, MI: Society of Manufacturing Engineers, November, 1977.
9. Ottinger, L. S, and Stauffer, R. N. Update on robotic standards development. *Robotics Today*, October, 1983, pp. 25–29.
10. Peck, R. W. Advantages of modularity in robot application software. *Proc. Robots 8 Conference*, Detroit, MI, 1984, pp. 20-77–20-93.
11. Tilove, R. B., Shapiro, V., and Pickett, M. S. Modeling and analysis of robot work cells in RoboTeach. *Computer Integrated Manufacturing and Robotics*, Proc. Winter Annual Meeting of the American Society of Mechanical Engineers, New Orleans, LA, 1984, PED-Vol. 13, pp. 33–52.
12. Gondert, S. Off-line programming increases robotic productivity. *Design News*, March 26, 1984, pp. 60–66.
13. Gevarter, W. B. An overview of artificial intelligence and robotics. NASA Technical Memorandum 85836, 1983.
14. Schreiber, R. R. How to teach a robot. *Robotics Today*, June 1984, pp. 51–56.

Chapter 11
ROBOT VISION

The generic term for robot vision is machine vision, which means the automatic acquisition and analysis of images in order to obtain data for interpreting an object and/or controlling an activity. Machine vision enables the robot to recognize shapes, sizes, distances, and sometimes colors, and to take action on the basis of these inputs. Machine vision comprises image formation, image preprocessing, image analysis, and image interpretation.

There is little or no comparison between machine vision and the human eye, which, being interfaced with the brain, is infinitely superior in perception and judgment. Machine vision extracts a minimum of information to guide the robot through its programmed routine. To call the robot with visual feedback "smart" or "intelligent" is permissible but, nevertheless, a gross exaggeration. This having been said, machine vision represents a viable alternative to human vision in the usage of robots. Within its limitations, it can be more reliable and more sensitive than the human eye, and it certainly is an undisputable necessity in developing to the fullest the usefulness of industrial robots.

As one of the many examples that illustrate the widening of robot potential by means of vision, consider the Partracker system by Automatix, Inc. It can locate a part or component when the object's position is not precisely known and then perform a task such as welding or assembly. Thus, Partracker enables manufacturers to automate operations in which accurate and repeatable placement of workpieces into fixtures is either impossible or uneconomical.

One application occurs in a General Motors assembly plant, where a robot is used to locate automotive body seams and fill them with silicon-bronze braze materials. The location of a point on the car body with respect to an assembly line can vary by as much as +2 in. The width of the gap to be filled with metal also varies from car to car. The use of the Partracker system enables the robot to locate the seam and apply the braze material necessary to fill the gap.

In operation, a laser projects a beam of light across the gap. Cameras sense the features in the reflected laser light pattern and feed this information into a processing unit built around two Motorola 68000 microprocessing chips. The processor determines the location of the gap and sends this information to an Automatix A132 robot controller that uses a pretaught program to guide a robot arm through the correct motions to braze the seam, in addition to setting welding parameters such as wire feed and arc current.

PRINCIPLES

Machine vision as considered here refers simply to some limited image formed of a workpiece that is to be handled by a robot. This requires illumination of the workpiece, its sensing by a camera, and the conversion of the image into digitized signals that are used for input to the robot control system. The light sources can be incandescent lights, laser beams, or fiber optics, the last used in combination with either of the other two. The camera or cameras belong either to the vidicon (see below) or the solid-state category.

Image preprocessing converts the voltage signals received from the camera into a digitized image. It implies windowing, that is, the conversion of selected portions of the image into binary or grey-scale values, and with this the restoration of essential parts of the image.

Image analysis determines position, orientation, and geometric configuration of the object on the basis of distribution of binary or grey-scale values.

Image interpretation consists of matching the image with a memory-stored model and making a decision as a result of the matching process as to whether or not the image corresponds to the model. The decision is converted into an output to the robot control system. The hardware used in this case usually combines floppy or hard discs as memory storage, microcomputers and host computer, possibly a console with CRT and keyboard, and an interface for connecting to the robot control system.

PIXELS AND PROCESSING

Pixel stands for picture element. The number of pixels in an image expresses the resolution of a machine vision system. Since the sensor elements of solid-state cameras are individual photodetectors, an image of 256 × 256 pixels means that it is composed of 256 picture elements in each horizontal row and 256 in each vertical column. All together, there are 65,536 pixels, which require the same quantity of photodetectors to produce the pixels. The greatest problem of vision technology is to find a cost-effective way to process this amount of information fast enough to initiate action of the robot in a minimum of time so that it can be guided by the vision system in real time.

Pictures represented by this kind of information are just beginning to be read by vision systems in real time.[1] An example of a real-time system is the APP system by Pattern Processing Corp., described later in this chapter.

One way of substantially improving the image processing speed of vision systems is being investigated at the National Bureau of Standards. The concept is called the pipelined image processing engine (PIPE). By pipelining is meant the moving of the image data through several consecutive processors, rather than their processing in a single parallel batch.

The first prototype will be a three-stage system that receives and processes images as they come from the camera at the rate of 60 frames/sec. As pictures move along from the first to the second to the third stage processor, each stage, directed by algorithms downloaded from a host computer, performs an operation on the image. These operations include edge or line finding, as described further below, motion detection, noise reduction, varying the picture resolution, and several others. Each stage can process 65,000 eight-bit pixels every 15 msec. The present three-stage approach will be expanded to an eight-stage PIPE system in the next phase of development.

CAMERAS

Vidicon Cameras

The vidicon camera is well known as a TV camera. It is based on the vidicon tube, the front of which is covered with a transparent, conducting film. A thin photoconductive layer is deposited on the film. Each cross-sectional element of the photoconductive layer is an insulator in the dark but becomes a conductor when light impinges on it. The amount of electric current that is capable of flowing through this conductive element is a function of the light intensity. The combination of the transparent, conducting film and the photoconductive layer with variously conducting elements allows a positive-potential replica of the image to be produced on the backside of the layer. By scanning the layer with an electron beam, a current can be produced that constitutes the video signal.

There are a number of disadvantages connected with vidicon cameras as compared with solid-state cameras. Among them are the fragility of the tube, the limited life expectancy, the sensitivity to electromagnetic interference (noise), and the considerably higher electric power requirements of vidicon cameras.

Solid-State Cameras

Because of the limited resolution offered by present-day solid-state cameras, robotic vision systems developed around them often provide the capacity for supporting more than one camera. Itran, for example, has developed a vision system that supports up to eight cameras. The Vidomet II from Penn Video can support up to 32 cameras.

An example of the solid-state camera itself is Honeywell's HDS vision system. It is shown in Fig. 11–1 mounted on the end effector of the Yamatake-Honeywell Robo-Ace SR1 robot. The sensor module of the cam-

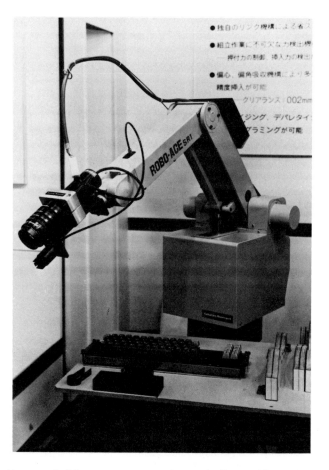

Fig. 11–1. Honeywell vision system mounted on end effector. (*Courtesy of Honeywell-Visitronic*)

era measures less than ½ × ½ × ⅛ in. The vision system is totally contained within the camera body and can communicate directly to the robot computer without further processing or hardware. The camera transmits information on object size, location, and distance to the robot. The feature of performing the vision processing in the camera is a particular advantage with multiple camera systems, since computation of the information takes no more time than with a single camera design.

At the heart of the camera is a charge-coupled device (CCD), which is an integrated circuit chip that stores localized packets of charges and transfers them to adjacent locations when stimulated by externally manipulated voltages. The charge in each packet can represent digital information. CCDs are extremely fast. Their upper operating limit lies close to the 5- to 7-MHz range.

During the last decade, great progress has been made in the development of CCDs for optical-sensor application. The resistance to shock and vibration, nonsensitivity to magnetic fields, and longevity of CCDs was evident from the beginning. However, initially they lacked resolution and light sensitivity. Advances in LSI design and manufacturing techniques have improved resolution as well as light sensitivity.

The CCD is part of an integrated sensor. The same chip contains not only the CCD but 96 photodetectors arranged in 48 pairs. Each pair is behind a tiny microlens measuring less than 0.008 in. in diameter. The microlenses focus an object's image on the 48 detector pairs. Analysis of the image allows a microprocessor to calculate the distance of the object.

With this system, the robot can detect the edge of an object, and then move to predetermined points on the object to perform a programmed task.

The vision system also allows a robot hand to approach and grip an object properly, regardless of the object's position. It can measure accurately to within 0.002 in. and can do so while the object is moving because the system updates its calculations 20 times a second. It is sensitive to light differences of one part in a hundred. By comparison, the human eye can only sense light differences of one part in 20.

A solid-state camera of the MC9000 Series from *EG&G Reticon* is shown in its actual size in Fig. 11-2. Its block diagram together with the camera connector are shown in Fig. 11-3. The low weight, less than 12 oz, of the camera facilitates its mounting on the end effector of a robot with a minimum of loading. It produces a square image format with either a 128 × 128 pixel image sensor (MC9128) or with a 256 × 256 pixel image sensor (MC9256).

The two-dimensional image sensor consists of individual photodiodes arranged in a square pattern. The active area of the 128 × 128 sensor measures 0.302 in. on each side, with the photodiode center-to-center spacing

Actual Size

Fig. 11–2. Solid-state camera. (*Courtesy of EG&G Reticon*)

equal to 0.0024 in. The 256 × 256 sensor measures 0.403 in. on each side
with the photodiodes on 0.0016-in. centers.

The photodiodes convert incident light energy to electrical output signals
whose amplitude is proportional to the light intensity integrated over the
exposure time. On-chip scanning of the photodiodes produces a series of
charge pulses corresponding to the pattern of light intensity to produce an
electrical analog of the image. Conventional photographic lenses are used
on the camera to form an image on the array of photodiodes. The camera
produces data corresponding to the object's size, position, shape, and
brightness (grey level).

Two hybrid microcircuits, in which several circuits are mounted on the
same substrate, are sandwiched on an aluminum mounting plate that also
functions as a heat sink. The image sensor is bonded to the front hybrid.

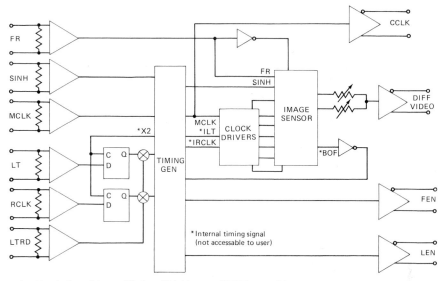

−Input termination resistors are 100 ohms −Digital inputs are RS422A compatible

Block Diagram

MCLK . . . . MASTER CLOCK INPUT		FR . . . . FRAME RESET INPUT
CCLK . . . . CAMERA CLOCK OUTPUT		FEN . . . . FRAME ENABLE OUTPUT
LT . . . . LINE TRANSFER INPUT		RCLK . . . . ROW CLOCK INPUT
ILT . . . . INTERNAL LINE TRANSFER		IRCLK . . . . INTERNAL ROW CLOCK
LEN . . . . LINE ENABLE OUTPUT		ϕX1 . . . . ARRAY CLOCK
SINH . . . . START INHIBIT INPUT		ϕX2 . . . . ARRAY CLOCK INVERTED
LTRD . . LINE TRANSFER TO ROW CLOCK DISABLE INPUT		BOF . . . . BEGINNING OF FRAME

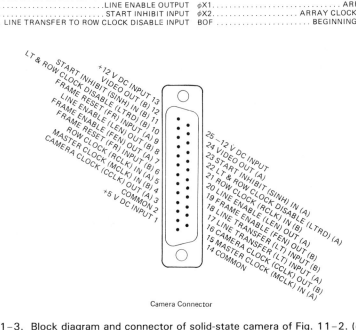

Camera Connector

Fig. 11−3. Block diagram and connector of solid-state camera of Fig. 11−2. (*Courtesy of EG&G Reticon*)

In the basic mode of operation, only a single clock input to the camera is required to scan the image sensor. Pixel rates up to 8 MHz and corresponding frame rates of up to 380 frames/sec for the MC 9128 and up to 105 frames/sec for the MC9256 can be achieved. The pixel rate is equal to the clock rate. Each row is scanned in sequence.

Not all solid-state image sensors are matrices. The linear sensor also plays an important role, as will be seen in the following. It has only a single row of photodiodes. Figure 11–4 shows the three different types of monolithic silicon integrated circuits manufactured for image sensing by Reticon. They are the matrix type, the linear type (of which three different lengths are shown), and the circular type. Cameras that use the linear type are known as line cameras or linear array cameras.

There are also chips under development that combine optical sensing operations and control functions, thus making vision increasingly cost-effective. Together with faster three-dimensional image-processing algorithms, they can be expected to develop robotic vision to its fullest potential.

N. V. Philips in the Netherlands appears to be in the forefront of this development. A single chip with a surface area of about 0.1 in.² contains 360,000 pixels. Half of these pixels register the image; the other half store

Fig. 11–4. Solid-state image sensors. (*Courtesy of EG&G Reticon*)

it. This allows a new image every 20 msec, while the previous image is read out.

STRUCTURED LIGHT

Structured light systems use an illumination technique that provides depth information in a two-dimensional image. The most representative of the structured light vision systems is probably CONSIGHT, which was developed by the General Motors Research Laboratories. CONSIGHT has the following characteristics:[2]

- It determines position and orientation of a wide cross section of manufactured parts, including complex curved objects.
- It provides easy reprogrammability by insertion of new part data.
- It works on visually noisy picture data typical of many plant environments.

CONSIGHT uses binary pictures, which are digitized images in which the brightness of the pixels can have only two different values, such as white and black or light and dark. The experimental set-up illustrated in Fig. 11–5 consists of a Reticon 256 × 1 line camera combined with a Digital Equipment Corporation PDP 11/34 computer to control a Stanford robot arm. The object is to recognize and pick up an object on a moving conveyor.

The computer coordinates and controls the operation of the system and

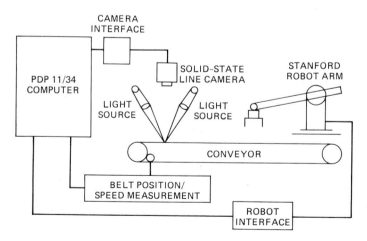

Fig. 11–5. Experimental set-up for CONSIGHT vision system. (*Courtesy of General Motors Research Laboratories*)

assists in calibration and programming for new parts. It stores data to access them in proper order, so that the system is capable of handling a continuous stream of parts on the belt.

A narrow line of light is focused on the belt at an angle, as shown in Fig. 11–6. This line, when seen from directly above, appears to be displaced by the object. Since the belt is moving, it is possible to build a conventional two-dimensional image of passing parts by taking pictures of a sequence of these image strips. Uniform spacing of these strips is achieved between sample points by use of the belt position detector which signals the computer the appropriate intervals to record the camera scans of the belt.

It is also permissible for several parts to be within the field of view simultaneously. Parts that overlap or touch each other are ignored and allowed to pass by the robot for subsequent recycling.

Object detection and position determination are thus accomplished, and it is then up to the robot to execute a previously taught program and transfer the part from the conveyor to a fixed position.

There are two light sources rather than one in Fig. 11–5 to avoid a particular problem: When the leading edge of an object enters the slanted light beam, it casts a shadow ahead of itself. This produces a distorted image of the object before it actually reaches the imaged line. The solution is to use two (or more) light sources that direct their lines of light so that they coincide on the belt. This approach essentially eliminates the shadowing effect.

GREY–SCALE VISION

The grey level is a measure of pixel brightness. No longer is the vision binary as before, but it can differentiate among many levels of grey. This is the concept of grey-scale vision.

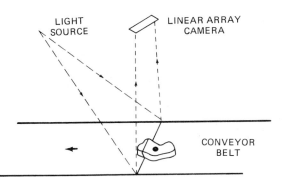

Fig. 11–6. Basic lighting principle. (*Courtesy of General Motors Research Laboratories*)

VS-100

The model VS-100 is a visual system that was developed by Machine Intelligence Corp. (MIC). It uses grey-scale vision and includes a Digital Equipment Corp. LSI-11/2 microcomputer, image processing circuitry, interfaces to four types of solid-state cameras, a computer interface, and software.

The VS-100 accepts images from the camera and determines type, location, and orientation of the object. This information is then fed to a robot controller via a standard interface.

Objects are recognized by comparing features of their silhouettes with those of samples or patterns shown to the camera beforehand and stored in the microcomputer's memory. As identifying features, the VS-100 uses about a dozen geometric properties, such as perimeter length, area, and geometric (two-dimensional) center of gravity of the objects to be recognized. They are chosen to facilitate recognition and, hence, are independent of the silhouette's location and orientation. Recognition involves a number of steps, as outlined in the following paragraphs.

1. The silhouette, shown in Fig. 11-7, is formed by thresholding. This means a threshold of light intensities has been programmed. All pixels that are above this threshold count as white. All pixels with less intensity count as black. Whether the objects appear as black or white depends on whether they are lit from the front or from the back.

As an aid for the initial threshold determination, the vision system can display a histogram, as shown in Fig. 11-8, of the intensities encountered in a typical scene. If the scene has sufficient contrast, the histogram will have two peaks, one dark and one bright, which are equivalent to background and objects. The threshold value will then be the bottom of the valley between the two peaks.

Fig. 11-7. Reduction of grey values to black-and-white images. (*Courtesy of Machine Intelligence Corp.*)

Fig. 11-8. Histogram of light intensities. (*Courtesy of Machine Intelligence Corp.*)

2. Silhouette tracing is next. This is done by scanning the image for silhouette edges, starting in the upper left corner and moving row by row to the bottom. To minimize scanning time, the system actually scans only a compressed version of the image. The compressed image contains the beginning and end locations of continuous pixel runs of either the black or the white category. This is, in fact, the only information necessary to form the silhouette, which then appears as in Fig. 11-9.

A transition from black to white followed by one from white to black signals the edges of a white silhouette on a black ground.

Whenever the computer encounters an edge point, it determines to what edge the point belongs. This is done by examining neighboring points on the pixel row above. If it shows that the edge is being continued, it enters the point's location in a memory list for that edge. However, if the point is isolated, the system assumes a new silhouette and creates a new list.

Because the computer always scans the image from top to bottom, it can

Fig. 11-9. Outline of silhouettes. (*Courtesy of Machine Intelligence Corp.*)

never be sure that the edges it encounters partway through the scan are not connected farther down in the image. For example, if the system first encounters the tines of a fork, it will treat the tines as separate objects until it reaches the fork's palm. For this reason, edge lists are always provisional until a scan is completed. Lists are consolidated when edges are found to be connected, which happens when the system discovers a point common to two or more edges.

3. Having completed the silhouette, the system computes its location and orientation by the method illustrated in Fig. 11-10. The silhouette's location is defined as its geometric center of gravity; its orientation, as the orientation of an ellipse that has the same area.

4. As a final step, the system attempts to match the silhouettes to the patterns stored in its memory. Close similarity signifies recognition. What is close is determined by scoring the number of individual features that are the same in image and pattern and then weighting and combining the individual scores to create a total score. By adjusting the weighting factors, it is possible to recognize objects with variable features.

Associative Pattern Processor

The question has been raised of whether grey-level analysis that creates a threshold to separate the pixels into black and white, as is done in the above-described system, is truly grey-level, or should instead be considered a binary system. In any case, the Associative Pattern Processor (APP) developed by Pattern Processing Technologies digitizes the image, analyzes it in live grey-level terms, and then reduces it to a statistical representation. This allows not only extremely high processing speeds of 20 to 30 msec, but also storage of up to 256 totally independent images in as little memory as 256K bytes.

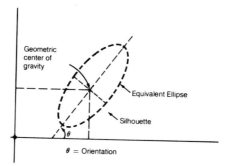

Fig. 11-10. Defining location and orientation of silhouettes.

The system has the capability to adjust to nonuniform light and has no shape restriction regarding images. Recognition of images is continuous, sensitivity can be adjusted from zero to 100%, and an image can be freeze-framed without stopping the process. A typical image produced by an APP is shown in Fig. 11-11.

Arne L. Watland, the director of sales and marketing of Pattern Processing Technologies, points out that most robot vision systems are based on some form of well-defined processes for the solution of an image acquisition algorithm developed by SRI International. Although the SRI program accomplishes the appropriate vision task, it has some drawbacks. First of all, to accurately implement the algorithm, a large amount of data storage is required, especially if more than two images are to be stored in the main memory. Processing this memory takes too long for many applications.

These are precisely the problems that the APP 100/200 from Pattern Pro-

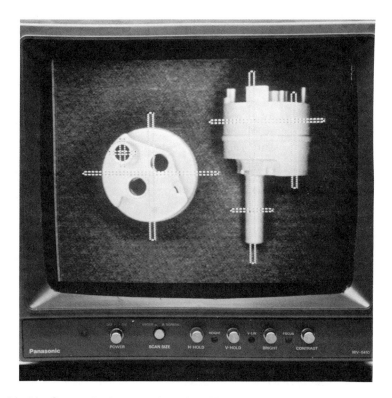

Fig. 11-11. Grey-scale picture produced by APP. (*Courtesy of Pattern Processing Technologies, Inc.*)

cessing Technologies has endeavored to overcome. It identifies images, accepts a standard video format from the camera, and communicates with almost any manufacturer's control system. As before, visual recognition implies a comparison of the camera image with a stored sample image. The system is taught by being shown the sample and storing its image in the memory. A label that identifies the sample image is entered at the controller's keyboard. A subsequent occurrence of any of the taught images causes the APP to generate the associated label for immediate access by the controller.

A block diagram of the APP system is shown in Fig. 11–12. The camera image is transmitted to the A/D converter, which converts the analog electrical signals transmitted from the camera into equivalent digital signals, and to the image memory, which stores the picture frame as a grey-scale representation. From there the signals may branch off to an optional CRT display, but their direct line goes to an encoder interface. The encoder, which in this case refers to a statistical encoding technique developed by Pattern Processing Technologies, receives the signal and passes it on to the response memory interface and associated components. Further details are as follows:

The A/D converter digitizes the video signal it receives from the camera so that the image can be stored in memory. The camera has a matrix of 320×484 photocells; that is, it produces a total of 154,880 signals for each image. The video signals are digitized into 64 levels of intensity at a speed of 40 to 50 image/sec.

The image memory stores the pixels produced by the A/D converter. This action is continuous even though the encoder is equally continuously sampling the image memory. Thus, high-speed recognition and real-time access to the image data become possible.

The encoder—and there may be several, in fact, up to 256—converts the large amount of image data to a relatively small set of repeating response memory addresses, amounting to some 500 in the typical case. A given image will generate a specific set of repetitious response memory addresses; a different image will generate a different set of addresses. Each different image can be considered to be encoded as a unique randomly selected set of repeating response addresses. The encoding process is continuous. That is, the encoder provides a continuous stream of sample addresses to the response memory, and the data at those addresses are used by the encoder to generate a continuous stream of response addresses. Typically the repeating response address set occurs within 20 msec, in which time the image memory is totally refreshed.

Associations between a label (in the form of an eight-bit code) and an image are formed in the APP using a special statistical encoding technique

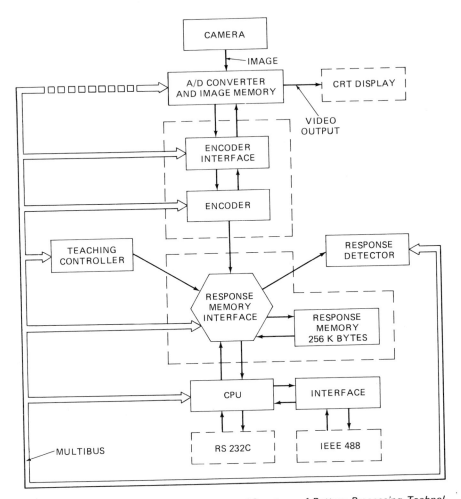

Fig. 11-12. Block diagram of APP systems. (*Courtesy of Pattern Processing Technologies, Inc.*)

developed by Pattern Processing Technologies. This technique is implemented directly in electronic hardware and requires no programming.

The response memory contains the image labels, also called response codes. The addresses for the labels are provided by the encoders. Since for a given image the encoders generate a unique random set of repeating response addresses, the response codes will be stored at a unique random set of locations. This means that different images can have some response memory addresses in common, a condition that is called overlap.

As stated above, up to 256 encoders can be operating simultaneously, each contributing independently to the total recognition task. Each encoder can be individually assigned a window within the field of view, with each window defined by X-Y coordinates.

One configuration could utilize 256 different windows distributed throughout the field of view. In total there are four pages of 256 independent windows (i.e., 1024 windows in all), which can be used to queue other window subsets. This type of configuration is useful in recognizing local features. The encoders can be taught to recognize an arbitrarily selected set of features that, by windowing, can be simultaneously applied to arbitrarily selected regions throughout the field of view. Even though different encoders may be windowing different regions, each will generate the same response address set if the images windowed are the same.

Consider, for example, encoders taught to recognize corners consisting of horizontal and vertical edges and their intersections. Any object in the field of view with such features would cause the encoders to generate a response address set corresponding to appropriate response codes for these features that were previously stored while teaching. A square, for example, placed anywhere in the field of view would produce a characteristic distribution of edge and corner responses. If the encoders were taught to recognize rotated edges as well, then the square could also be recognized even though it was rotated in the field of view.

This ability to teach arbitrary features and then to apply them to arbitrarily selected windows and locations offers great flexibility. Some additional examples are:

- Windows are placed over the perimeter of an object to inspect the quality of its edges with regard to burrs or nicks. The features taught are the edges of a good object, and the sensitivity to detect is controlled by the size of the windows.
- Windows locate a particular type of object within the field of view of multiple-type parts. The window size is adjusted to the size of the part and then redistributed throughout the field of view until one of the encoders responds with the part's response code and thus locates the parts.

Such a system, as described here, offers the possibility of being expanded to include color recognition, parallax, acoustics, and touch, all in a single system.

STEREO VISION

Triangulation

To obtain maximum visibility of objects, a three-dimensional image is required. A frequently used method for this purpose is triangulation. It determines the depth of an object point by directing toward it the light from two sources that are a known distance apart. The object point will be at the intersection of the two light lines. Triangulation methods concentrate on the third dimension, that is, the Z-axis. The X- and Y-axes can be obtained by conventional grey-scale methods. Once all three dimensions are known, the stereo image of the object can be computed.

National Bureau of Standards.   Figure 11–13 illustrates an arrangement that was developed at the National Bureau of Standards.[3] A camera and a structured light source are fastened to the wrist of a robot arm. The structured light source is a stroboscopic light that emits a plane of light through a cylindrical lens.

The camera is oriented to position the columns of the image pixels perpendicular to the flash plane, so that each column has one intersection with the plane of light. By mounting the camera and the light source at a fixed angle, it is possible for the computer to determine the distance from the robot to each pixel by triangulation.

A camera interface is attached to an eight-bit microprocessor and provides data reduction as well as control of the stroboscopic flash duration, camera clock rate, and iris adjustment. In addition, the interface receives

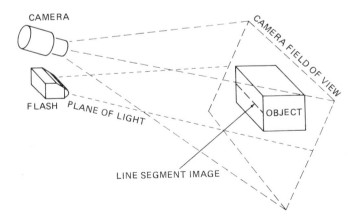

Fig. 11–13. Object recognition. (*Courtesy of National Bureau of Standards*)

a threshold value signal from the computer via the microprocessor and produces the resultant binary image.

The vision system looks for the object on the basis of a given maximum distance. It then sends the control system a status signal, the endpoint of the closest part edge (in the coordinate system of the wrist), and an edge characteristic.

Operation of the system is divided into two primary phases. In the first phase, the system acquires an image and evaluates its quality. In the second phase, the system analyzes the image to produce the data necessary for the control system.

The reason for the first phase is to obtain an image that can readily be analyzed. Complex procedures to connect lines that have not been clearly defined in the image are avoided, for example. The concept is rather to adjust the flash, iris setting, and threshold parameters on the basis of the quality evaluation of phase one.

The image is broken up into segments corresponding to surface configurations. The analysis phase of the vision distinguishes three types of such segments: straight lines, angles, and curves. Thus, if convex parts are examined, these segments are a complete description of the possible images.

There are, furthermore, four types of border points: left and right endpoints, minimum points and vertices, as shown in Fig. 11–14. All segments have left and right endpoints corresponding to the start and end of the segment. A curve has a minimum point that is defined to be the lowest row value in the curve. An angle segment has a vertex that is defined to be the corner point of the angle. Thus, line segments have two border points, while curves and angle segments have three.

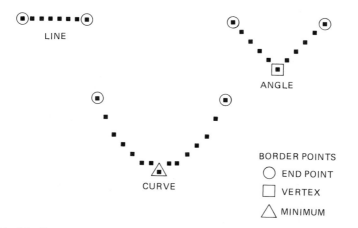

Fig. 11–14. Segment border points. (*Courtesy of National Bureau of Standards*)

Three conditions are required to identify a segment as a curve:

- Enough points must have been found for the slope to be computed.
- The range between maximum and minimum slopes must be greater than a specified parameter.
- The percentage of the curve between the last occurrence of the maximum and the first occurrence of the minimum must be greater than a specified parameter.

The first test ensures that there are enough data to be significant. The second prevents a line that is noisy at the ends from being considered a curve. The third separates curves and angles, the difference being that curves begin gradually changing their slope at the endpoints, while angles retain their maximum slope until very near the middle of the segment.

If the segment passes the curve test, any possible vertices found are discarded, and it is labeled a curve. If the segment is not found to be a curve, it is considered to be an angle or a line. If no vertices are found, the segment is classified as a line.

Although the vision system does not know what type of part it is, the line, angle, and curve data from the vision system are used by the control system to identify the part side it is facing. Thus, in Fig. 11–15 the grey-scale image of a cylinder is obtained by the vision system hardware. The bright line, which is caused by the plane of light, will be all that remains of the image after it is thresholded. The control system can measure the width of the object from coordinates supplied by the vision system. It moves the robot with the camera on its wrist to face the selected side of the object head on, and can then make it proceed to pick up the randomly located object and place it in a fixed orientation in a machine tool, vise, or some equivalent.

Laser-Based Triangulation System.   The development of solid-state laser diodes and the ease of interfacing the optical sensor to a microcomputer have led to the design and implementation of compact and accurate optical rangefinders that can determine the third dimension for three-dimensional vision systems. Developmental work at the Case Western Reserve University[4] was based on the concept that an ideal robot rangefinder sensor should meet the following requirements:

1. The sensor must be capable of gauging distances from 1 cm to 1 m with a resolution of 1 mm or better.
2. Ideally, measurement should be continuous, but a rate of 10 measurements/sec may be adequate.

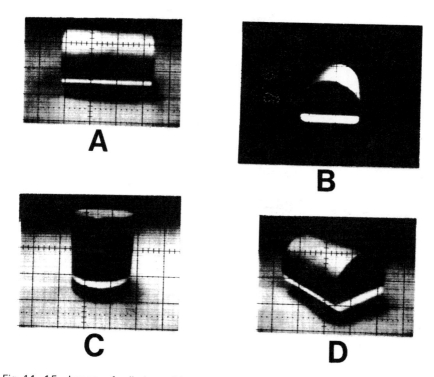

Fig. 11–15. Images of cylinders with structured light lines. (*Courtesy of National Bureau of Standards*)

3. The size should be on the order of a few cubic inches maximum, so that the sensor can be mounted directly on the manipulator's hand.
4. The sensor should be relatively insensitive to the degree of reflectivity of the object being monitored.

Figure 11–16 illustrates the geometrical configuration of the laser-based triangulation rangefinder based on these concepts. One or several lasers generate a narrow collimated beam either directly, as in the case of a single-mode HeNe laser, or through a set of collimating lenses, as in the case of a continuous-wave (CW) laser diode. The beam impinges upon the center of a mirror that rotates at a fixed rate, so that a given plane is continuously scanned. A position-sensitive photodetector is placed in the scanned plane. The distance between the axis of rotation of the mirror and the center of the photodetector constitutes the base line of the rangefinder.

A focusing lens rotates around the photodetector. It is synchronized with the mirror in such a way that the angle between the mirror-reflected beam

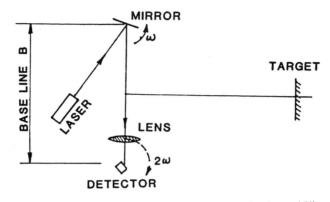

Fig. 11–16. Laser-based triangulation rangefinder. (*From Mergler and Nimrod, Ref. 4*)

and the base line is always equal to the angle between the optical axis of the focusing lens and the base line. This is achieved by mechanically aligning both components and having the lens rotate at twice the rate of the mirror but in the opposite direction.

Consequently, the line along which the range is measured lies in the plane defined by the scanning beam. It is the perpendicular bisector of the base line. If an object is located along the axis of measurement within the scanning range, two optical center-crossing signals per mirror rotation cycle are generated by the photodetector. The first one occurs when the mirror reflects the incident beam directly along the base line and through the lens onto the photodetector. The second one occurs when the mirror-reflected beam illuminates the object at the intersection point with the measurement axis. At that moment, the synchronized lens focuses the image of the illuminated spot on the center of the photodetector. This is the triangulation instant when the parameters are defined that permit calculation of the distance of the object point.

Experiments were made with a base line of 50 cm and a mirror cycle time of 4.5 sec. A HeNe laser was used to generate a 1-mW beam 2.1 mm in diameter, and the optical signals were detected by a dual-axis, linear, position-sensitive photodetector. Measurements were taken visually from a CRT, and the distance to the object was calculated manually.

Later, to minimize the size of the rangefinder, two modifications were made. First, the HeNe laser was replaced by a CW laser emitting a 7-mW beam at a wavelength of 860 nm. A focal lens collimated the diverging radiation of the diode into a beam of 1 to 2 mm diameter. Then, as illustrated in Fig. 11–17, the rotating lens was replaced by an arrangement of

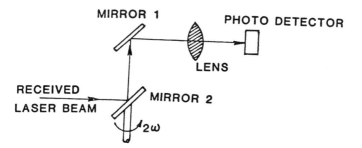

Fig. 11–17. Modified configuration of Fig. 11–16. (*From Mergler and Nimrod, Ref. 4*)

a rotating mirror at 45° inclination, a second fixed mirror at the same inclination, and a fixed lens.

Automatix, Inc.   The purpose of the stereo vision systems developed by Automatix, such as the Partracker system mentioned above, is primarily to locate objects whose location is not accurately known or which are imprecisely placed in their corresponding fixtures. The system can use either structural light or stereo cameras. If multiple cameras are used, they can be two rigidly fixed cameras, two cameras mounted on a robot arm, or even a single camera mounted on a robot arm and moved to two positions. A point imaged by the cameras determines a pair of lines in space. These lines intersect in space at the object point, as illustrated in Fig. 11–18.

Triangulation in the Automatix concept consists of processing certain

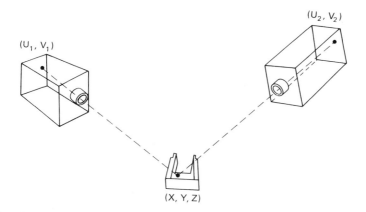

Fig. 11–18. Stereo projection of a point into focal planes. (*Courtesy of Automatix, Inc.*)

camera calibration coefficients that essentially characterize the cameras in space.[5] Together with the focal-plane coordinates, they can be used to locate the points in space.

The camera calibration procedure, as developed by Automatix, does not require measuring any of the parameters with a measuring rule. Instead, "calibration target" locations are stored in the memory of the system, and a target block is shown to the system. For this purpose, the cameras initially take pictures of the calibration block with the relative features on it in known positions. Camera images can then be processed to extract these relevant features correctly.

After the features have been extracted, correspondence between the same features in the two images must be found. For example, in Fig. 11-19 the object has two holes. The software, which is based on the RAIL® programming language, must recognize that holes A and B in picture 1 correspond to holes B and A, respectively, in picture 2.

The information from the stereo vision system is presented to the robot, which is manually taught to move along a three-dimensional path to perform the programmed operation at the point indicated by the stereo vision system. Figure 11-20 shows a typical application. Here, parts have been stacked on a pallet in a partially ordered manner. The location of each part

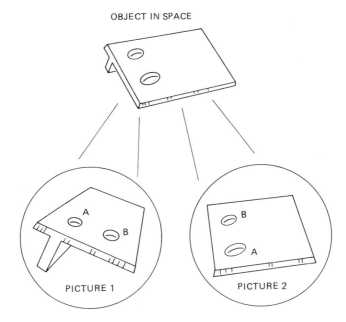

Fig. 11-19. Stereo view of an object in space. (*Courtesy of Automatix, Inc.*)

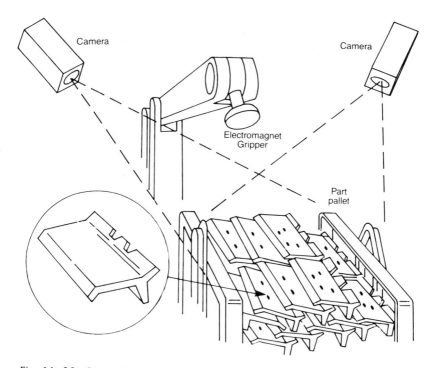

Fig. 11–20. Bin-picking of prearranged objects. (*Courtesy of Automatix, Inc.*)

is determined using stereo vision, and a robot is directed to pick it up. This process is repeated until the pallet is unloaded.

Bin-Picking

In bin-picking, the problem becomes more complex than in the preceding example. Here the placement of objects is completely random; they may be touching or overlapping and in any conceivable position. SRI International and others are working on a general-purpose technique for locating parts under such undefined conditions. The initial goal of SRI was to locate cylinders in a jumble, given their diameter and range data taken from above them by a visual system.[6]

The sensor computes the three-dimensional positions of points along the intersection of a light plane and the object. The intersection forms the typical line of structured light. In the case of cylinders, the series of intersecting lines are ellipses, as shown in Fig. 11–21. Therefore, clusters of similar ellipses can be assumed to represent cylinders. The difficulty, however, is that at most half of each ellipse is visible, while the rest is hidden. In reality,

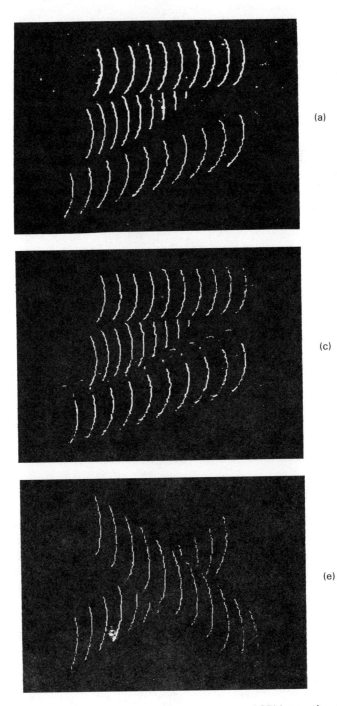

(a)

(c)

(e)

Fig. 11-21. Ellipse-clustering program. (*Courtesy of SRI International*)

(b)

(d)

(f)

401

only about 30% is visible, even with an unoccluded cylinder, since the camera and the light source do not focus on exactly the same parts of the scene, and because the apparent width of the light line gets narrower as it curves down the cylinder.

The program that uses clusters of ellipses for recognizing cylinders operates as follows: Using an elliptical segment, it tries to find a circle that corresponds to the diameter of the cylinder. If it cannot find such a circle, then it tries to find an ellipse with a minor diameter that is equal to the cylinder diameter. Groups of similar ellipses are formed by finding peaks of a histogram of the orientation of the ellipses. At the end, the program divides the group of circles and the group of ellipses into subgroups that correspond to individual cylinders. A RANSAC fitting technique is used for this purpose.[7]

Figure 11–21 shows results of this program as follows:

(a) A composite figure formed from ten slices along three cylinders.
(b) and (c) Circles and ellipses fitted to curve segments appearing as ellipses because they have been transformed back into the image plane of the camera that took the data. Groups such as this group of circles have been partitioned into subgroups corresponding to individual cylinders, by means of the RANSAC fitting technique.
(d) Hypothesized cylinders.
(e) A set of slices from two crossed cylinders.
(f) The hypothesized cylinders that result from (e). Here, the program could not fit circles or ellipses to the curve segments in the upper right corner of the picture, and therefore could not include them in the hypotheses.

Symbolic representations of complex objects often involve a hierarchy where an object is represented as a structure of volumes, each volume is represented as a set of surfaces, and so on. SRI International has proposed a representation with the following geometric features of objects:

Volumes: block, sheet, cylinder, cone, snake, hole, notch, groove.
Surfaces: plane, cylinder, cone.
Surface patches: planar, convex, concave.
Edges: line, circle, curve.
Edge elements: convex, concave, tangential, jump, sheet, picture.
Curve segments: line, convex, concave.
Segment ends: convex, concave, tangential, jump, sheet, picture, shadow.

Most of these terms are self-explanatory. Sheets are broader than they are thick. Snakes are long thin volumes. The exact shape of a snake's cross section is generally less important than the shape of its spine. Surface patches are similar to segments, and edge elements are essentially the same as segment ends. Figure 11–22 shows examples of different types of (a) edge elements and (b) segment ends.

The BinVision system developed by General Electric, which includes GE's own software, follows the general principles of the method laid out by SRI International. The digitized image information is analyzed for structural features, and the system calculates a set of programmed parameters, such as areas, distances from one point to another, perimeters, and so on. The gripper selects the object that, on the basis of linear and rotary motion required by the robot, is the closest to reach. If the gripper misses the object, it withdraws, the camera surveys the scene again, and the process is repeated.

A somewhat different approach is used by the i-bot 1, a vision system produced by i-bot Vision Systems, Inc., a subsidiary of Object Recognition Systems, Inc. The i-bot 1 vision system was described by Nello Zuech and Rajarshi Ray.[8] It is based on research that was done by the University of Rhode Island[9] (URI).

The method developed by URI consists of computing the location and orientation of objects by viewing them from the top. The data are fed to a robot that then proceeds to acquire a random object from the field of view by means of a parallel jaw gripper or a vacuum cup gripper. The gripper then moves with the object to present it to a camera that provides the image for computing the orientation of the object.

Object Recognition Systems combined URI approaches with its own grey-scale capability. The sequence starts with the robot arm moving to a home position, where the camera has an unobstructed view of the entire scene. In this position, the camera acquires an image of the scene and passes it on to the i-bot-1 vision controller for processing.

The basic unit consists of a vision controller, a gripper controller, and an imaging system that includes a solid-state camera. The vision controller comprises a high-speed frame grabber that stores an entire picture frame, a 16-bit Intel 8086 microprocessor, a dynamic RAM of 256 K bytes, a non-volatile RAM of 2K bytes plus 64K bytes of image processing firmware, hardware, math processor, I/O expander, and several other components.

The video imaging system provides images at the rate of 30/sec to the frame grabber, which stores a given image of 320×256 pixels with 64 shades of grey. Grey-scale vision that uses thresholding by means of a histogram is similar to the systems described earlier.

Pixels associated with the defined edges are grouped into multiple clusters, and for the three biggest clusters (i.e., one cluster for each of up to three objects selected for removal) the centroid and axis of least moment of inertia are computed for two coordinate systems. One set of coordinates

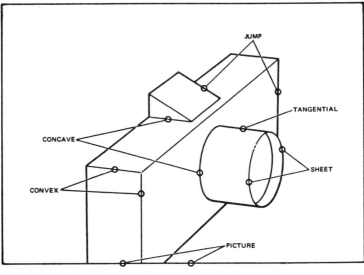

(a)

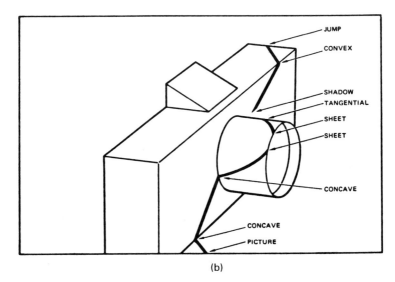

(b)

Fig. 11-22. Examples of elements (a) Edge element. (b) Segment ends. (*Courtesy of SRI International*)

represents the object at the top level of a bin, the other set of coordinates at the bottom level. The robot is then directed to go to the top object location and move from there to the bottom object location, thereby moving along the line of sight of the camera and providing a three-dimensional representation of the object location.

The total cycle time to process the image and make estimates of location and orientation of up to three objects is less than 3 sec.

SIX-DIMENSIONAL VISION SYSTEM

As previously pointed out, there are six degrees of freedom that define the position and orientation of any object relative to a robot. It would be desirable to have all six visually determined to enable the robot to grasp the object in a uniquely specified manner. Such a visual system was designed by the National Bureau of Standards. It measures all six degrees of freedom using two frames of video data taken sequentially from the same camera position.[10] Structured light techniques are being used.

The reference described the system operating with planar surfaces, but it is stressed that the principle can be extended to curved surfaces as well.

The structured light projector and the camera are mounted on the robot hand. Two frames of video data are used. The first frame is taken while the scene is illuminated by two parallel planes of light provided by two slit projectors. In the second frame, the same scene is illuminated by the flood-light of a point source located between the slit projectors.

The first image, which is produced by the two planes of light, makes it possible to compute the range as well as pitch and yaw orientations of any simple geometrical surface. The second image, produced by the point light source, can then be analyzed to obtain the elevation, horizontal angle, and roll orientation of the illuminated surface. The combined information determines the six degrees of freedom of the surface relative to the vision system.

EYE-IN-HAND VISION

Camera placement must be decided. So far, two possibilities have been considered. One is the use of one or more fixed cameras that are directed on the object. The other is using a camera mounted on the end effector. The latter has the advantage that the robot arm cannot move between the object and the camera, that the camera gets closer to the object, and that immediate feedback between the end effector and the camera, and vice versa, is assured. Its disadvantages, however, are that even small solid-state cameras are often too bulky for this purpose, and that evaluation of the images requires considerable computer power.

Another solution, which was proposed by Arun Agrawal and Max Epstein,[11] is "eye-in-hand" robot vision that utilizes (a) coherent fiber-optic bundles for carrying light from an object to be imaged, and (b) photodiodes to convert this light into electrical signals for processing. The advantages that could be obtained are:

- The physical separation of electronics and optics.
- The smallness and light weight of the sensing head.
- The immunity of optical fibers to electrical noise.
- The possibility of geometric preprocessing by arranging the optical fibers in a suitable pattern in the sensing head, while the other end of the aligned fibers is matched to the photodiode array.

A schematic of the proposed system is shown in Fig. 11–23. It consists of a fiber-optic imaging cable, photodiodes in the form of a CCD array, array support circuitry, computer interface, microcomputer, and display.

The light inputs conveyed by the fibers of the imaging cable are applied to the corresponding photodiodes. The CCD array converts incident light into an electric charge that is integrated and stored in a shift register. In addition, MOS transistor switches are used to multiplex photodiode outputs to a single serial output line via a shift register. The diode array, shift register, and multiplexing switches are integrated into a single silicon chip.

Support circuitry is needed to process the output signals from the array via a sample and hold circuit. A continuous analog signal is generated, corresponding to outputs from the photodiodes. The typical maximum signal is about 3 V at saturation. This signal is converted into digital form for computer processing. For binary images, a threshold is used to differentiate between light and dark regions. For grey-scale images, an appropriate A/D converter can be used.

To reduce computation time, one can take advantage of several prop-

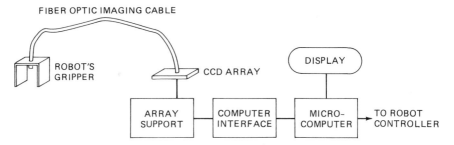

Fig. 11–23. Fiber-optic eye-in-hand system. (*From Agrawal and Epstein, Ref. 11*)

erties of the system described. Since the object is scanned line by line and there is continuity of image features, one needs only to examine changes and compare them against certain criteria to interpret the significance of these changes. Since there is a large amount of redundant information in the image, a large fraction of these changes can be discarded, since they do not contribute to the feature information that is being sought. This approach also reduces data storage space, since at any given time only the current data line is stored, and possibly a few items from a previous scan.

Another possibility, as mentioned above, is to arrange the optical fibers in a suitable pattern according to the object features that need to be detected. This concept can be expanded further when several fiber-optic scanning heads can be positioned in a gripper and be interfaced to a single CCD array. This is analogous to having several cameras but shared electronics, which is under common control. These "cameras" are independent of each other while coupled to the same feedback information for servo control.

REFERENCES

1. Keller, E. L. Clever robots set to enter industry en masse. *Electronics,* November 17, 1983, pp. 116–129.
2. Ward, M. R., Rossol, L., Holland, S. W., and Dewar R. CONSIGHT: a practical vision-based robot guidance system. *Proc. Ninth International Symposium on Industrial Robots,* Washington, DC, 1979, pp. 195–211.
3. Nagel, R. N., VanderBrug, G. J., Albus, J. S., and Lowenfeld, E. Experiments in part acquisition using robot vision, in *Industrial Robots,* Vol. 1, W. R. Tanner (ed.), 2nd ed. Robotics International of SME: Dearborn, MI 1981.
4. Mergler, H. W., and Nimrod, N. A triangulation laser-based scanning rangefinder. Eighth NSF Grantees' Conference on Production Research and Technology, Stanford, CA, January 27–29, 1981, pp. F1–F4.
5. Lees, D. E. B., and Trepagnier, P. Stereo vision guided robotics. *Electronic Imaging,* February, 1984, pp. 62–64.
6. Bolles, R. C. Three-dimensional locating of industrial parts. *Proc. Ninth Conference on Production Research and Technology,* November, 1981.
7. Bolles, R. C., and Fischler, M. A. A RANSAC-based approach to model fitting and its application to finding cylinders in range data. *Proc. Seventh International Joint Conference on Artificial Intelligence,* Vancouver, British Columbia, Canada, 1981.
8. Zuech, N., and Ray, R. Vision guided robotic arm control for part acquisition. Object Recognition Systems Inc., Princeton, NJ, no year.
9. Birk, J., et al. General methods to enable robots with vision to acquire, orient and transport workpieces. Seventh Report, University of Rhode Island, December, 1981.
10. Albus, J., et al. Six-dimensional vision system. *SPIE Vol. 336—Robot Vision.* Society of Photo-Optical Instrumentation Engineers, Bellingham, WA, 1982, pp. 142–153.
11. Agrawal, A., and Epstein, M. Robot eye-in-hand using fiber optics. *Proceedings of SPIE* (The International Society for Optical Engineering), *Robotics and Robot Sensing Systems,* San Diego, CA, August, 1983, pp. 257–262.

Chapter 12
PROXIMITY AND TACTILE SENSORS

PROXIMITY SENSORS

Gripper-Integrated Proximity Sensors

For the purposes of the bin picking system that was developed at University of Rhode Island (see Chapter 11), a parallel jaw gripper was developed.[1] A sectional view of the gripper is shown in Fig. 12-1. The tachometer as part of the compliant overload structure was necessary for the position control system. An optical limit switch established a reference opening position for the fingers of the gripper. The fingertips, however, were equipped with a proximity sensor. An infrared transmitter–receiver pair provided a light beam between the two tips. To obtain reliable performance, the transmitter was pulsed at a 10% duty cycle, and the receiver was synchronized with the pulses. Interruption of the beam would detect the instant when a piece passed between the fingertips. It would also verify that a piece was present when the fingers were closed.

This simple and reliable form of proximity sensor is frequently used in similar ways. Infrared light is usually preferred to white light, since infrared light limits the influence of surrounding light sources. Other forms of sensing, such as air jets or eddy-current devices, are often substituted for the light-beam proximity sensor.

Three Sensors Compared

International Harvester investigated three types of proximity sensors to provide assistance in the unloading of forgings from pallets.[2] A Cincinnati Milacron T^3 robot was used. A vision system was ruled out on the basis that it was then too expensive to be cost-effective, and that the requirements of controlled lighting and high contrast between object and background could not be met.

The purpose of the proximity sensor was to sense the edges of a gear blank forging accurately enough to locate the center of the forging by com-

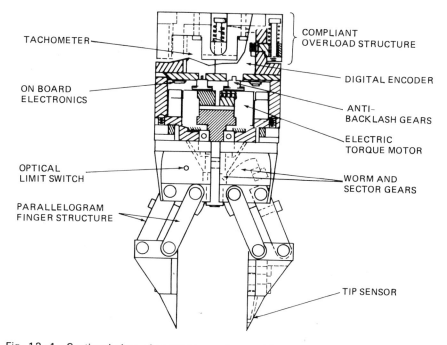

Fig. 12-1. Sectional view of parallel jaw gripper with optical tip sensor. (*Courtesy of University of Rhode Island*)

putation. The sensor had to be capable of meeting these conditions in spite of variations of $\pm \frac{3}{16}$ in. in the vertical distance from the sensor to the forging. The output from the sensor initiated the action of the robot. Thus, only an on–off signal was required.

Three sensors were investigated: an optical proximity sensor with an incandescent light source, a pulsed light-emitting diode (LED) sensor, and an eddy-current sensor.

The incandescent sensor contains the light source, a focusing lens, and a phototransistor in a package that measures $0.88 \times 0.50 \times 0.16$ in. It has a maximum range of 3.0 in., an optimum range of 0.16 in., a field of view that is 0.11 in. wide, and an object definition of 0.015 in. The chief advantage of the incandescent sensor over the other sensors is its small size, which enables it to fit into a minimum space including the tip of the gripper.

The pulsed LED sensor contains an infrared LED, a phototransistor, and two focusing lenses in a package that measures $3.25 \times 3.10 \times 1.00$ in. It has a detection range of 2.5 ± 0.25 in. and a spot size of $\frac{1}{8}$ in. The chief advantage of this sensor is its long sensing distance, which makes it less likely to be accidentally rammed into a forging.

The eddy-current sensor is contained in a cylindrical package with a diameter of 0.65 in. and a length of 3.2 in. It has a maximum sensing distance of 0.4 in. and 4% typical hysteresis. Its main advantage is that reflections from the forging do not interfere with its accuracy.

Before the incandescent and LED sensors could be used for detecting forgings, their sensitivity controls had to be adjusted to have as much sensitivity to the forging as possible without being sensitive to the pallet surface.

The LED sensor was adjusted for maximum sensitivity and was still insensitive to the pallet surface. The incandescent sensor, on the other hand, could not be set at maximum sensitivity without its detecting the pallet and thereby causing wrong signals. Since the forging is 1.8 in. high, the pallet will be at least that far away from the sensor. Consequently, its sensitivity was set just below the point where it would detect the pallet at a distance of 1.8 in.

Eddy-Current Sensors

Cutler-Hammer E54 sensors consist of an oscillator with sensing coil, detector, and output switch and two LEDs. Units with short-circuit protection, which open the switching circuit when the output is shorted, use three instead of two LEDs, as shown in Fig. 12-2. The standard two LEDs indicate "Power On" (left) and "Target Present" (right). The optional LED in the center indicates when the protection circuit has been activated.

Figure 12-2 illustrates the principle of the E54. When a metallic object is brought within the sensing field, eddy currents are induced in the target, causing a voltage drop in the oscillator. The voltage drop is sensed by the detector, which triggers the output signal.

The sensing positions are, depending on the model, on the top, on the front, on the left, or on the right, as illustrated in Fig. 12-3. The dimensions

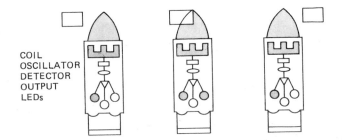

COIL
OSCILLATOR
DETECTOR
OUTPUT
LEDs

Fig. 12-2. Schematic of eddy-current sensor. (*Courtesy of Eaton Corporation, Cutler-Hammer Products*)

Fig. 12–3. Various models of eddy-current proximity sensor. (*Courtesy of Eaton Corporation, Cutler-Hammer Products*)

of these sensors are only 1.36 × 1.36 × 3.72 in. They are often referred to as welding sensors and are used with welding robots because they are resistant to the electromagnetic fields generated by resistance welders.

Optical Position Indicators

Optocator. The Optocator is an optical proximity sensor manufactured by Selcom Selective Electronic Inc. It has many applications for sensing positions, dimensions, contours, vibrations, thicknesses, and so on, one particular one being seam finding in ASEA's robotic arc welding system. It is available in measuring ranges of 0.3, 0.6, 1.25, 2.5, 5.0, and 10.0 in.

Its frequency response is up to 2 kHz, and its resolution capability 0.025% of measuring range.

As Fig. 12–4 illustrates, an infrared light source—either an LED or a laser diode—illuminates a spot on the surface to be measured. In the majority of applications, the light is beamed perpendicular to the surface. Most materials—whether hot or cold, hard or soft—will scatter a portion of this light. The Optocator "sees" a portion of this scattered light on its detector. Any variations in the position of the measured surface result in a change of the location of the focused spot's image on the detector. In this manner, the Optocator measures the point of origin of the scattered light—and hence the distance to the surface being measured.

The optoelectronic detector measures minute changes in position. At a measuring range of 5 in., for example, it will have an accuracy of ±0.005 in. and a resolution of slightly more than 0.0001 in. It does this by responding to the location of a spot focused on its surface and detecting the center of the light image, generating output signals that are converted into precise position information.

The light source is controlled to maintain constant intensity on the detector surface. This permits wide variations in the reflectivity of the measured surface.

One additional advantage of the Optocator is that the reference plane need not be fixed. Thus, if the height of a part on a moving, vibrating conveyor were being measured, a second Optocator head would measure the position of the conveyor belt surface, while the first would locate the top of the part. The signal processor would compare the signals from the two heads and give an output that would be the measurement of the height of the part.

The signal from the Optocator is in a serial digital format. It can be converted into both an analog dc voltage signal and a TTL compatible, digital signal for computer interfacing.

Fig. 12–4. Principle of Optocator.

Op-Eye.   Op-Eye is a product of United Detector Technology. One of its applications is to calibrate the position in space of robot arms, as shown in Fig. 12–5. A photodiode with lateral effect, as described below, is built into each of the Op-Eye optical heads. The location of the arm is signaled by the LED attached to the arm. The light is focused by a 28-mm lens on the photodiode system. The field of view of the system is a circle of 6.7 in. diameter at a distance of 3.3 ft. with a resolution of 0.0033 in.

The Op-Eye system can be used either in this dual-axis mode or as a single-axis device, where only one sensor head measures changes in displacement or angular position. It may also serve as a discrete sensor that only detects the presence or absence of a light beam. An Op-Eye system can accommodate up to 4 dual-axis position-sensing photodetectors, 8 single-axis sensors, 16 discrete sensors, or a combination thereof.

The lateral effect of the photodiode was discovered by Schottky in 1930. It is capable of resolving the position of a light spot over a single active element. The impinging light generates electrons within the substrate of the semiconductor photodiode. These electrons are divided according to position by applying multiple ohmic contacts on the back layer of the device. Two back contacts are made at opposite ends of the sensor for single-axis versions, while dual-axis units utilize four contacts coinciding with the extremes of the Cartesian axes. The back contacts are connected to the ohmic contacts. The light-spot position is measured by calculating the differential in photocurrent contribution to each ohmic contact. The photoinduced current flowing through the lateral contacts is the result of all incident light flux, based upon relative location and intensity over the active area of the diode. If it is difficult to locate the light source on the object, as in the case

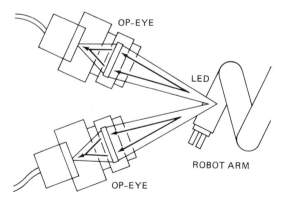

Fig. 12–5. Op-Eye. (*Courtesy of United Detector Technology*)

of the robot arm, a secondary source is used. A small white diffuser is then situated on the object, and the movement of the diffuser is in turn monitored by the detector.

Sensing of Position and Orientation

The Robotics Institute of Carnegie-Mellon University developed an optical proximity sensor for robot manipulation[3] that measures surface position and orientation in a range of 1.6 to 2 in. Schematics of the sensor are shown in Fig. 12–6. The sensor is useful for such applications as the tracing of an object surface by a robot arm at a specified distance and orientation relative to the surface.

The sensor uses six infrared LEDs and a pin-diode sensor chip. The latter has four outer electrodes (X_1, X_2, Y_1, Y_2), as shown in the illustration. When light impinges on a spot of the sensor chip face, a photocurrent is generated at this spot. The magnitude of the currents that then appear at the four electrodes will depend on their distances from this point, and a total current can be obtained that is a function of these four currents. It will be proportional to the intensity of the light impinging on the chip spot and will define the center of the light distribution at this spot.

The six LEDs are provided with optics for collimating the beam and are mounted at the sensor head. The directions of the beams are aligned to form a cone of light converging at a distance of 1.77 in. from the sensor head. As each LED is sequentially pulsed, the sensor chip detects the position of the spot projected by the LED light beam on the object surface. The 3-D location of the spot on the surface can be computed by triangu-

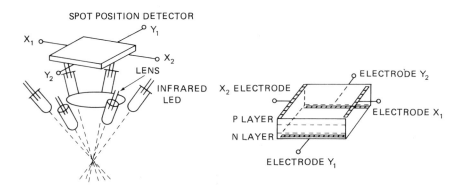

Fig. 12–6. Schematics of proximity sensor with pin-diode sensor. (*From Kanade and Sommer, Ref. 3*)

lation. By doing this for six LEDs, a set of six three-dimensional points is obtained. Then, by fitting a plane to these points, the distance and the orientation of a small portion of the object surface are calculated.

The sensor can give approximately 1000 measurements of distance and orientation per second with a precision of 0.003 in. for distance and 1.5° for surface orientation.

TACTILE SENSORS

Tactility is capability of an object to be felt or touched. For the purposes of robotics, tactility is the feeling of the presence of force, torque, and slip. Tactile sensing, thus defined, may give information about presence, proximity, shape, orientation, weight, or unbalance of an object. In many cases, it can replace vision, particularly today when vision technology is barely out of the laboratory. Thus, tactile sensing has gained great importance, and it is surprising how many different approaches exist for this purpose.

One of the most frequently used types of tactile sensor responds to force exerted by manipulating an object. Force is exerted (a) by the gripper when it picks up an object, and (b) by the robot arm when it lifts the object. Such forces can be made evident by gripping force, incipient slipping of the load, or torque exerted on the gripper, wrist, or joints.

Many robotic assembly tasks require only relatively simple tactile sensors. Dr. Leon D. Harmon, of the department of biomedical engineering at Case Western University, surveyed six manufacturing plants where automobile and electronic components and appliances are assembled. Of hundreds of manual assembly tasks studied in these plants, 41 were chosen as likely applications for robots, and 80% of those tasks were judged to be suited for robots with two-jawed grippers or three-fingered hands equipped with tactile sensor arrays with no more than eight-pressure sensitive points on a side. These sensors would have to be capable of sensing slip or shear.

When a gripper picks up an object, it is frequently desirable to measure the force that the gripper exerts. It is particularly important when delicate objects are being handled to assure that the force exerted is sufficient to lift the object but not so large as to damage it. A simple approach, as used occasionally by Cincinnati Milacron, is to monitor the air pressure that is applied to pneumatic actuated grippers. The air pressure applied is related to the force exerted.

Another form of force sensing is often desirable to protect the robot from overload. This may be done directly by limiting the torque exerted by the actuator, or by torque sensors in the joints of the robot. The sensors described in the following paragraphs, however, are primarily those used with the gripper of the robot.

Strain Gages

Either tension or compression forces applied to a fixed object produce a change in length. Such changes can be measured by a strain gage down to one-millionth of an inch per inch. The conventional strain gage consists of a grid of fine alloy wire or foil, typically bonded to paper and covered for protection with a felt pad. Ideally the wire changes its electrical resistance in proportion to the amount of deformation under strain. Frequently, strain gages are arranged as a Wheatstone bridge. They should also include temperature compensation, since a strain gage is sensitive to temperature as well as strain. The compensation is generally effected by a resistance unit on a nonstrained substrate adjacent to the active gage.

The Astek Accommodator remote-center compliance device from the Barry Wright Corporation was described in Chapter 3 (see Fig. 3-26). This was a passive approach, since the device itself yields to external forces and thereby positions the peg into the hole even if slight offsets are encountered. An extension of this principle is an active force sensor that produces an electric feedback signal to a control system in response to external forces. This concept has been designed into a six-axis force sensor with onboard microprocessor, shown in Fig. 12-7. It includes transducer structure, strain gages, and instrumentation electronics to completely measure the six components of force that may be applied to the sensor body by an arbitrary load. A schematic of the circuitry, complete with six foil-type strain gages—one for each force component—is shown in Fig. 12-8.

The sensor is provided with two printed circuit boards. One is for multiplexing by scanning the six strain-gage bridges, the other for converting the analog signals from the first board into digital signals. The output from this board is an RS-232 compatible serial-bit stream that contains measurement of instantaneous forces acting on the sensors. Also provided is an analog output port for chart recording and other analog devices. The discrete digital port serves to signal overlimit conditions produced by forces exerted on the sensor.

The unit is 4.73 in. in diameter and 2.37 in. high. The full-scale ranges are 25, 50, or 150 lb, depending on the model. The corresponding moments that can be sensed are 18, 36, and 108 in.-lb, respectively. Resolution of the 50-lb model is 0.025 lb force, or 0.018 in.-lb.

Silicon Strain Gages. Silicon is not only a semi-conductor, as described earlier; it is also piezoresistive, and hence changes its resistance when strained. Furthermore, it physically moves in response to pressure or strain. Using its piezoelectric and solid-state characteristics, a silicon semiconductor can thus become the basis of a strain gage.

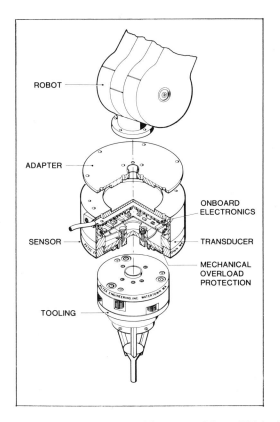

Fig. 12-7. Six-axis force sensor. (*Courtesy of Barry Wright Corp.*)

The rotary joint sensor, designed at Purdue University, consists of four p-type semiconductor strain gages of 350 ohms each. In their work on the advancement of industrial robot systems, Purdue University researchers postulated that robots without feedback do not provide precise position control and hence require feel, force, and touch sensors to enhance their precision.[4] They reported as follows:

A good force sensing technique is imperative to increase the robot's adaptabilities, especially in the area of batch assembly. A fast, reliable force servo system based on the proper sensing techniques can increase the speed of robot operations. However, the existing force servo systems are differential approximations or are computationally difficult, making the compliance either slow or approximate. To overcome these disad-

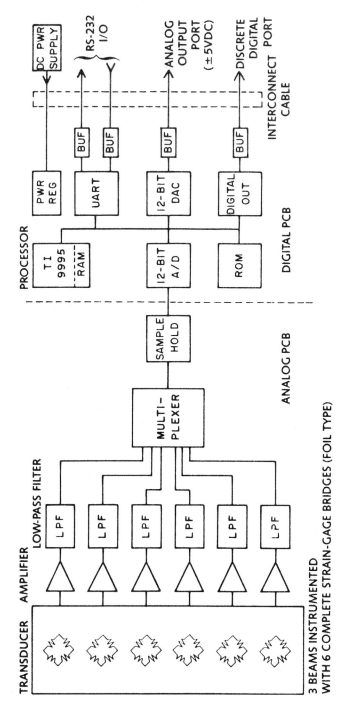

Fig. 12–8. Schematic of circuit for six axis force sensor. (*Courtesy of Barry Wright Corp.*)

418

vantages a simple, high gain, wide bandwidth joint torque servo system has been developed which contains no differential approximations and provides a fast response with no computation problems. The resolution of the sensed torque into control information is much simpler than with a wrist sensor. We have tested a single joint manipulator with the joint sensor to verify the experimental results of the theoretical analysis.

Based on this experiment, two joints of an industrial robot have been redesigned and fabricated to include torque sensing capability by means of strain gages. The resulting control systems reduced the effective frictional torques of the joints from 1077 oz-inches to 33.5 oz-inches.

The above-mentioned four p-type semiconductor strain gages of 350 ohms each, which Purdue designed in following these concepts, are connected as a Wheatstone bridge and provided with temperature compensation. The gages are cemented to a thin short section of the circular aluminum joint shaft. Two strain gages are mounted at 45° and the other two at 135° with respect to the central axis of the shaft, to sense the tension and compression strain. The output is 0.329 mV/oz-in. of torque.

Skins

The concept of robotic skin is to provide a robot hand with the equivalent of the skin of a human hand, in particular with respect to its capability for using its tactile sense. There are about 20,000 sensory-nerve endings in a human fingertip, which, however, has a resolution of 0.004 in., making it comparable to state-of-the-art sensor technology. The following describes some of the robot "skins" being used or being developed.

Conductive Membranes.   The Artificial Intelligence Laboratory of the Massachusetts Institute of Technology has been developing such a hand, and as William Hillis, a researcher at that institution, pointed out:[5] "The mechanical hand of the future will roll a screw between its fingers and sense, by touch, which end is which." The hand he developed toward this purpose incorporates an artificial skin with 256 tactile sensors in a space the size of a fingertip. Each sensor has an area of less than 0.016 in.^{2} The array of sensors is scanned one column at a time to minimize the number of connecting wires.

Figure 12–9 shows the concept of the skin. It is constructed of three layers. The top layer is a sheet of anisotropically conductive silicone rubber (ACS). It is electrically conductive only vertically to its plane, since it consists of vertical slices of silicone rubber impregnated with either graphite or

NYLON MESH OR FILM OF SPRAY LACQUER

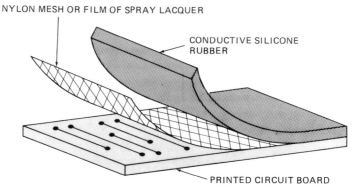

CONDUCTIVE SILICONE
RUBBER

PRINTED CIRCUIT BOARD

Fig. 12–9. Skin with conductive elastomer. (*Courtesy of Massachusetts Institute of Technology*)

silver, alternating with similar nonconductive slices. Each slice is approximately 0.01 in. thick.

The bottom layer is a flexible printed circuit board. It is etched into fine parallel lines, so that it too conducts in only one dimension, but one that is at right angles to the conductivity of the ACS. The contact points at each intersection form the pressure sensor.

Sandwiched between the ACS and the board is an isolating layer that tends to spread them apart. The sensitivity and range of the sensor depend largely on the construction of this layer, which may consist of either a fine-mesh nylon weaving for a large pressure range or a fine mist of nonconductive paint where high sensitivity is more desirable. The ACS and the board can meet only at those points where pressure is applied on the top side of the ACS and compresses the isolating layer sufficiently to establish contact between the outer layers.

Wires are connected to the board at the edges of the array. The array is scanned by applying a voltage to one column at a time and measuring the current flowing in each row.

The Barry Wright Corporation developed the Sensoflex Tactile System using a similar compliant tactile pad of a conductive elastomer and an electronic interface device. The system provides data that can be used to determine gripping force, part position, and orientation. At the same time, its surface is durable and compliant to serve as a gripper pad for picking up workpieces.

The Sensoflex sensor pad and the image produced by it are shown in Fig. 12–10. The pad contains an array of either 16 × 16 sensor points with distances between centers of 0.1 in., or 8 × 16 points with 0.05-in. centers.

Fig. 12–10. Typical image produced by Sensoflex tactile system. (*Courtesy of Barry Wright Corp.*)

The former has overall measurements of 1.77 × 2.04 × 0.33 in., while the latter measures 0.54 × 1.04 × 0.26 in.

To drive the sensor, the array must be excited, and the data must be obtained from each sensor point. The data are read by exciting a row with a constant 5-V signal while reading 16 columns.

LSI Sensors.   NASA's Jet Propulsion Laboratory has proposed a large-scale integrated (LSI) circuit placed on the contact surface of end effectors as touch sensors. It would combine the functions of transduction, computing, and communicating.

The concept is shown in Fig. 12–11. The entire surface of an LSI wafer is used as an array of tactile sensors with sensing material connected directly to computation elements in the wafer. These elements process and reduce the raw sensory data and send the resulting signals to a control computer. A number of functions are thus combined, and signal bandwidth, number of connecting wires, weight, and power consumption are reduced, while high tactile resolution is achieved.

A key feature of the concept is that computation at the site of transduc-

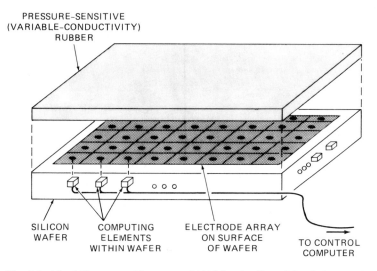

PRESSURE-SENSITIVE
(VARIABLE-CONDUCTIVITY)
RUBBER

SILICON
WAFER

COMPUTING
ELEMENTS
WITHIN WAFER

ELECTRODE ARRAY
ON SURFACE
OF WAFER

TO CONTROL
COMPUTER

Fig. 12-11. LSI sensor. (*Courtesy of NASA—Jet Propulsion Laboratory*)

tion permits useful information to be extracted from the raw tactile image and selectively communicated to a central controller. Thus, grasping force, touch pattern, contact area, and slippage can be extracted readily.

The exposed surface of the LSI wafer contains an array of pairs of electrodes covered by a sheet of electrically conductive rubber. When the rubber sheet is deformed by touch, its resistivity varies locally. Current passing through the sheet between a pair of electrodes indicates the local contact pressure.

Associated with each electrode pair is a computation element consisting of an analog comparator, a data register, and an accumulator. Each element reads the local conductivity data from its electrode pair and performs simple calculations. In addition, each element communicates directly with its nearest neighbors and indirectly with others. All elements in the array execute the same sequence of instructions simultaneously, as directed by a controller that is also in the LSI wafer. Together the elements form a tactile-pattern parallel processor.

Silicon Touch Sensors. As mentioned above, silicon has piezoelectric characteristics and moves physically in response to pressure. Furthermore, the development of micromechanics[6] permits shaping silicon not in two but in three dimensions. It also permits making very thin silicon membranes that are strained by the slightest touch, thereby producing electrical signals. Transensory Devices, Inc. uses this process in making silicon touch sensors.

It permits mass production of minute structures with extremely high precision. By combining these structures with an onboard microprocessor, signal processing can be performed on the touch sensor itself, increasing the overall system accuracy, control, and reliability.

The TP4011 Tactile Sensor Array as produced by Transensory Devices consists of minute silicon touch sensors arranged in a three-by-three matrix with a spacing between elements of 0.08 in. Its normal force range is zero to 2 lb. The complete dimensions of the unit are 0.64 × 0.72 × 0.08 in., and its weight is 0.1 oz. The built-in addressing logic permits ready expansion to larger arrays and interfacing to controllers.

In addition, an electronic interface is available for signal processing and analog-to-digital conversion along with an internal microprocessor and RS232 interface. To obtain information, the host computer sends a command to the interface module and gets back compensated binary data.

Magnetoresistive Skin. Magnetoresistivity is the capability of a material to change its electrical conductivity under the influence of a magnetic field. This change is a function of the magnetic field intensity. John Vranish, of the National Bureau of Standards, has used this effect for the magnetoresistive skin of a robot,[7] and Fig. 12–12 shows the construction of such a skin. The magnetoresistive material is permalloy, an alloy of nickel and iron. In this case it consists of 81% nickel and 19% iron.

Permalloy resistors of 50 ohms each are etched on a substrate of aluminum oxide. Connected by gold shorts and thick-film edge conductors, they form an array of magnetoresistive elements. Over this array lies a thin film of rubber, typically an open-cell sponge rubber that is about 0.01 in. thick, and is in turn covered by a thin Mylar film with a pattern of copper

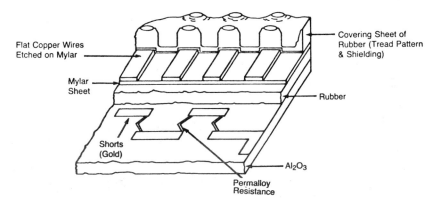

Fig. 12–12. Construction of magnetoresistive skin. (*From Vranish, Ref. 7*)

strips that are about 0.00025 in. wide. The copper strips provide the magnetic fields that are to affect the magnetoresistive permalloy elements located on the substrate underneath. Figure 12–13 shows a photograph of an array with the magnetoresistive elements spaced 0.1 in. apart, forming an array of 64 elements on a 1 × 1 in. square.

The skin is constructed so that each permalloy element has a copper strip directly over it. The permalloy element is exposed to a magnetic field when a current passes through the corresponding copper strip above it. When this occurs, the permalloy element changes its resistivity in inverse proportion to the distance from the strip. The multiplexing circuit scans the different elements continuously.

If 30 mA is pulsed through a copper strip 0.01 in. above a permalloy element, that element will be subject to a magnetic field intensity of about 0.26 Oe (= oersted or gauss, which both are used to express the unit of the magnetic field intensity). If the separating rubber pad is compressed so that the distance is decreased by 33%, the magnetizing force will increase to about 0.39 Oe. Within this 1/3 compression the force/compression relationship is linear to within 10%. Tests with the rubber pad have shown that,

Fig. 12–13. Magnetoresistive array. (*From Vranish, Ref. 7*)

within this range, the rubber compresses 0.05 in./psi. Thus, the sensor can be used for tactile pressures of up to 0.56 psi at a sensitivity of 0.004 psi. The minimum detectable pressure on the tactile skin would be 0.006 psi.

Ultrasonic Force Sensors

To measure the change in thickness of a compliant skin material, Grahn and Astle of Bonneville Scientific developed an ultrasonic method.[8] These sensors are capable of measuring pressure or force over a 2000:1 range. Distances between centers of sensing points can be made as small as 0.02 in. Response time of the sensor is on the order of 5 μsec.

One particular advantage claimed is that the elastic element does not need electrical conductivity, so that the material may be chosen strictly on the basis of mechanical considerations and not as a compromise between mechanical and electrical behavior. However, as shown below, the desirable characteristics of compliance and linearity may also make the requirements for the elastic element more stringent.

The elastic element is the pad of the sensor and is made of silicone rubber. The external force is applied to the pad, causing a change of its thickness that is measured by an ultrasonic pulse-echo technique. The time it takes for an ultrasonic pulse to traverse the pad and return to the transducer is a direct measurement of its thickness. Since the thickness of the pad in its relaxed condition is 0.12 in., it takes about 6 μsec for the pulse to traverse it and return through it. If the expected maximum force compresses the pad to 80% of its original thickness and it is necessary to resolve this force to one part out of 50, then the ultrasonic pulse transit time needs to be resolved within 24 nsec. Though this is within the capabilities of the electronic system, a thicker or more compliant pad or reduced force resolution requirements would somewhat lessen the demands on the system.

An ultrasonic pulse-echo system requires a transducer that produces and receives the ultrasonic waves.

A piezoelectric material—which generates a voltage when stressed and, inversely, gets stressed when a voltage is applied—is generally used for this purpose. One of the most common piezoelectric materials is quartz, which has the best characteristics. However, there are other materials that, while not quite as powerful as transducers, have the advantage of being more adaptable than quartz to the packaging of a tactile sensor.

It was discovered in 1969 that polyvinylidene fluoride (PVDF) has a strong piezoelectric effect. There are other plastics that have the same effect, but in PVDF, provided that it is prestretched uniaxially or biaxially, it is particularly high. It was this material that was chosen by Grahn and Astle as transducer for their ultrasonic force sensor.

Thus, the ultrasonic transducer consists of a PVDF foil that is 1.1 μin. thick and is covered with aluminum-over-tin electrodes of 3 to 4 μin. thickness. The unit is connected to a pulser, producing 5-V pulses of 200 nsec duration. These pulses produce stress changes in the PVDF that are transmitted through the silicone rubber pad and reflected. The reflected ultrasonic waves produce voltages in the PVDF, with the time difference between the pulse and its reflected echo proportional to the thickness of the pad.

The surface of the silicone rubber pad was indented with the flat end of a $\frac{1}{8}$-in.-diameter brass rod, which detected the echos. It was found that an object of 0.0088 in.^{2} area and 0.022 oz weight that rested on the pad would readily signal its presence by this method. This corresponds to 0.15 psi as the lower end of the sensitivity range, while the pressure at the upper end is 300 psi. Thus, the pressure range for the sensor is the above-mentioned 2000:1.

The array construction and the principle of operation are illustrated in Fig. 12–14. The PVDF is metallized on both sides. On the top, the metal is selectively etched lengthwise and crosswise to leave a number of element electrodes, totaling 16 elements in this prototype. Further development would use a thin-film hybrid circuit for the sensor assembly, together with leads with integrated circuit chips bonded to the substrate and connected to the array elements for multiplexing. The hybrid circuit technique would use integrated circuit chips mounted on a ceramic substrate.

Magnetoelastic Sensors

When a magnetic material changes its flux density as a result of an externally applied force, it is termed a "magnetoelastic material exhibiting the villari effect." When it changes its length, it is a "magnetoelastic material

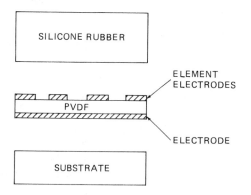

Fig. 12–14. Array construction of ultrasonic sensor. (*From Grahn and Astle, Ref. 8*)

exhibiting the magnetostrictive effect.'' There are two specific magnetoelastic materials that exhibit both the villari and the magnetostrictive effect. These are: (a) amorphous ribbons produced from an iron–cobalt–boron alloy with small amounts of silicon and cobalt, and (b) terfenol rods that contain terbium, dysprosium, and iron. Amorphous ribbons are well suited for measuring stretching forces because of their strong villari effect under such conditions. Terfenol is superior in magnetostriction but also performs well under compression, where it shows a high villari response.

Magnetoelastic materials research has been in process at the Naval Surface Weapons Center since the 1960s, and terfenol was discovered there by Dr. A. E. Clark as a result of this work. Progress in more recent times revealed the potential of these materials as force sensors for robotic applications.[9]

The materials, whether amorphous ribbons or terfenol, are magnetized so that each internal magnetic particle or domain lines up with its net magnetic moment perpendicular to the long axis of the material. At the same time, the net magnetic moment of each domain is pointed in a direction opposite to the adjacent domain. This leaves the material with a total net magnetic moment of zero, thus reducing spurious effects and increasing its sensitivity.

Thereafter, any external magnetic field to which the material is exposed rotates the magnetic moments toward the direction of the field.

Furthermore, the net magnetic moment of each of the domains of the amorphous ribbon will rotate toward the axis of the ribbon when it is stretched. The rate of rotation is proportional to the tensile force. The effect is that the flux density of the ribbon changes in proportion to the applied tensile force per unit area.

The same thing happens in reverse with a terfenol rod when it is compressed by an external force. In this case, the flux density increases in proportion to the compression force per unit area.

A second magnetic field is produced by wire coils that are wound around the amorphous ribbons or terfenol rods. Current flowing through the wire will expand or contract the magnetoelastic material to an extent that is dependent on the magnetostrictive properties of the material and the product of current and number of coils. As already stated, the magnetostrictive effect is most pronounced with terfenol rods.

Magnetoelastic materials can be used to measure either the total force directly or the rate of change of the force acting on the sensor.

The ability of magnetic material to carry magnetic flux is its permeability. It can be compared with the inverse of resistance (i.e., the conductance) of an electric wire. An amorphous ribbon, when used for direct tensile force measurements, has a permeability that is about 40,000 times greater than that of a terfenol rod. It is the obvious choice for such measurements.

To build a direct force sensor, for example, a coil of insulated wire is wrapped around an amorphous ribbon. This becomes the drive circuit, which is connected to an oscillator. A second coil on the same ribbon provides feedback to control the drive circuit oscillating frequency. A third coil generates a voltage output signal that is proportional to the magnetoelastic effect produced in the ribbon.

The driving circuit drives the amorphous ribbon to saturation from one polarity to the other. Each time the wave form drops to zero, the output signal is maximum. This maximum, and thus the voltage amplitude of the output oscillation, is determined by the change of permeability due to the tensile force applied to the sensor.

The reasons why these materials are considered such good force feedback and tactile sensors are essentially the following:

- Sensitivity: The ratio of output voltage to the voltage equivalent of the applied force is 0.7 and expected to climb to 0.8. A ratio of 0.040 was previously considered to be excellent.
- Range: Measurements with amorphous ribbons are linear within a range from 0 to 1000 psi. Practical design considerations often require that the ribbons be prestressed to 500 psi. This still leaves a linear measuring range of 500 psi. With terfenol rods the range is 0 to 4000 psi, with prestressing to 2000 psi.
- Resolution and hysteresis: Experiments have shown that it is possible to resolve a change of 0.004 psi at 266 psi. There is no observable mechanical hysteresis.

Other favorable characteristics are the simple circuitry, as well as durability, corrosion resistance, and flux density. It needs only an essentially passive sensor with one set of coils and no drive circuit. Generally, the sensor provides a relatively low-voltage, high-power output.

There are also some problems. One is sensitivity to stray fields and voltage changes. This especially affects the amorphous ribbons because they have such high magnetic permeability, and they operate with an oscillator. Several techniques can alleviate this problem. Among them are magnetic shielding around the amorphous ribbon, second sets of symmetrical windings to cancel errors, filtering, and digital signal processing techniques. Another problem is the brittleness of terfenol rods, but recent experiments have already resulted in tensile strengths of 2000 psi, while the compressive strength remains extremely high.

Figure 12–15 shows the gripper of a robot equipped with two torque sensors in the gripper base combined with two slip sensors in the gripper faceplates. Torque as well as slip sensors are magnetoelastic. They are designed for a large industrial robot that has a load capacity of 400 lbs.

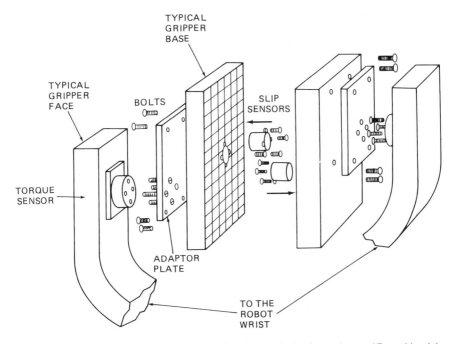

TYPICAL
GRIPPER
BASE

TYPICAL
GRIPPER
FACE

BOLTS

SLIP
SENSORS

TORQUE
SENSOR

ADAPTOR
PLATE

TO THE
ROBOT
WRIST

Fig. 12-15. Magnetoelastic sensor retrofitted to typical robot gripper. (*From Vranish, Ref. 9*)

The torque sensors can each measure a torque of up to 100 ft-lb in a package 2.5 in. in diameter and 0.8 in. high. The top plate or turret of the torque sensor in Fig. 12-16 is mounted to the gripper face. The lower portion of the sensor is bolted to the robot gripper base, which is rigidly connected to the wrist.

Torque exerted by an unbalanced load on the gripper would tend to rotate the turret of the torque sensor against its lower portion, which contains the amorphous ribbons.

Since the inner radius of the ring is 1 in., the maximum design torque of 100 ft-lb will produce a force of 1200 lb. However, the ribbon itself will only require 3 lb to produce its full output. The remainder will be taken up by steel rods, which are shown in the illustration.

A direct force measurement with oscillator is used. The ring of the torque sensor as shown consists of two identical halves. Each half is composed of 15 laminated layers of amorphous ribbons, with coils wound around them. Depending on the direction of torque, one or the other of the ring halves will be stressed, thus producing a voltage signal that is proportional to the torque.

The slip sensor is shown in Fig. 12-17. It is designed to be sensitive to

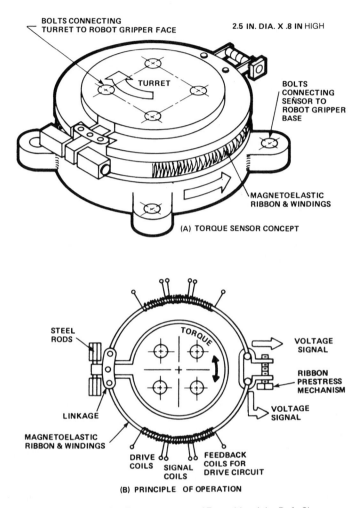

Fig. 12-16. Torque sensor. (*From Vranish, Ref. 9*)

0.2 oz slipping force and to yield 30 m V at 0.01 in./sec slip. Its diameter is 1.1 in., and its height, 0.8 in. While the torque sensor measures the force directly, the slip sensor is intended to measure its rate of change. This means the oscillator can be dispensed with.

A constant normal force, F_n, of 1 lb is produced by a helical spring. The slip force, F_s, is at right angles to it. The direction of the net force thus produced is between the two. It stretches the amorphous ribbon laminate, which in turn produces the output signal.

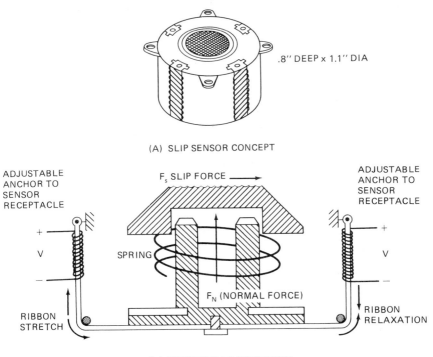

(A) SLIP SENSOR CONCEPT

.8" DEEP x 1.1" DIA

ADJUSTABLE ANCHOR TO SENSOR RECEPTACLE

F_s SLIP FORCE ⟶

ADJUSTABLE ANCHOR TO SENSOR RECEPTACLE

SPRING

F_N (NORMAL FORCE)

RIBBON STRETCH

RIBBON RELAXATION

(B) PRINCIPLE OF OPERATION

Fig. 12-17. Slip sensor. (From Vranish, Ref. 9)

The laminate consists of eight ribbons, each 0.002 in. thick, 0.1 in. wide, and 1.44 in. long. Two such ribbon laminates are used, at right angles to each other, in order to respond to slip in any direction.

REFERENCES

1. Birk, J., et al. General methods to enable robots with vision to acquire, orient and transport workpieces. Sixth Report prepared for National Science Foundation, University of Rhode Island, August, 1980.
2. Clark, L., Webber N., and Sutton, T. Adaptive control with the T^3 robot. *Proc. Thirteenth International Symposium on Industrial Robots,* Chicago, April, 1983, pp. 13-1/13-12.
3. Kanade, T., and Sommer, T. M. An optical proximity sensor for measuring surface position and orientation for robot manipulation. *Proc. Symposium on Robotics and Robot Sensing Systems*, San Diego, The International Society for Optical Engineers, August, 1983, pp. 301-307.
4. Paul, R. P., Luh, J. Y. S., and Nof, S. Y. Advanced industrial robot control systems. *Proc. Ninth Conference on Production Research and Technology*, Ann Arbor, November, 1981.

5. Hillis, W. D. Active touch sensing. Report AD-A099 255, National Technical Information Service, April, 1981.
6. Petersen, K. E. Silicon as a mechanical material. *Proceedings of the IEEE,* Vol. 70, No. 5, May, 1982, pp. 420–457.
7. Vranish, J. M., Magnetoresistive skin for robots. *Proc. Conference on Robotics Research: The Next Five Years and Beyond,* Bethlehem, PA, August, 1984. SME Paper MS84-506.
8. Grahn, A. R., and Astle, L. Robotic ultrasonic force sensor arrays. *Proc. Robots 8 Conference,* Detroit, June, 1984, pp. 21-1/21-18.
9. Vranish, J. M. Magnetoelastic force feedback sensors for robots and machine tools. *Proc. Robots VI Conference,* Detroit, March, 1982, pp. 492–522.

Chapter 13
SAFETY

It is not the purpose of this chapter to discuss the many safety procedures and warnings that robotic manufacturers attach to their equipment. They should be carefully observed. The following discussion considers only some concepts of (a) perimeter protection and (b) collision protection. Another approach to increased safety would be to provide enough resiliency in a robot arm so that even in case of collision, damage could be avoided. An example is the robot with pneumatic actuators that was described on p. 217.

PERIMETER PROTECTION

A straightforward means to protect people from the hazards of the robot's work area is a solid enclosure with a door that only authorized persons can use. Standard with all robots from Prab Robots, Inc., for example, is a chainlink fence containing an interlocking gate. The interlocking gate automatically shuts down the robot when a worker opens it to enter the area. The only way the robot can be restarted is to shut the gate and restart its controller.

Figure 13-1 shows a robot guard made by Wire Crafters Inc., protecting the work area of a robot from GMF Robotics Inc. It is made of modular woven wire panels and square steel posts. The panel used in front of the guard in the illustration, for example, is a stock size 6 ft wide and 4 ft high. This provides a highly visible barrier that prevents personnel from walking into the robot working area. In addition, the barrier keeps the robot from ejecting objects from its working area.

Where the work area requires some limited access and it is only necessary to prevent accidental entering of the work area, handrails may suffice, though their protection is decidedly less reliable than that of a solid enclosure.

Another form of perimeter protection is a light-beam curtain. An example is the SafeScan Perimeter Guard, produced by Dolan-Jenner. Here, a curtain of low-level, near-infrared light beams surrounds the work area

Fig. 13-1. Wire panel robot guard. (*Courtesy of Wire Crafters Inc., Louisville, KY*)

434

of the robot, as shown in Fig. 13–2. The light is invisible to the eye, but any crossing of the light curtain shuts down the robot. Heights of either 35 in. or 70 in. can be protected. The system consists of a number of posts. One post contains the transmitter that sends out the light beams. Corner posts are equipped with mirrors that pass the light beams around the perimeter to an end post that contains the receiver.

The light beams are produced by light emitting diodes (LEDs). These are solid-state devices that have proved to be dependable over long periods of time in high shock, vibration, and dirty industrial environments. They are part of the transmitter and are rapidly switched on and off to emit brief flashes of near-infrared light, which is practically invisible to the human eye. The receiver is tuned to the repetition rate and duration of the pulses. Thus, effects of extraneous light are eliminated, and the system becomes virtually immune to light sources other than those of the transmitter. The LED pulse modulation technique permits reliable operating ranges up to 65 ft.

The response time of the system in case of an obstruction in the light beam is 70 msec maximum, 40 msec average for the 35-in.-height protection, and 58 msec maximum, 24 msec average for the 70-in.-height protection.

A SafeScan perimeter guard for 70 in. protected height is provided with

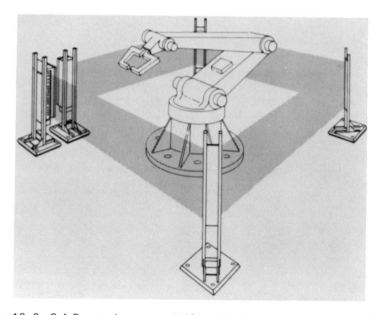

Fig. 13–2. SafeScan perimeter guard. (*Courtesy of Dolan-Jenner Industries, Inc.*)

a vertical array of 20 LEDs that emit the near-infrared light beams. For approximately a 12-in. distance about the center, the space between LEDs is reduced so that the minimum resolution is $1\frac{5}{8}$ in. to allow for hand intrusion, stock entry, and so on, in the event this is necessary. The spacing of emitters and detectors is gradually increased on either side of this area to a minimum resolution of approximately 9 in. at the ends.

The top channel in the receiver unit is a coded LED that sends synchronized modulated information to a detector in the top of the transmitter. The transmitter uses the coded information to actuate infrared emission from each transmitter beam at the proper time.

Two relays provide the output signal from the perimeter guard—two being used for redundancy, so that if one fails the other will still give a signal in case of unauthorized entry into the work space. The outputs consist of one set of normally open and one set of normally closed contacts to interface with the robot controller. Generally, the output is connected to the emergency stop circuit of the robot controller.

There are three modes of operation selectable by a key-operated switch:

1. The "latch off" mode, which automatically restores the system to normal after an obstruction has passed through the light curtain.
2. The "latch reset" mode, which keeps the output relays activated until a reset pushbutton has been depressed.
3. The "latch on" mode, which permits return to normal operation only after a supervisory key switch has reconditioned the system.

The SafeScan perimeter guard allows partial permanent obstruction of anywhere between 6% and 89% of the protected zone while providing full protection of the remaining safety coverage area. Authorized personnel simply unlock the receiver enclosure and set the proper channel deselection switches for the area to be blocked by conveyor, material handlers, or other obstructions. If a deselected channel sees light, the robot will be shut down.

Another approach was described by Roger D. Kilmer.[1] His report concerns a prototype safety system that was developed by the National Bureau of Standards. The area to be protected is divided into two different zones. One is the working volume of the robot, which corresponds to the region that the robot can reach with its arm at maximum extension. The other corresponds to a larger perimeter identified as the workstation of the robot.

A set of pressure-sensitive floor mats is positioned around the robot. A foot pressure of 30 lb or approximately 5 lb/in.^{2} is required to activate the mat to turn on a warning light. Pressure-sensitive mats are used because they do not create any obstacles to isolate the workstation from the surrounding areas.

The floor area that corresponds to the working volume is protected by another such mat. However, in this case stepping on the mat produces a signal that stops the robot.

Thus, when an intruder enters the workstation area, a visual and/or an audio signal is broadcast. When he enters the working volume, the robot stops at its current position. The robot remains in this position until the intruder leaves the working volume. Once the intruder has left this area, the robot resumes its programmed task from the point at which it stopped. For cases where personnel must be within the working volume, a hand-operated emergency stop switch is used to halt the robot.

COLLISION PROTECTION

Together with the floor mat system described above, the National Bureau of Standards[1] developed what it called the Level II system. Here, it is not so much the intrusion but the incipient collision against which the system is to protect. The Level II consists of an array of five ultrasonic sensors, mounted on critical points of the robot. Their basic operation involves:

- Transmission of an ultrasonic pulse from the electrostatic transducer of the sensor.
- Reception of any reflected signals using the transducer as the receiver.
- Measurement of the time the ultrasonic pulse requires to cover the distance from transducer to target and back to transducer.
- Measurement of the separation distance on the basis of the time signal.

The ultrasonic system is used to determine whether an intruder gets closer than some predetermined minimum distance from the robot. If this is the case, a signal is sent to the robot controller to halt the robot. The operating range of the Level II sensors is 0.9 to 35 ft.

Another collision protection system is Roboguard, which was developed at General Motors' Technical Center. A multiple-branched antenna is mounted rigidly on the robot's arm. A weak electromagnetic field is projected around the antenna in about a 12-in. envelope. The antenna is connected to a capacitance-type sensor that detects any conductive or dielectric object, including a person, that intrudes within the envelope. The sensor output is a voltage that varies between 2 and 14 V. The voltage is then converted into a digital signal that is fed into the computer.

The robot is initially guided through its work routine. During this operation, the sensor output responds to any metal workpieces or other conductive or dielectric objects it encounters and that will be present during its normal work routines. These inputs are stored in the computer memory as the "signature" of the operating path of the robot. Once the robot is applied to normal operating conditions, the computer compares the sensor's input with the initially stored signature every 0.1 sec. Any deviation will bring the robot to an immediate stop.

A sensor of this sort could also be modified for use with a collision avoidance system such as the one designed by Gerald J. Agin and John K. Myers of the Robotics Institute of Carnegie-Mellon University. Here, not only are obstacles in the path of the manipulator detected, but alternate routes are automatically chosen to avoid collisions.

The system uses simulation to test a given path of the manipulator. If the proposed path results in a predicted collision, a number of intermediate points in the vicinity of the obstacle are specified. The routines that select these points account for the nature, size, and shape of the obstacle and the portion of the manipulator involved in the predicted collision.

Since the program is capable of successively trying various approaches to find the subpaths, search techniques can be used to find a pair of collision-free paths from the starting point to the intermediate point, and from there to the goal. An alternate path is then drawn on the basis of least deviation from the originally proposed path.

Experiments with a six-degrees-of-freedom robot have confirmed the versatility of the system, which can fulfill its role in a number of difficult situations. It can avoid collisions resulting from movements in any direction and with any part of the robot. The system is capable of proposing as many intermediate positions along the path as the situation may require.

With the development of off-line programming, the avoidance of collision becomes part of the programming. Thus, Alan de Pennington, M. Susan Bloor, and Maxin Balila, all of the The University of Leeds (United Kingdom),[2] have described a geometric modeling system based on constructive solid geometry. It includes robot selection, determination of the actual work ranges, choices of the best robot, verifying the robot commands on the screeen, and safe trajectory planning. Collision detection and avoidance are based on swept volumes.

REFERENCES

1. Kilmer, R. D. Safety sensor systems for industrial robots. *Conference Proc. Robots VI,* Detroit, MI. March, 1982, pp. 479–491.
2. de Pennington, A., Bloor, M. S., and Balila M. Geometric modelling: a contribution towards intelligent robots. *Conference Proc. 13th International Symposium on Industrial Robots and Robots VII,* Chicago, IL, April, 1983, pp. 7-35/7-42.

Chapter 14
SELECTED APPLICATION TOPICS

ARC WELDING

Figure 14-1 shows the Motoman arc welding robot system from Hobart Brothers Company. The Motoman L-10WX robot (A) is controlled by dc servomotors, moving it in five or six degrees of freedom. The robot has a reach of 56 in. without and 73.8 in. with torch, and a volume of 140 ft.^{3} at an accuracy of ±0.008 in. A point-to-point teaching system is used. The controller (B^1) plays back the weld trajectory in continuous-path operation at both welding and high speed. It has a standard data capacity of 2200 positions and 1200 instructions, and controls the servomotors and synchronizes the overall operation of the welding robot system with signals received from external sources, including a through-the-arc-tracker as described below. The teach pendant (B^2) hangs on the side of the controller cabinet, from where it can be lifted and carried to the point of manipulating and teaching the robot trajectory. The power source (C) supplies the power required for welding.

For feeding electrode wire to the welding torch (E), a Hobart H4S/40 solid-state controlled wired feeder (D) with tachometer feedback for precise speed control is provided. The Mega-Con 114-X interface controller (F) allows the wire feeder and power supply to be controlled by the robot computer. The computer starts the welding operation, specifies the welding voltage and the wire feed speed, changes them according to the program, and stops the welding arc. The interface takes the specified values and maintains the power source voltage and wire feed speed through the closed loop system. A work positioner (G^1) permits positioning the workpiece in advance into its fixture and feeding it to the robot, while the welded piece is removed. The operator station (G^2) provides manual pushbutton control over the position movement.

Welding systems of this sort may be simply programmed by the teach pendant, and then will follow the stored routine in actual operation within the accuracies of the system. There are, however, many cases where this

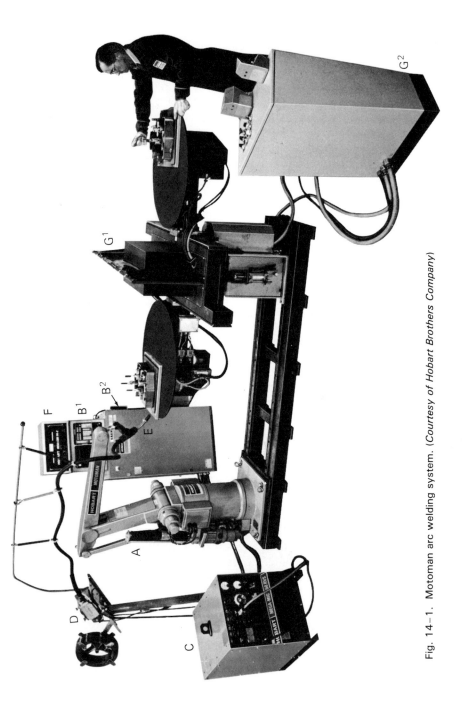

Fig. 14–1. Motoman arc welding system. (*Courtesy of Hobart Brothers Company*)

440

method needs refinement by sensory feedback. Often the part to be welded is a casting, or a formed sheet metal part, or a flame-cut part, or a worn, used part that is to be repaired, and so forth. In all such cases, tolerance variations impede the use of preprogrammed weld path trajectories, even disregarding tolerances of fixtures, distortions due to welding heat, and limited accuracies of robots. Continuous adaptation to changing parameters is required under such conditions, and the development of cost-effective sensors for this purpose is the major challenge.

The following discussion on sensors for robotic welding is based on a paper by Richard E. Hohn and John G. Holmes.[1] Its use is acknowledged with thanks, but inaccuracies in the presentation of these concepts are the sole responsibility of the author.

Contact and Non-contact Sensors

The most basic sensor for guiding an arc welding torch along the groove to be welded is a fingerlike extension attached to the robot wrist. The finger runs in the weld joint immediately ahead of the molten pool produced by the arc. Two axes can be measured and controlled by this method. In the case of a horizontal weld, they would be the up-and-down axis and the left-and-right axis.

Feedback information from the sensor system is used either to control two positioning slides, one for each of the two axes, or to integrate the sensors with the robot controls. The latter method eliminates the need for the slides, since the robot controls accept the sensor information directly, process this information, and make real-time positional changes of the end effector. In either case, the result is a relatively simple and inexpensive control system.

Disadvantages of such a contacting sensor are:

- The difficulty of keeping the sensor in the groove.
- Its relative bulkiness.
- The need to mount the sensor in front of the weld, and the resulting time delays between measurements and correction.
- The directional constraint due to the fixed relationship between torch, sensor, and weld joint, which prevents welding into corners or other restrictive areas of the weldment.

With all this, these control systems are well suited and frequently used for long continuous and essentially straight welds, such as welding plates or circumferential pipe welds.

Use of a noncontacting, instead of the above-described contacting sensor is quite conceivable. It could follow the welding groove without the problem of keeping the sensor in the groove. However, the difficulties of maintaining a sensor close to the heat and contamination of a welding seam have so far precluded a generally acceptable, commercial version.

Inductive, capacitive, and eddy-current sensors have been designed for this purpose. Their disadvantage is that they are too bulky and too sensitive to temperature extremes. To minimize thermally induced errors, they have to be mounted 2 to 4 in. from the arc, and this produces a serious limitation to the flexibility of the robot. Under certain limited applications they do perform well. A universal solution, however, apparently has not been found.

Visual systems certainly have many potential advantages. Thus, the location and orientation of a randomly located part can be determined prior to welding, which solves the important problem of finding the starting point of the seam to be welded.

In using visual systems, one of the following two methods may be used to track the total seam trajectory:

1. The welding torch is taught a certain part program that implies the welding path. It is then cycled through the program without welding. In this phase, the camera is used to view the seam and compute corrections to the part program. The corrected program is then used for welding the seam without the camera. Drawbacks of this method are: (a) the loss in overall cycle time due to the initial preview pass, and (b) the possibility of uncorrected changes of the welding seam due to the distortions while welding.
2. The camera is used during welding. The conditions for doing this are not very favorable, considering the intense light, smoke, heat, and spatter that interfere with the vision signal. Although these problems have been solved to a certain extent, the mounting location of the camera remains a serious limitation. It becomes extremely difficult, for example, to weld inside boxlike assemblies. Furthermore, the vision systems must be custom-tailored to welding applications, which adds to cost. Hardly less costly is the high level of computing power that is required to process the data from the camera.

Another sensing method for tracking a welding seam is through-the-arc sensing; that is, use of the welding arc to determine the position of the welding torch relative to the workpiece at all times. The arc can be oscillated within a small amplitude, as shown in Fig. 14–2. Under this condition, voltage and current, as measured at the arc, are related to its position relative

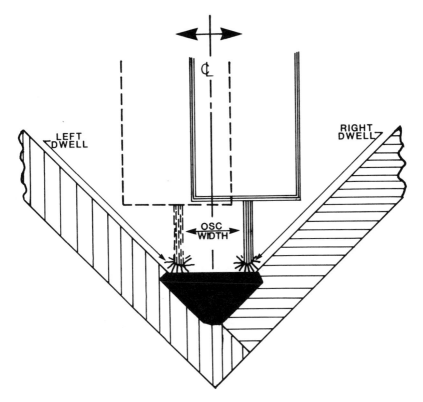

Fig. 14–2. Oscillating arc. (*From Hohn and Holmes, Ref. 1*)

to the groove. While Fig. 14–2 refers to a fillet weld, the same would apply to butt, lap, and other forms of welds.

The application of this method by Cincinnati Milacron is described further below. By way of introduction, however, some of the underlying concepts are discussed next.

Through-the-Arc Sensing

The possibilities of using electric signals for seam tracking were investigated by Merrick Corp. in a study supported by the National Science Foundation.[2] The Project Director was Dr. George E. Cook. Two of the essential objectives of this study were:

- The correlation of electrical signals obtained from the welding arc with joint geometry.

- The development of control algorithms to demonstrate that electrical signals of the welding arc can be used for both lateral and vertical positioning control of the welding arc with respect to the weld joint.

Preliminary experiments showed the existence of a definite relationship between the distance of the welding torch to the workpiece and either arc current in the case of gas metal arc welding (GMAW) or arc voltage in the case of gas tungsten arc welding (GTAW).

Limitations were considered with respect to practical welding joint geometries, welding travel speeds, time delays inherent in the process, magnitudes of signals, signal-to-noise ratios, and the ability to measure these signals without substantially interfering with the welding process.

In developing the control algorithm, the objective was to demonstrate the feasibility of a method of control using both analog and digital signal processing, with the digital processing and control being done by microcomputer.

The microcomputer was also to be used for control of the sampling of the electrical arc signals, the digital filtering of the signals, the arithmetic computations required for comparing signals at each torch oscillation extremity (see Fig. 14–2), the output of suitable correction signals, and the multiplexing of the lateral and perpendicular control actions.

The investigations showed that for GTAW the sensitivity of the arc potential to arc length lies between roughly 0.5 and 0.75 V/mm, and that for GMAW the corresponding change amounted to approximately 1 to 1.5% of the average welding current per millimeter change in arc length.

It was demonstrated that a high-performance dc servo can be used to automatically maintain the arc voltage in the GTAW process and the arc current in the GMAW process to better than 1% accuracy.

Control algorithms were developed to use the information obtained from the arc signals for lateral and vertical control by multiplexing the two control actions. Limited experiments conducted with the arc positioning system showed accuracies of +0.25 mm in lateral as well as vertical directions. The tests were conducted at relatively slow welding speeds for the GMAW tests. However, it was also established that welding speeds of 9 to 13 mm/sec are quite feasible for both lateral and vertical control. For lateral control only, frequently the only type needed with robotic welders, increasing the welding speed beyond 9 to 13 mm/sec appeared quite feasible, if the required correction were small enough.

A number of commercial systems have adopted through-the-arc tracking. Among them are the Unimate 7000 by Unimation, and the systems by Advanced Robotics, CRC, Cybotech, Hitachi, and Yaskawa-Motoman. Thus, the above-described Motoman arc welding robot system is also available with through-the-arc tracking. For systems used in this fashion, welding speeds of 24 in./min are said to be possible for heavy welds. A quick-search function is provided to determine the start of the seam. Low current through the welding electrode is used in the search mode. The system switches automatically to welding current, once welding begins. As a somewhat more detailed example of through-the-arc tracking, the Cincinnati T^3 welding robot is described in the following section.

The Cincinnati T^3 Welding Robot.[1]   One particular characteristic of the Cincinnati T^3 welding robot is the provision of the robot with the necessary control capabilities to oscillate the torch directly, without the use of additional mechanical or magnetic oscillators.

Cincinnati recognized that even through-the-arc tracking has its limitations; for example:

1. A minimum joint wall thickness of $\frac{3}{16}$ in. must be available for measurement purposes.
2. The maximum welding speed is limited to 30 to 50 in./min.
3. High-conductivity materials, such as aluminum, may limit the effectiveness of through-the-arc sensing.

However, Cincinnati researchers investigated the practical requirements of the arc-welding market and its applications and found that these limitations were not significant compared to the advantages of such a system. Thus, the robot-controlled oscillations of through-the-arc sensing became standard for the Cincinnati welding robots.

The through-the-arc system requires correlation of welding current and voltage relative to the distance between the electrode and the workpiece. Various types of sensors can be used to measure the welding parameters. Thus, in gas metal arc welding (GMAW) a clamp-on current sensor or shunt can be used, as shown in the block diagram of Fig. 14–3. The analog signal from the current sensor is applied to the sensor electronics where it is filtered, sampled, and processed through a measurement algorithm. This algorithm is dependent upon the particular process being used.

Positional information is transmitted to the robot control, where the necessary position correction is computed. The corrections are then automatically made in the cross-seam and perpendicular directions for the next weave cycle. The accuracy of the system is ±0.010 in. in the cross-seam direction and ±1% of nominal stickout in the vertical direction.

The through-the-arc system developed for the T^3 welding robot has three basic modes of operation: cross-seam, perpendicular, and dual. All three function with the weld gun being oscillated over the weld joint and can be

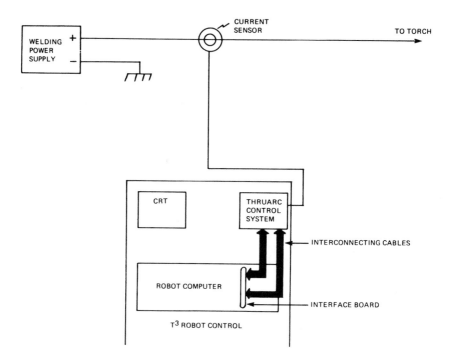

Fig. 14-3. Seam system tracking diagram. (*From Hohn and Holmes, Ref. 1*)

used in any welding position. At the start of welding, the wire may be manually located in the weld seam, if the weld start location is not repeatable.

Cross-seam Mode.   When only cross-seam correction is required, the welding current is sampled on each side of the weld joint. The exact weave pattern desired is selected from a previously programmed weave schedule table. This table contains the number of weave cycles per inch, the amplitude of the weave on each side of the weld centerline, and the dwell at the end of each weave amplitude as a percentage of one complete weave cycle.

For GMAW, when the centerline is not in the center of the groove, the distance between the electrode tip and the workpiece is different on each side of the joint. Hence the measured current is also different. In this case, the sampled current values for each side of the joint are processed by the measurement algorithm and communicated to the robot control, which applies a correction value to balance the system.

Perpendicular Mode.   In the perpendicular mode, the arc current is sampled as the weld gun traverses the weld joint during a weave cycle. The sampled data are processed by a vertical measurement algorithm and communicated to the robot control to be compared to the vertical-distance reference value previously programmed by the user. The difference between reference value and measured value is then used to obtain the position correction value normal to the weld plane. In the perpendicular mode, it is also possible to track a seam normal to a weld plane without weaving.

Dual Mode.   The dual mode is the one used most frequently. It is merely a combination of the cross-seam and perpendicular modes. Figure 14–4 shows the weave motion. During one weave cycle, the measurements for the cross-seam mode are made, and during the second cycle, the measurements for the perpendicular mode are made. The two-directional correction of the weave centerline is made at the end of the second cycle. Thus, two complete weave cycles are required for measurement and correction in the dual mode.

Further developments.   Hohn and Holmes[1] discuss the continuation of their work in the following paragraphs:

> Adaptive positional seam tracking is a reality. The next major hurdle is adaptive process control. Integration of a basic data collection system with a computer-controlled robot allows for additions and expansions to the current algorithm to enhance not only the adaptive position control capabilities, but also to generate adaptive process control algorithms. This technology is feasible by combining the information gleaned by the through-the-arc sensory system with process control algorithms which will allow for changes in welding parameters when the seam geometry changes.
> When systems with both positional and process control exist, we will

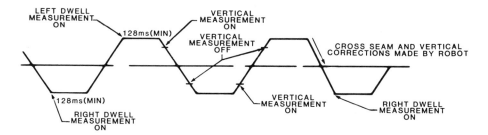

Fig. 14–4. Weave pattern for dual mode. (*From Hohn and Holmes, Ref. 1*)

begin to approach the capabilities of the human weldor. Much of the required information is in the arc; it is just a matter of interpreting it and using it effectively.

CRC's M-1000.   The CRC Automatic Welding Co. also uses through-the-arc tracking in its M-1000 welding unit. This is a fully automated portable GMAW system that is being used particularly in shipbuilding. Here, a condition of irregular fits due to a buildup of fabrication tolerances and workmanship factors has created particular problems in automatic welding, which the M-1000 is capable of correcting.

The entire welding process is controlled by an RCA eight-bit microprocessor controller system via the through-the-arc sensing technique. The unit is designed as a portable robotic-type system with four-axis control. Weighing about 45 lb, it is capable of welding the outside of a curve with a minimum external diameter of 14 in. and the inside of a curve with a minimum internal diameter of 100 in.

The welding parameters that are controlled include heat input, fill height, torch tracking, centerline tracking, and width of weld. Other parameters, such as bar feed speeds, travel speeds, dwell times, and oscillator rates, are automatically modified as the joint widens or narrows.

When the operator first turns on the system, it goes through a routine where it finds the plate surface, then backs up and extends the proper electrode stickout lengths. It then backs up again and moves across the weld joint.

Then the operator turns the system off momentarily while it is more or less over the bevel of the joint. He then turns it on again. The unit now goes through a seek routine. It moves down through the joint below the surface of the plate and seeks to the right and left of the joint, indexing the centerline. Next, it drives to the bottom of the joint to find the depth of the well bead or where the next pass will be placed.

By this time the computer possesses the data of a geometric model of the joint profile at the start of the weld. All the necessary welding parameters can be calculated, and the system is ready to take over the automatic welding of the joint.

Optical Sensing

While the use of cameras in connection with tracking the weld seam has problems, as mentioned above, it also has the advantage that the camera can be removed from the aggressive atmosphere of the welding action and that it can sense configuration and position of the weld with great detail. Thus, General Electric opted for an electro-optic system. Their WeldVision

System is intended for GTAW. A series of lenses that are an integral part of the welding torch are focused on the weld puddle and the joint. The image concentrated by the lenses is picked up by a fiber-optic bundle and transmitted to a solid-state camera. Analysis of the image then proceeds in the usual way as described in Chapter 11.

In addition, the joint is tracked by two parallel laser stripes that are projected across the weld joint 0.30 in. ahead of the welding electrode. This structured-light technique indicates the position and width of the welding groove and permits the robot to follow any deviations, curves, and so on.

At the same time, since the camera has picked up the geometry of the weld puddle, this information can be fed back to the welding process controller to make any required adjustments for optimal puddle geometry, so that optimal quality of the welded seam results.

The Univision system from Unimation uses a somewhat similar lens–fiber-optic-camera system. However, here the joint shape is detected by viewing at an angle the light projected on the workpiece.

The system first scans the joint to be welded at a speed of up to 3 ft/sec. During this first run, which takes about 10% of the welding time, the system is programmed for the welding operation that follows. During the welding, the robot makes additional corrections to minimize any deviations from the preprogrammed trajectory.

Robovision II is the optical seam tracking system developed by Automatix Inc. It does not require the preparatory run. Thus, some time is saved, and it is also claimed that thermal distortions produced by the welding heat while the welding proceeds are compensated for. Robovision II uses a linear array camera with structured light projected onto the workpiece about 1 in. ahead of the weld. From the image thus obtained, the system determines the location of the center of the seam, its width, and the distance of the workpiece from the sensor.

Figure 14–5 shows the laser-operated arc welding seam finder system from ASEA Robotics Inc. The system consists of an optical laser sensor and microcomputer interfaced with ASEA's S2 controller and its IRB 6AW/2, IRB L6AW/2, or IRB 60/2 robot.

The ASEA seam finder system can be used in most arc welding, as well as resistance welding and plasma-arc cutting applications. It is particularly suited to the welding of thin sheets measuring as little as 0.030 in. in thickness, and to situations requiring many short welds and quick cycle times.

The optical sensor has a resolution of 0.002 in., a measurement distance of 6.9 in., and a measuring range of 1.3 in. The search location accuracy is +0.016 in.

During the search process, the start and stop points along the joint are defined in either two or three dimensions, and the welding gun is posi-

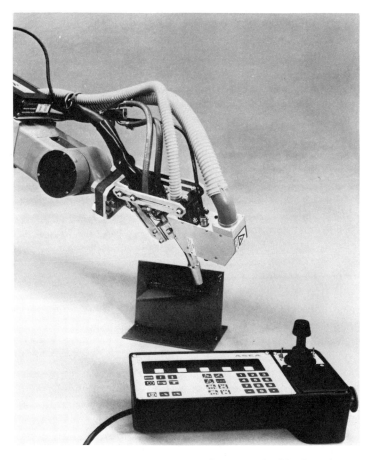

Fig. 14-5. Laser-operated seam finder. (*Courtesy of ASEA Robotics Inc.*)

tioned. Most often, a two-dimensional search requiring less than a second is sufficient to set up the welding gun properly. Three-dimensional searches typically require approximately 1.5 sec. All searches are accomplished without the arc being actuated.

The process enables the robot to quickly position the weld torch at the actual seam location, as opposed to the theoretical starting point of the weld, and to move the welding gun along the joint to the desired stop point.

For long weld seams with a three-dimensional weld path, the seam finder system can search out as many intermediate points along a joint as required. It can also be used to automatically adjust weld parameters to accommodate variations in the weld seam.

LASERS

Lasers can produce very high energy per unit area. They require no physical contact with the workpiece, a favorable characteristic, particularly when they are used as a tool on abrasive substances, because no wear results. Lasers can operate on almost any material and can be aimed at small, precisely defined areas, thus making selective work hardening and similar operations possible. As the heat-affected zone is extremely small, metallurgical or other changes in the vicinity of operation are minimized. It is also possible to direct the laser beam by means of mirrors and refractive lenses to almost any desired location, and even to split it up and use it for a variety of different operations. Thus the extreme versatility of the laser beam is obvious.

Computer control makes the laser respond to processing information at the speed of electronic switching. Power density and dwell time can be regulated, and the beam may be used for welding, cutting, drilling, or surface treating.

Combining lasers with robots fully utilizes the versatility of lasers in industry. Two basic concepts are used in this combination: one is to have the robot move the workpiece into the laser beam; the other is to have the robot move the beam. The latter is particularly useful when the workpiece is too large or unwieldy to be easily handled by the robot, particularly when exact contouring becomes necessary.

Today, the generally preferred method is to have the laser beam controlled by the robot. There are two options. One is to use a floor-mounted robot and transmit the beam along the robot arm either by beam-bending optics or by flexible optic fibers. The other is to mount the laser system on the robot arm. The latter has become more feasible with the introduction of smaller, lightweight CO_2 lasers with an output power of up to 1000 W.

The difficulty in using fiber optics with laser beams is the requirement that the intense laser energy be channeled into the small-diameter fibers. General Electric's Research and Development Center developed a special input coupler for this purpose. It is capable of channeling the 0.75-in.-diameter beam of a neodymium–yttrium aluminum garnet (ND-YAG) laser into a glass fiber of 0.04 in. diameter. The fiber-optic cable is capable of carrying power from the laser, which has a peak value in excess of 10,000 W and an average value of 400 W, for 25 yd from the laser to the workpiece. At the end of the cable, a lens assembly focuses the laser energy onto a tiny spot on the workpiece. In initial tests, intricate patterns were cut in steel, titanium, and nickel-based alloys (Inconel). At 100-W power, the results were as follows: Carbon steel of 0.048 in. thickness was cut at 6 in./min, and titanium 6-4 of 0.062 in. thickness and Inconel 718 of 0.090 in. thickness were both cut at 3 in./min. It is anticipated that a single laser could supply energy for a number of robots with different tasks, such as cutting, welding, drilling, and heat treating. According to Dr. Marshall G. Jones, the manager of GE's Laser Technology Program, this laser/robot system can "cut, weld, and drill with a dexterity never before possible."

The Laserflex 100 from Spectra-Physics uses a combination of rotating mirrors and a telescoping tube to precisely direct the laser beam along a contoured path within its working envelope. The system uses a CO_2 laser that is capable of higher output power than ND-YAG lasers. The Laserflex 100 will deliver up to 1500 W of power. It is designed for contoured cutting and trimming of steels, aluminum, plastics, rubber, wood, ceramics, glass, and cloth, as well as for continuous welding and spot welding of contoured parts.

The Laserflex 100 is illustrated in Fig. 14-6, mounted on a robot and in its functional details. As shown in the illustration, the Laserflex 100 contains three main elements: the optical shoulder, the telescoping tube, and the optical wrist. The incoming beam enters the shoulder at a fixed inlet. The two-mirror shoulder is attached to an adjustable base that can be mounted in any position, depending on the desired routing of the laser beam. An optional third mirror helps to direct the beam into the shoulder if necessary. Mirror adjustment screws allow quick set-up and accurate alignment. Two rotating mirror-blocks in the shoulder maintain the beam precisely in the center of the telescoping tube over the full range of motion.

The telescoping tube provides the link between the optical shoulder and the wrist. It can be extended from 45 in. to 90 in. while maintaining the laser beam alignment through the system. The tube is airtight and filled with dry nitrogen gas at a positive pressure to prevent contaminants from entering it.

The wrist is composed of modular mirror blocks. One, two, or three blocks are used, depending on the geometry of the part being processed. The wrist adds further flexibility to the system while keeping the laser beam in alignment through the exit aperture and lens/nozzle assembly. A bracket attached to the final block of the wrist allows the robot to control the exact location of the laser power.

WATER JET CUTTING

Water jet cutting has developed into a feasible technique for contour cutting of plastics, textiles, paper, and so on. Where it is necessary to follow contours within exact tolerances and where the ability to rapidly change from one contour to another is important, a combination of water jet and robot is the ideal solution. Kevin Ostby[3] has described such a system, used in the

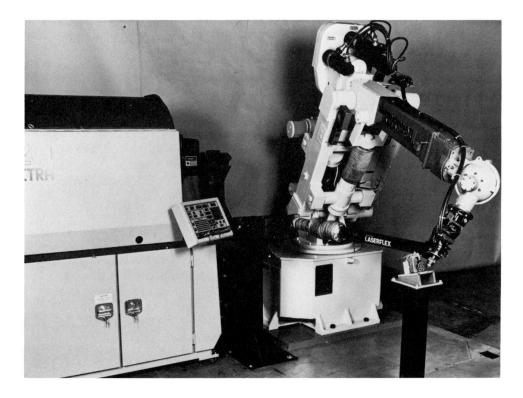

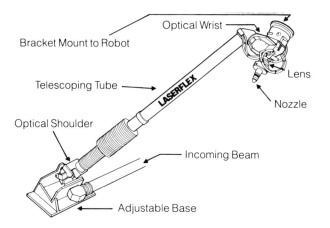

Fig. 14-6. Laserflex 100. (*Courtesy of Spectra Physics*)

manufacture of polyethylene fuel tank shields at the General Motors Corp. Chevrolet Division plant. Two identical production lines are used, each line including four S-360R robots from GMF Robotics Corp.

The fuel tank shields are made in two sizes, each in both right- and left-hand styles. The flexibility obtained by the programmability of the robots facilitates response to a variety of orders that often include part specification changes.

Polyethylene blanks of 0.195 in. thickness are formed and then moved to the robot piercing and trimming line. On each production line, two robots are equipped with water jet tooling to cut 12 holes and slots. At the next station, the formed polyethylene is trimmed at the edges and divided into two identical parts by another two robots equipped with water jets.

Figure 14-7 shows a water jet nozzle manipulated by a GMF robot, as it cuts the periphery of a plastic form consisting of two fuel tank shields. Very high pressure is used for these operations. The pump, piping, and nozzles are built by Flow Systems, Inc. An electrically driven oil-hydraulic pump drives a reciprocating plunger pump that serves as an intensifier. Intensification is obtained by applying the hydraulic oil pressure to a large piston, causing it to reciprocate and to drive plungers with a smaller diameter. The plungers pump the water and produce a pressure that is proportional to the relative areas of the piston and the plunger.

Each water jet is applied through a sapphire nozzle of 0.008 in. diameter at a pressure of 50,000 psi. The maximum cutting rate is 10 in./sec. Before being used, the water is passed through softeners to prevent damage to the sapphire nozzles by minerals in the water.

The robots keep the nozzles 0.5 to 1 in. from the part surface, depending on what portion of the program is active. The distance is important in that it affects cutting efficiency. After the water has cut through the polyethylene, it is collected by a catcher that dissipates the jet energy, thus reducing noise levels and safety hazards.

The four robots in each line work in such close proximity to one another that their work envelopes overlap. Collisions are avoided by special software interlocks and a programmable controller that continuously compares positional data from each robot. If a collision is incipient, all robots in the system stop. A special "pullout" program keeps the robots from colliding while returning to their home positions.

The GMF-360 robots are driven by ac servos. They operate with six degrees of freedom, and all axes move simultaneously to guide the nozzles through the cutting trajectory. The robot's R Model C controller from GMF incorporates a self-diagnostics control system, magnetic bubble memory with 6000 points, and an auxiliary bubble cassette for an extra 2700 points.

Fig. 14–7.   GMF Robotics S-360 robot in water jet cutting application. (*Courtesy of GMF Robotics Corp.*)

The bubble cassette permits simple changing of programs in accordance with production schedules.

Another example of robotic water-jet systems is the water-blasting system developed at the Marshall Space Flight Center.[4] It removes hard, dense, extraneous material from surfaces. A pump forces water at pressures of up to 20,000 psi at supersonic speed through a nozzle manipulated by a robot, as shown in Fig. 14–8. The impact of the water blasts away unwanted material from the workpiece, which is rotated on an airbearing turntable. Designed for removing thermal-protection material from the Space Shuttle during postflight refurbishment, the system is adaptable to such industrial processes as cleaning iron or steel castings.

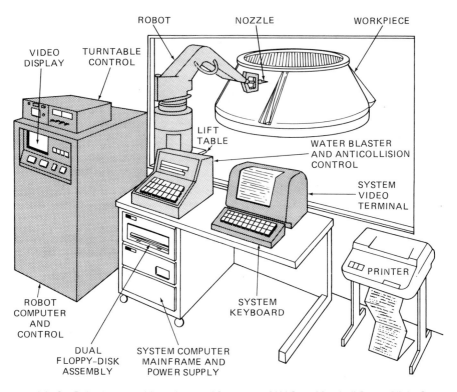

Fig. 14–8. Robotic water-blast cleaner. (*Courtesy of NASA — Marshall Space Flight Center*)

Two 200-hp electric motors drive two five-plunger positive-displacement pumps. The pump speed is continuously variable between 50 and 400 rpm. The robot is a 25-hp waterproof industrial machine that can withstand 225 lb of reaction force and still maintain position with an accuracy of 0.050 in. The robot can move in six axes, and the turntable provides a seventh axis of rotation. A computer subsystem monitors and controls such parameters as turntable location, position, and speed; pump speed, pressure, and flow rate; robot program; and anti-collision signals.

The robot can manipulate the blast nozzle in any of four distinct modes as follows:

1. The workpiece is rotated continuously on the turntable while the robot moves the nozzle vertically, removing material in a spiral pattern. This mode ordinarily allows the fastest removal.
2. The workpiece is rotated back and forth on the turntable while the

robot moves the nozzle vertically, removing material completely from a section. This mode allows the removal of material in horizontal strips without using excessive amounts of robot memory.

3. The workpiece is held stationary at a predetermined point while the robot moves the nozzle over a section of the surface. This mode removes material in difficult areas, such as around posts.

4. The workpiece is rotated incrementally in fractions of a degree while the robot sweeps the nozzle vertically. This mode is effective in open areas between protrusions.

Air bearings support the turntable during rotation and raise it above the floor when it is moved into or away from the blast area. Powered by two counterrotating dc electric motors, the turntable can be positioned with an angular accuracy of 0.05°.

The automatic removal system replaces manual water-blast equipment in which an operator must make split-second decisions in manipulating a blast nozzle under adverse conditions of noise, poor visibility, and wet, restrictive clothing. The new system removes the operator from the hazardous blast area and greatly improves the safety of personnel and equipment.

ELECTRONICS ASSEMBLY

Wafer Processing

Wafers are thin semiconductor slices on which matrices of microcircuits or individual semiconductors can be formed. After processing, the wafers are separated into dice or chips containing individual circuits. Besides their transportation from work station to work station, the wafers must be handled during their processing, such as at a wet sink. They must also be loaded into and removed from automated machines.

Robots are being used for this purpose because they can be adapted to the clean-room requirements of some of the processes, and they can safely handle minute parts, contact hazardous chemicals in acid dip tanks, assure more consistent processing, and minimize human errors.

SMI developed a wafer processing system[5] as illustrated in Fig. 14–9. An IBM 7535 Manufacturing System was chosen. The robot is of the scara design (see Chapter 2) and has four degrees of freedom. They cover the two arm swivels, a roll axis, and an up-and-down Z-axis of 3 in. movement. SMI reconfigured the robot as follows:

- Higher positioning resolution on all four axes was provided.
- The stepper motor for the roll axis was replaced with a closed-loop dc servo.

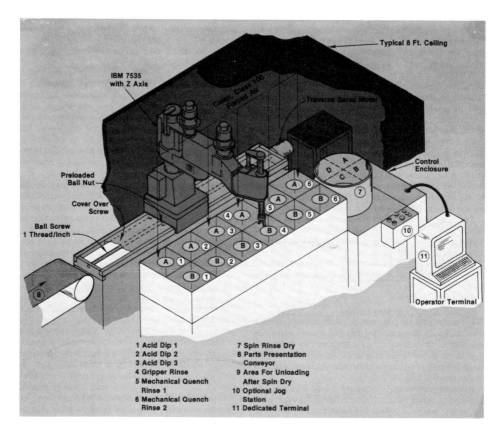

Fig. 14-9. Wafer processing system. (*Courtesy of SMI*)

- The programmable Z-axis was extended to 18 in.
- An additional degree of freedom was provided by mounting the robot on a way system and moving it with a ball screw.
- The outer joints of the robot were sealed and booted to minimize any contamination into the tank area.

An advantage for fast and accurate positioning along the additional degree of freedom was that the IBM 7535 robot arm has a nominal weight of only 132 lb. Another contributing factor for fast and accurate positioning was the use of a ball screw with ground threads and preloaded ball nut, as well as a way system utilizing 2.5 in.-diameter ball bushing guides with wipers.

Although clean-room conditions are not necessary for this operation, so-called class 100 requirements are observed. Screws and guides are sealed to prevent lubricant particles from being thrown into the atmosphere during operation. All air surrounding the robot system is forced downward so that contamination is maintained at floor level and not at the work surface.

Control of the work cell is handled by an ICC 3200 Flexible Automation Controller. The controller provides eight axes of closed-loop servo drive capability using resolver feedback. This permits the controller to handle other axes in the work cell environment in addition to the robot system.

The robot is taught by a hand-held pendant or through a keyboard terminal. Processes can be programmed and stored for later call-up by part number or code number.

Parts arrive in baskets, and without being removed from the baskets are processed through the various tanks of the system. During normal operation, the time required for each process is sufficient to allow continual movement of the wafer baskets. Ideally, and for maximum throughput, baskets should be in process in each tank.

Soldering Systems

Figure 14-10 shows the ECS 201 robot soldering system from Chad Industries. It permits random soldering of discrete points on printed-circuit board (pc board) by program control. The user can solder a wide variety

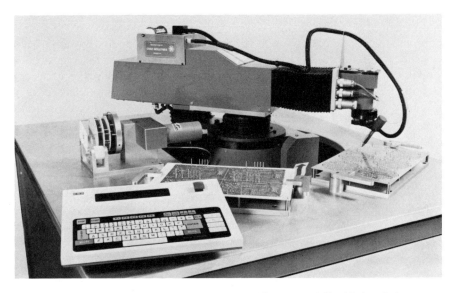

Fig. 14-10. Robotic soldering system. (*Courtesy of Chad Industries*)

of component junctions on the top and bottom of a pc board, before and after wave soldering is completed. The system is fully automatic and will solder standoff pins, component leads, and other types of junctions.

The ECS 201 is under the control of a Seiko RT-3000 robot that is actuated by dc servos. Solid, single, or multicore solder is fed automatically to the soldering head from a reel. Built-in compliance of the robot protects the solder head against damage. A fluxing attachment permits additional fluxing. The solder iron temperature is controllable with an accuracy of ±10°F within a range of 500 to 850°F. The control permits programming of 400 different positions. Feed time is programmable with 0.001 sec resolution. The typical soldering cycle time is 1.5 to 3 sec, and movement to the next point is at speeds up to 55 in./sec.

An even more elaborate soldering system has been developed by Toshiba, combining two of its TSR-500 robots into one system that tries to simulate the two-handed actions and eyesight of a human operator. One of the robots holds and manipulates the piece, and the other solders it to its counterpart. A solid-state camera is attached to one of the two robots and controls the correct position of the soldering joint. More than ten microcomputers are involved to process the data and the visual information, to control the motions of the robot, and to coordinate the soldering functions. By switching end effectors, it is possible also to use the system for multiple assembly tasks.

Assembly of Printed-Circuit Boards

The tedious insertion of many components in a pc board by hand lends itself to automation. In particular, where short-volume production runs are involved, robots can increase productivity while providing the necessary flexibility to be reprogrammed for frequent production-line changeovers. William L. Huck[6] described a three-axis robotic assembly with a four-position turret head for multiple end effectors that was developed for such applications by Anorad Corporation. It is illustrated in Fig. 14–11.

Special emphasis was put on flexibility because of the variety of components the system must handle. These components include resistors, capacitors, transistors, and so on, together with various sizes of multi-pin DIPs (such as shown in Fig. 9–15), sockets, potentiometers, switches, connectors, and so forth. At the same time, high speed is required in picking up and placing all these components with great accuracy on a densely populated board.

The position and orientation of the printed wiring pattern and hole locations vary from board to board. This requires the boards to be positioned in their fixture and aligned with the help of an electro-optical vision system

Fig. 14-11. ANOROBOT for PCB assembly. (*Courtesy of Anorad Corporation*)

attached to the arm of the robot. The optical system can also be used to redefine certain hole locations on the board.

A programmable microprocessor controller operates the system, including optical alignment, component feed, robotic tool selection, and component positioning and placement.

The entire system is mounted on a granite base plate that has inherently good vibration damping characteristics and a low coefficient of expansion. It provides stability for the high degree of positional accuracy and repeatability that operation of the system requires.

An $X-Y-Z$ travel of $10 \times 10 \times 2$ in. is adequate for assembling pc boards up to 6×9 in. Other travel dimensions could also be furnished. Linear optical encoders are used to provide closed loop servo control of the three axes. These are high-precision feedback devices that are used with robots only in special cases where the highest degree of accuracy is required. The glass scales have gratings of 500 lines/in., which when used with a 20 times enlarging controller logic provides a resolution of 0.0001 in.

The various axes of the system can be positioned to an accuracy of ± 0.0005 in./ft with a repeatability of ± 0.0002 in./ft. Depending on the length of the required move, velocities of up to 40 in./sec can be attained with acceleration rates on the order of 200 to 400 in./sec^2. At a velocity of 30 in./sec, the X-axis can traverse the length of a 6 $\times$ 9 in. board in 455 msec, while the Y-axis can traverse the width in 394 msec. For a travel distance of 4.5 in., the average "pick-and-place" move time is about 350 msec.

A four-position turret head is provided as shown. It is programmed so that on command any of the four devices can be moved in position. The turret positions can be equipped with vision and a variety of special tools, such as rotary wrenches and screwdrivers, grippers, solder tip, epoxy applicator, and so on. There is a wide choice of gripping actions: mechanical, vacuum, or magnetic, with the shapes of jaws, magnets, or vacuum cups corresponding to the parts handled. Force sensors can be incorporated as a programmable function to prevent bent leads or to detect misaligned leads or board holes.

The turret may also be provided with rotary tools using servo drivers with torque sensing for control of speed and position, and with remote-center compliance devices allowing the tool to locate its working point if it is within 0.06 in. of its final position.

An electro-optical auto-centering system can be used where initial location is defined only to within ± 0.25 in. For applications requiring a more precise vision capability, a solid-state camera can be incorporated with an image analyzer.

Four programming options are available:

1. Teach mode by means of a pendant.
2. Manual self-programming by a built-in editor.
3. Off-line program development with program entry by magnetic tape, floppy disc, or PROM.
4. Program download from a host computer.

INSPECTION SYSTEMS

Optical System

The Robo Sensor System Model 210 Series from Robotic Vision Systems Inc. is a 3-D vision system. It consists of a vision module and a processor unit. The vision module can be mounted either on a robot's tool plate or on an electro-mechanical translation equipment which is programmable and capable of going through the same motion as a robot. This is illustrated in

Fig. 14–12 which shows the application as an Automated Component Optical Measurement System (ACOMS).

The system is a noncontact, automated inspection machine for engine block castings. It is an interesting example for the fact that a manipulator does not have to "look like a robot" but yet corresponds with the definition of page X of the Preface. It provides programmable motions in the X, Y, and Z axes and also in a rotary C-motion. An additional degree of freedom is obtained by the X-axis motion of the measurement table. It is used at Cummins Engine Co. where it checks incoming castings and reduced the time needed for the inspection from an average of 35 hr to 40 min. ACOMS gives Cummins a large volume of measurement data that can be used to analyze for trends. By feeding this information back to its suppliers, Cummins is able to improve the quality of its engine blocks.

The measurement system consists of a main column supporting an array of four sensors, which face the specimen casting. The sensors use structured

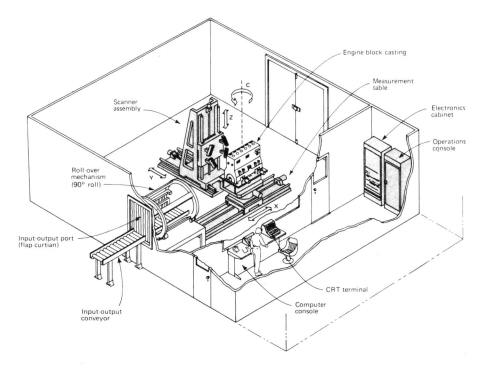

Fig. 14–12. Automated component optical measurement system—ACOMS. (*Courtesy of Robotic Vision Systems Inc.*)

laser light with an optical triangulation technique, together with the corresponding software to acquire data of 1250 dimensions with measuring accuracy of better than 0.010 in.

The incoming casting rolls into the inspection room through flap curtains and into the rollover barrel, where it remains until the operator enters its serial number into the control console via a series of thumbwheels. From the barrel, it moves onto the workpiece measurement table, where clamps grab the casting and hold it in position. The stage moves forward in X while the inspection column moves forward in Y and up and down in Z until it is approximately 3 ft from the closest face of the workpiece. With the four sensors focused on a small portion of the casting, they scan this portion and transmit the necessary data. This is followed by the next portion until the entire face is scanned by the sensors. Then, the operating program directs the workpiece table to rotate 90 degrees showing a new face to the laser array. When all four sides of the casting have been scanned, the workpiece returns to the roll-over barrel, where it is turned 90 degrees about its longitudinal axis. The work returns to the table, which again moves it through scanning, this time on the top and bottom and, for reference, on one of the previously scanned end faces.

Ultrasonic System

Martin-Marietta Aerospace has developed a fully automated system for nondestructive quality control testing of composite structures. The name of the system is Russ, which stands for Robotic Ultrasonic Scanning System. It combines robots, computers, and digital ultrasonic instruments to examine large or complex components manufactured from epoxy resins reinforced with graphite fibers or other materials used in such composites.

Russ is programmed to send ultrasonic signals through a specially designed water jet that is scanned by the robot on either side of a test component. Ultrasonic signals, both transmitted and reflected, are used to evaluate the integrity of the composite component.

Russ was built for Martin Marietta by MatEval, Ltd. It includes Unimation Puma robots, MatEval micropulse ultrasonic instrumentation, and Hewlett-Packard computers with data storage and color graphics output. Its reproducibility and sensitivity are said to exceed the levels previously attainable in electronic scan measurements.

ROBOTIC MANUFACTURING SYSTEMS

Anderson, Greenwood & Co. is a precision valve manufacturer. In a manufacturing cell, it produces valve stems in three shifts. Operators are available during one shift, whereas during the other two shifts the cell is unattended. The installation includes two Unimation Puma robots, two Reed thread rollers, and a special vertical turning lathe designed and built by Centennial Tool Co. Automation of this production unit increased production from 43 units per standard hour to 187 units.

The cell machines valve stems that were previously blanked out of $\frac{3}{8}$-in. rods. Most of them are stainless steel. Two diameters are ground by a centerless grinder, after which an operator manually loads a vibratory bowl that stages and prepositions the stems for the first robot operation.

The robot picks up a stem and moves it over a continuously cycling vertical thread roller. When signaled, the robot drops the stem in place. After the thread is rolled, the gripper takes the ejected part and places it in a buffer queue that orients the part for pickup by the next robot.

The second robot picks up the stem and loads it into the vertical turning lathe which drills a 0.317-in.-diameter hole 0.150 in. deep in the stem end, inserts a ball of $\frac{5}{16}$-in. diameter into the hole, and swages the stem over to hold the free-rolling ball in place.

After the ball assembly, the robot moves the part to the second thread roller for burnishing. The finished stem is loaded, ejected, and caught, to be placed by the robot in an inspection unit for gaging. From there it is selectively placed in a finished part basket. Human interference is reduced to a minimum.

The above discussion describes the beginning of a trend. How will it continue? The National Bureau of Standards (NBS) has instituted an Automated Manufacturing Research Facility (AMRF), and it is the goal of this operation to combine computer-controlled robots and machine tools of any type and producer into flexible manufacturing systems, and to make them suitable for machine shops that produce only small numbers of a variety of metal parts. Computer software developed at NBS will tie the individual parts of the manufacturing system into a closed network, sending appropriate commands while continuously collecting data from sensors that include highly developed vision capabilities. Programs will continuously correct themselves by the input of experimental data gathered during the operation. The user can change the part being manufactured simply by changing the description of the part in the database. This sort of flexible manufacturing system is a step into the future of robotic technology.

The concept of hierarchical control shown in Fig. 14–13 was designed for such a system.[7] A more detailed diagram of a part of this concept is represented in Fig. 14–14. The following material is a summary of the important paper by James S. Albus, Anthony J. Barbera, M. L. Fitzgerald, and others.[7] For more detailed information, Reference 7 should be consulted directly.

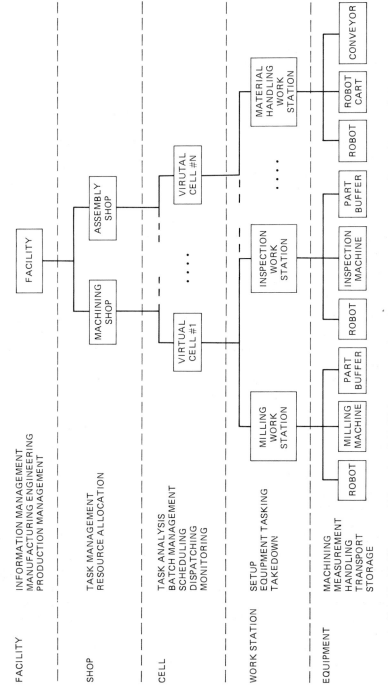

FACILITY   INFORMATION MANAGEMENT
                MANUFACTURING ENGINEERING
                PRODUCTION MANAGEMENT

SHOP   TASK MANAGEMENT
                RESOURCE ALLOCATION

CELL   TASK ANALYSIS
                BATCH MANAGEMENT
                SCHEDULING
                DISPATCHING
                MONITORING

WORK STATION   SETUP
                EQUIPMENT TASKING
                TAKEDOWN

EQUIPMENT   MACHINING
                MEASUREMENT
                HANDLING
                TRANSPORT
                STORAGE

Fig. 14–13. The control hierarchy of the NBS Automated Manufacturing Research Facility. *(From Albus et al., Ref. 7)*

466

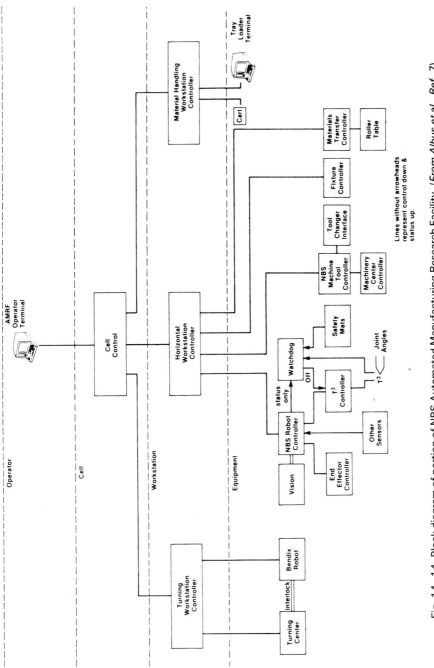

Fig. 14–14. Block diagram of portion of NBS Automated Manufacturing Research Facility. (*From Albus et al., Ref. 7*)

467

The turning workstation consists of a workstation controller, a turning center, a turning center controller, a robot, a robot controller, and a roller table materials loader/unloader. The workstation controller provides closed-loop control over all the elements in the turning center workstation.

The workstation controller receives commands from the cell controller. These commands define which items are to be made and how many of each. The workstation controller returns status information indicating that the commands have been received and the commanded process either is being executed, is completed, or has failed. The workstation controller can also request part programs for the turning center from the AMRF database.

It is also possible for the workstation controller to send commands to the turning center controller through a keyboard input simulator, which was designed and constructed at NBS. This allows the workstation controller to perform all the machine tool keyboard entry functions that a human operator must otherwise perform to load and unload parts and set tooling.

The materials handling workstation consists of an inventory request terminal, a robot cart controller, a cart radio frequency communication modem, and a robot cart with an onboard microcomputer.

The materials handling workstation controller receives commands from the cell controller of the form: STARTUP, REPLENISH, RESET, and SHUTDOWN. It responds with status reports of EXECUTING, DONE, or FAIL.

When the materials handling workstation controller receives a REPLENISH command from the cell, it accesses the AMRF database to retrieve a kitting order, verifies that the order can be filled with the stock currently in inventory, and displays to the human operator at the inventory terminal the required layout of the tray to be loaded. The operator uses this display to determine which raw material blanks to place in what sectors of the parts tray.

The horizontal machining workstation consists of a workstation controller, a horizontal machining center, an industrial robot, a robot control system interactive with a number of sensors, and other parts shown in the illustration.

This book could have no better conclusion then this quotation from the Albus et al. paper:[7]

Much additional work is yet to be done, but the progress to date suggests that a system of interface standards may be possible for interconnecting computer-aided design, process and production planning, scheduling, materials transport, and control systems, with machine tools, robots, sensors and sensory processing, databases, modeling, and communications systems. If so, it may become practical to implement computer-integrated manufacturing systems incrementally using components from a variety of vendors.

REFERENCES

1. Hohn, R. E., and Holmes, J. G. Robotic arc welding—adding science to the art. *Proc. Robots VI Conference,* Detroit, 1982, pp. 303–317.
2. Cook, G. E. Research on adaptive arc tracking for welding. Final Report, September 1, 1980 to February 28, 1981. National Science Foundation, Report NSF/MEA-81004.
3. Ostby, K. Robotic water jet cutting. *Proc. Robots 8 Conference*, Vol. 1, Detroit, June 1984, pp. 5-26/5-38.
4. Robotic water-blast cleaner. *NASA Tech Briefs*, Spring 1983, pp. 329–330.
5. Shaum, L. E. Robot transfer system for wafer processing. *Solid State Technology*, November, 1984, pp. 155–158.
6. Huck, W. L. A small experimental robotic PCB assembly system. *Solid State Technology*, September, 1984, pp. 303–305.
7. Albus, J. A., et al. A control system for an automated manufacturing research facility. *Proc. Robots 8 Conference*, Vol. 2, Detroit, June, 1984, pp. 13-28/13-44.

APPENDIX: GRAPHIC SYMBOLS FOR HYDRAULIC SYSTEMS

BASIC SYMBOLS

Line :		1)
– continuous – long dashes – short dashes	} flow lines	$L > 10E$ $L < 5E$
– double	Mechanical connections (shafts, levers, piston-rods)	$D < 5E$
– long chain thin (optional use)	Enclosure for several components assembled in one unit (see 5.5.8)	— · — · —
Circle		
	As a rule, energy conversion units (pump, compressor, motor)	◯
	Measuring instruments	◯
	Non-return valve, rotary connection, etc.	○
	Mechanical link, roller etc.	∘
Square, rectangle	As a rule, control valves (valve) except for non-return valves	☐ ▭ ☐☐ ▭▭▭ ▭¦¦▭

BASIC SYMBOLS—(*Cont.*)

Diamond	Conditioning apparatus (filter, separator, lubricator, heat exchanger)	◇
Miscellaneous symbols	Flow line connection	$d \approx 5E$ *E* - Thickness of line
	Spring	⋀⋀⋁
	Restriction :	
	– affected by viscosity	≍
	– unaffected by viscosity	∨ ∧

FUNCTIONAL SYMBOLS

Triangle :	The direction of flow and the nature of the fluid	
– solid	Hydraulic flow	▼
– in outline only	Pneumatic flow or exhaust to atmosphere	▽
Arrow	Indication of :	
	– direction	↑ ↕ ↓
	– direction of rotation	↰ ↱
	– path and direction of flow through valves.	↓↘ ⌐↑ ↓↲ ↳↧
Sloping arrow	Indication of the possibility of a regulation or of a progressive variability	↗

PUMPS AND COMPRESSORS

Fixed capacity hydraulic pump : – with one direction of flow – with two directions of flow		
Variable capacity hydraulic pump : – with one direction of flow – with two directions of flow		
Fixed capacity compressor (always one direction of flow)		

MOTORS

Fixed capacity hydraulic motor : – with one direction of flow – with two directions of flow	
Variable capacity hydraulic motor : – with one direction of flow – with two directions of flow	
Fixed capacity pneumatic motor : – with one direction of flow – with two directions of flow	
Variable capacity pneumatic motor : – with one direction of flow with two directions of flow	
Oscillating motor : * - hydraulic - pneumatic	

*Rotary Actuator

PUMP/MOTOR UNITS

Fixed capacity pump/motor unit : – with reversal of the direction of flow – with one single direction of flow – with two directions of flow	
Variable capacity pump/motor unit : – with reversal of the direction of flow – with one single direction of flow – with two directions of flow	

VARIABLE-SPEED DRIVE UNITS

Variable speed drive units	

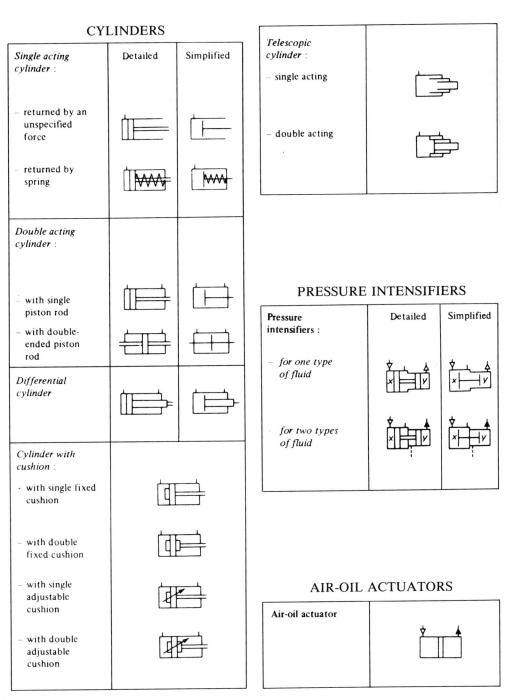

CYLINDERS

Single acting cylinder :	Detailed	Simplified
– returned by an unspecified force		
– returned by spring		
Double acting cylinder :		
– with single piston rod		
– with double-ended piston rod		
Differential cylinder		
Cylinder with cushion :		
– with single fixed cushion		
– with double fixed cushion		
– with single adjustable cushion		
– with double adjustable cushion		

Telescopic cylinder :	
– single acting	
– double acting	

PRESSURE INTENSIFIERS

Pressure intensifiers :	Detailed	Simplified
– for one type of fluid		
– for two types of fluid		

AIR-OIL ACTUATORS

Air-oil actuator	

MEASURING INSTRUMENTS

Pressure measurement : – Pressure gauge	
Temperature measurement : – Thermometer	
Measurement of flow : – Flow meter – Integrating flow meter	

OTHER APPARATUS

Pressure electric switch	

DIRECTIONAL CONTROL VALVES

Flow paths : – one flow path – two closed ports – two flow paths	

– two flow paths and one closed port	
– two flow paths with cross connection	
– one flow path in a by-pass position, two closed ports	
Non-throttling directional control valve	
Basic symbol for 3-position directional control valve A transitory but significant condition between two distinct positions is optionally represented by a square with dashed ends A basic symbol for a directional control valve with two distinct positions and one transitory intermediate condition	
Designation : The first figure in the *designation* shows the number of ports (excluding pilot ports) and the second figure the number of distinct positions	

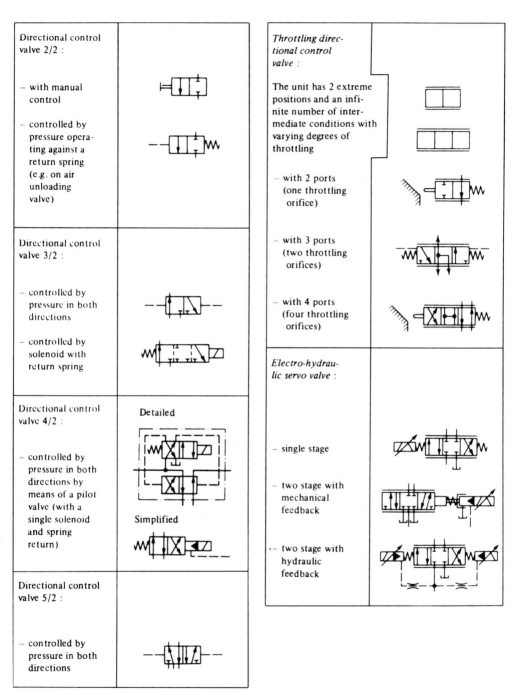

MECHANICAL FEEDBACK

NON-RETURN VALVES, SHUTTLE VALVES, RAPID EXHAUST VALVES

PRESSURE CONTROL VALVES

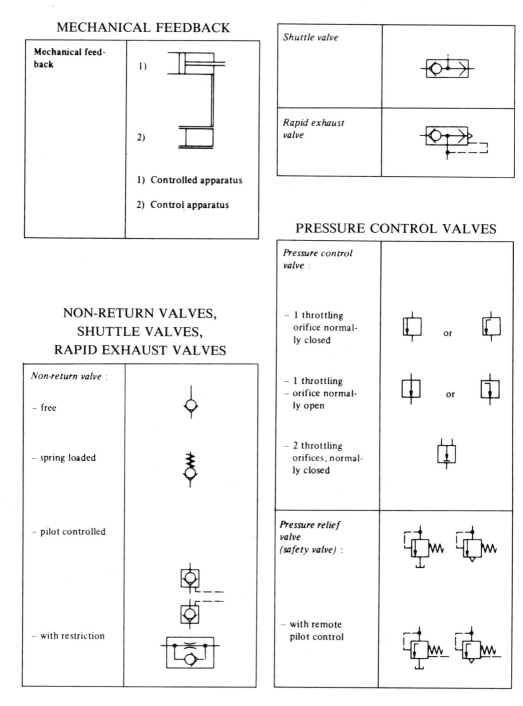

Mechanical feedback	1) 2) 1) Controlled apparatus 2) Control apparatus

Shuttle valve	
Rapid exhaust valve	

Non-return valve : – free	
– spring loaded	
– pilot controlled	
– with restriction	

Pressure control valve : – 1 throttling orifice normally closed	or
– 1 throttling – orifice normally open	or
– 2 throttling orifices, normally closed	
Pressure relief valve (safety valve) :	
– with remote pilot control	

FLOW CONTROL VALVES

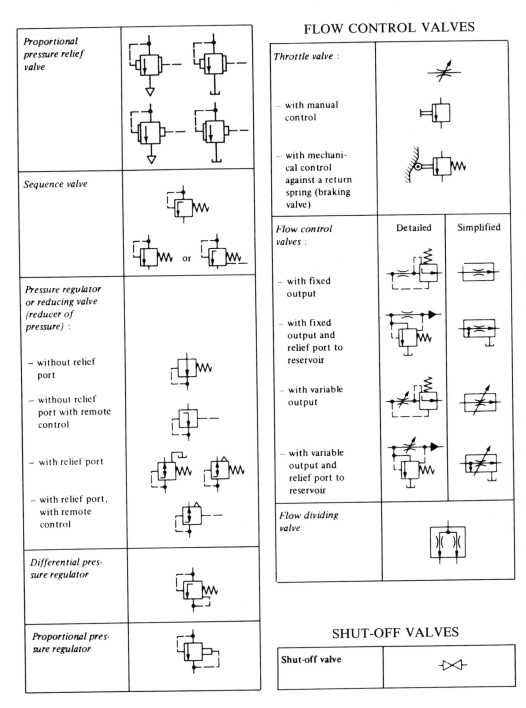

	Detailed	Simplified
Proportional pressure relief valve		
Sequence valve		
Pressure regulator or reducing valve (reducer of pressure) :		
– without relief port		
– without relief port with remote control		
– with relief port		
– with relief port, with remote control		
Differential pressure regulator		
Proportional pressure regulator		
Throttle valve :		
– with manual control		
– with mechanical control against a return spring (braking valve)		
Flow control valves :	Detailed	Simplified
– with fixed output		
– with fixed output and relief port to reservoir		
– with variable output		
– with variable output and relief port to reservoir		
Flow dividing valve		

SHUT-OFF VALVES

Shut-off valve	

SOURCES OF ENERGY

Pressure source	⊙−
Hydraulic press-ure source	⊙▶
Pneumatic press-ure source	⊙▷
Electric motor	Ⓜ=
Heat engine	[M]=

FLOW LINES AND CONNECTIONS

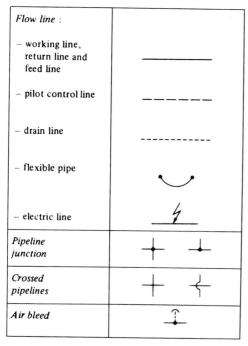

Flow line :	
– working line, return line and feed line	————
– pilot control line	– – – – –
– drain line	- - - - - -
– flexible pipe	⌣
– electric line	⚡
Pipeline junction	┿ ┷
Crossed pipelines	┼ ┼
Air bleed	↑

Exhaust port :	
– plain with no provision for connection	⊔
– threaded for connection	⊔̌
Power take-off :	
– plugged	→✕
– with take-off line	→✕←
Quick-acting coupling :	
– connected, without mechanically opened non-return valve	→•←
– connected, with mechanically opened non-return valves	⊶•⊷
– uncoupled, with open end	→⊢
– uncoupled, closed by free non-return valve	⊸⊢
Rotary connection :	
– one way	⊖
– three way	⊜
Silencer	▭▷

RESERVOIRS

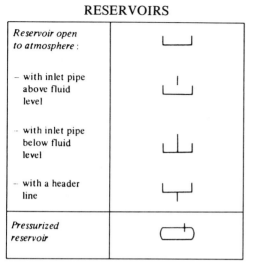

Reservoir open to atmosphere :	
– with inlet pipe above fluid level	
– with inlet pipe below fluid level	
– with a header line	
Pressurized reservoir	

ACCUMULATORS

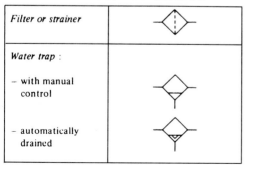

Accumulators	

FILTERS, WATER TRAPS, LUBRICATORS AND MISC. APPARATUS

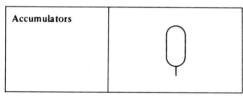

Filter or strainer	
Water trap :	
– with manual control	
– automatically drained	

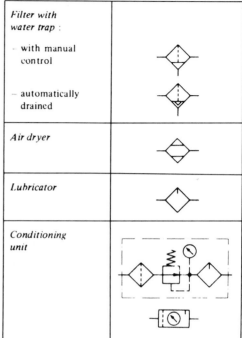

Filter with water trap :	
– with manual control	
– automatically drained	
Air dryer	
Lubricator	
Conditioning unit	

HEAT EXCHANGERS

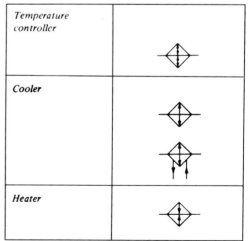

Temperature controller	
Cooler	
Heater	

MECHANICAL COMPONENTS

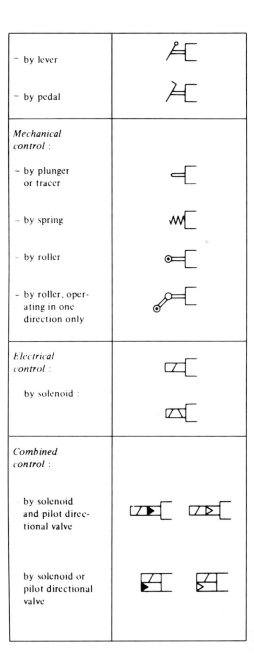

Rotating shaft :	
– in one direction	
– in either direction	
Detent	
Locking device	
Over-center device	
Pivoting devices :	
– simple	
– with traversing lever	
– with fixed fulcrum	

– by lever	
– by pedal	
Mechanical control :	
– by plunger or tracer	
– by spring	
– by roller	
– by roller, operating in one direction only	
Electrical control : by solenoid :	
Combined control :	
by solenoid and pilot directional valve	
by solenoid or pilot directional valve	

CONTROL METHODS

Muscular control :	
– by push button	

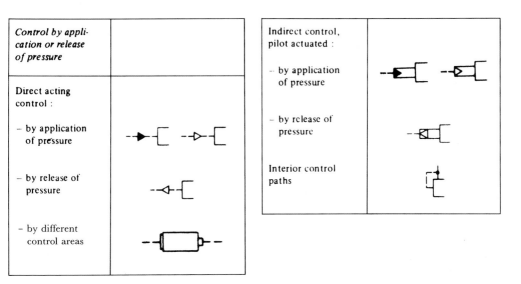

Control by appli-cation or release of pressure		Indirect control, pilot actuated :	
Direct acting control :		– by application of pressure	
– by application of pressure		– by release of pressure	
– by release of pressure		Interior control paths	
– by different control areas			

INDEX